OT 54
Operator Theory: Advances and Applications
Vol. 54

Birkhäuser Verlag
Basel · Boston · Berlin

Israel Gohberg
Naum Krupnik

One-Dimensional Linear Singular Integral Equations

Volume II
General Theory and Applications

1992

Birkhäuser Verlag
Basel · Boston · Berlin

Originally published in 1973 under the title “Vvedenie v teorijuodnomernych singuljarnych integralnych operatorov” by Stiinca, Kisinev.
German translation published in 1979 under the title »Einführung in die Theorie der eindimensionalen singulären Integraloperatoren« by Birkhäuser Verlag, Basel.

Authors’ addresses:

I. Gohberg
School of Mathematical Sciences
Raymond and Beverly Sackler
Faculty of Exact Sciences
Tel Aviv University
69978 Tel Aviv
Israel

N. Krupnik
Department of Mathematics
Bar Ilan University
Ramat Gan
Israel

A CIP catalogue record for this book is available from the Library of Congress, Washington D.C., USA

Deutsche Bibliothek Cataloging-in-Publication Data

Gochberg, Izrail’:
One-dimensional linear singular integral equations / Israel Gohberg ; Naum Krupnik. – Basel ; Boston ; Berlin : Birkhäuser
Einheitssacht.: Vvedenie v teoriju odnomernych singuljarnych integral’nych operatorov <engl.>

NE: Krupnik, Naum Y.:

Vol. 2. General theory and applications. – 1992
(Operator theory ; Vol. 54)
ISBN 3-7643-2796-0 (Basel …)
ISBN 0-8176-2796-0 (Boston …)
NE: GT

Printed in Germany on acid-free paper, directly from the authors’ camera-ready manuscript
ISBN 3-7643-2792-8
ISBN 0-8176-2792-8

Contents

Preface

This monograph is the second volume of a graduate text book on the modern theory of linear one-dimensional singular integral equations. Both volumes may be regarded as unique graduate text books.

Singular integral equations attract more and more attention since this class of equations appears in numerous applications, and also because they form one of the few classes of equations which can be solved explicitly.

The present book is to a great extent based upon material contained in the second part of the authors' monograph [6] which appeared in 1973 in Russian, and in 1979 in German translation. The present text includes a large number of additions and complementary material, essentially changing the character, structure and contents of the book, and making it accessible to a wider audience.

Our main subject in the first volume was the case of closed curves and continuous coefficients. Here, in the second volume, we turn to general curves and discontinuous coefficients.

We are deeply grateful to the editor Professor G. Heinig, to the translator Dr. S. Roch, and to the typist Mr. G. Lillack, for their patient work.

The authors

Ramat-Aviv, Ramat-Gan, May 26, 1991

Introduction

This book is the second volume of an introduction to the theory of linear one-dimensional singular integral operators. The main topics of both parts of the book are the invertibility and Fredholmness of these operators. Special attention is paid to inversion methods.

In the first volume of the book, the authors considered the case of integral equations on closed curves with continuous coefficients. The present second volume is concerned with general curves and discontinuous coefficients. As in the first part, the general abstract approach will be combined with concrete methods for the solution of singular integral equations. Numerous applications, examples, and exercises are given.

A singular integral operator is an operator of the form $A+T$ or $B+T$ where T is a compact operator, $A = aI + bS_\Gamma$, $B = aI + S_\Gamma bI$, I is the identity operator, $a \in L_\infty(\Gamma)$, $b \in L_\infty(\Gamma)$, and

$$(S_\Gamma \varphi)(t) = \frac{1}{\pi i} \int\limits_\Gamma \frac{\varphi(\tau) d\tau}{\tau - t} . \tag{0.1}$$

The integral in (0.1) is understood in the principal value sense. Functions a and b are called coefficients of operators A and B . Essentially, operators A and B are considered in the spaces $L_p(\Gamma, \rho)$ with weight

$$\rho(t) = \prod_{k=1}^{n} |t - t_k|^{\beta_k} , \tag{0.2}$$

where $t_1, \cdots, t_n$ are different points on Γ and

$$1 < p < \infty, \quad -1 < \beta_k < p - 1 .$$

Unlike the case of a closed curve and continuous coefficients, the conditions

$$a(t) - b(t) \neq 0 \quad \text{and} \quad a(t) + b(t) \neq 0 \quad (t \in \Gamma) \tag{0.3}$$

are no longer sufficient for the one-sided invertibility (or Fredholmness) of operators A and B . Let, for example, explain how the conditions for the one-sided invertibility of the operators A and B look if $\Gamma = [0,1]$, $\rho(t) = t^{\beta_1}(1-t)^{\beta_2}$, and the coefficients a and b are continuous on Γ . In this situation, operator A (or B) proves to be invertible if and only if the condition (0.3) and the following additional conditions (0.4) and (0.5) are satisfied:

$$(1 + \alpha^2 z^2) a(0) + (z(1+\alpha^2) - i(1-z^2)\alpha) b(0) \neq 0 \quad (-1 \leq z \leq 1) \tag{0.4}$$

$$(1 + \beta^2 z^2) a(1) + (z(1+\beta^2) + i(1-z^2)\beta) b(1) \neq 0 \quad (-1 \leq z \leq 1) , \tag{0.5}$$

where

$$\alpha = \cot \frac{\pi(1+\beta_1)}{p}, \qquad \beta = \cot \frac{\pi(1+\beta_2)}{p}.$$

If these conditions are fulfilled then

$$\operatorname{Ind} A = \operatorname{Ind} B = -\kappa,$$

where

$$\begin{aligned}\kappa \;=\; & \frac{1}{2\pi}\left[\arg \frac{a(t)+b(t)}{a(t)-b(t)}\right]_{t=0}^{1} \\ & -\frac{1}{2\pi}\left[\arg\left(a(0)+b(0)\,\frac{z(1+\alpha^2)-i(1-z^2)\alpha}{1+\alpha^2 z^2}\right)\right]_{z=-1}^{1} \\ & +\frac{1}{2\pi}\left[\arg\left(a(1)+b(1)\,\frac{z(1+\beta^2)+i(1-z^2)\beta}{1+\beta^2 z^2}\right)\right]_{z=-1}^{1},\end{aligned}$$

and where $[\arg f(x)]_{x=0}^{1}$ refers to the increase of the function $\arg f(x)$ along the interval [0,1]. Further, if conditions (0.3)-(0.5) are satisfied then each of the operators A and B is left invertible, right invertible, or two-sided invertible depending on whether the number κ is positive, negative, or equal to zero.

In the special case considered both conditions for the one-sided invertibility of operators A and B as well as the indices of these operators depend on the number p and the weight ρ. An analogous phenomenona takes place in the case of discontinuous coefficients on closed curves, and also in the general context.

As it was shown in the first volume, every continuous and non-vanishing function on a closed curve Γ admits a generalized factorization. For discontinuous coefficients, things turn out to be essentially more involved. Therefore one introduces the notion of p,ρ-factorization (that is a factorization which depends on the space $L_p(\Gamma,\rho)$) :

$$g(t) = g_-(t)t^{\kappa}g_+(t) \tag{0.6}$$

where $g_-^{\pm 1} \in L_1^-(\Gamma)$, $g_+^{\pm 1} \in L_1^+(\Gamma)$ and where additionally the operator

$$g_+^{-1} S_\Gamma g_+ I$$

is assumed to be bounded in the space $L_p(\Gamma,\rho)$.

The operator A (and B) is one-sided invertible in the space $L_p(\Gamma,\rho)$ if and only if

$$\operatorname*{ess\,inf}_{t\in\Gamma} |a(t)-b(t)| \neq 0,$$

$$\operatorname*{ess\,inf}_{t\in\Gamma} |a(t)+b(t)| \neq 0\,, \tag{0.7}$$

and the function $g=(a+b)(a-b)^{-1}$ admits a p,ρ-factorization.

The p, ρ–factorization plays the same role for the inversion of singular operators with discontinuous coefficients on closed curves as the generalized factorization of continuous functions did for the inversion of singular operators with continuous coefficients (see Chapter 3). If the operator A (or B) is one–sided invertible in $L_p(\Gamma, \rho)$ then conditions (0.7) are satisfied, and the function $g = (a+b)(a-b)^{-1}$ admits the factorization (0.6). The one–sided inverse operators A^{-1} (resp. B^{-1}) of A (resp. B) are of the form

$$A^{-1} = \frac{a}{a^2-b^2} I - \frac{b}{a^2-b^2} X S_\Gamma X^{-1} I ,$$

$$B^{-1} = \frac{a}{a^2-b^2} I - Y^{-1} S_\Gamma Y \frac{b}{a^2-b^2} I ,$$

where $X = (a-b)g_-$ and $Y = (a-b)g_+$. Moreover, if $\kappa < 0$ then the functions $\varphi_m(t) = g_+^{-1}(t)(1-g(t))t^m$ $(m = 0, 1, \cdots, |\kappa|-1)$ form a basis of $\ker A$, and the functions $\psi_m(t) = g_-(t)(a(t)+b(t))^{-1}t^{-m}$ $(m = 1, \cdots, |\kappa|)$ provide us with a basis of $\ker B$.

By means of the factors $g_\pm$ in the factorization (0.6) one can establish criteria for the solvability of the equations $Ax = y$ and $Bx = y$. Setting up conditions for the p, ρ–factorizability of a function $g \in L_\infty(\Gamma)$ as well as the explicit realization of the factorization are some of the fundamental problems in the theory of singular integral equations. So, special attention is paid to these problems throughout this book.

Let us now give a short description of the contents of this book. The numeration of the chapters in the first volume will be continued.

In the sixth chapter we summarize some basic facts about singular integral operators which are needed in this part of the book, and which were explained in detail in the first volume.

The seventh chapter contains the general theory of singular integral operators with coefficients in $L_\infty(\Gamma)$.

In the eighth chapter we give the definition of the p, ρ–factorizability of a function, establish criteria for the p, ρ–factorizability, and consider examples.

In the ninth chapter we present conditions for the one–sided invertibility and Fredholmness of singular integral operators on simple closed and open curves with coefficients having discontinuities of the first kind. We also study Toeplitz and paired operators. A special attention is devoted to applications of singular integral equations to boundary value problems with piecewise continuous coefficients, to problems in mathematical physics, to integral equations with Hilbert kernel, and to Wiener–Hopf equations with discontinuous symbols.

The tenth chapter is devoted to generalizations of the results of Chapter 9 to curves with intersections.

Criteria for the one–sided invertibility of singular integral operators with coefficients having discontinuities of almost periodic type is derived in the eleventh chapter, and in the twelfth chapter we treat the same questions for general coefficients in $L_\infty(\Gamma)$.

In the concluding chapter we compute the norms and the essential norms of singular integral operators in the spaces $L_p(\Gamma, \rho)$, and we point out some applications of these results to the invertibility and Fredholm theory of singular integral operators.

Each chapter ends with a series of exercises.

Chapter 6

Preliminaries

The present chapter is of a supporting nature. It contains for the reader's convenience it contains the basic material on singular operators which will be used in this book. A systematic treatment of the subject was given in the first volume GOHBERG/KRUPNIK [11], which we refer to throughout this volume as [GK1].

6.1 The operator of singular integration

A *simple arc* is an oriented, bounded and non–closed curve Γ without multiple points which satisfies a *Lyapunov* condition. The latter means that the angle $\theta_\Gamma(t)$ $(t \in \Gamma)$ formed by the tangent of Γ at the point t and the positive real axis satisfies a Hölder condition $|\theta_\Gamma(t_1) - \theta_\Gamma(t_2)| \leq C|t_1 - t_2|^\alpha$ for some $0 < \alpha < 1$ and a constant C. A *non–simple* curve is a curve Γ which consists of a finite number of oriented simple arcs $\Gamma_1, \Gamma_2, \ldots, \Gamma_n$ having only finitely many points in common. These points will be called *singular*. Further we assume that for Γ_i and Γ_k with a common point, the curve $\Gamma_i \cup \Gamma_k$ is either Lyapunov or the tangents of Γ_i and Γ_k at this point do not coincide. A non–simple curve Γ is said to be *closed* if it divides the extended complex plane $\mathbf{C}^\infty (= \mathbf{C} \cup \{\infty\})$ into two open components, F_Γ^+ and F_Γ^-, both having Γ as its boundary. Thereby, F_Γ^+ denotes the part which is located on the left of the curve Γ. Further we suppose for simplicity that $0 \in F_\Gamma^+$ and $\infty \in F_\Gamma^-$.

By $R(\Gamma)$ we denote the set of all rational functions possessing no poles on the curve Γ, and by $R^\pm(\Gamma)$ the set of all rational functions the poles of which lie in $F_\Gamma^\pm$. The *singular integral* or *operator of singular integration* along the non–simple curve Γ is the integral

$$\frac{1}{\pi i} \int_\Gamma \frac{\varphi(\tau)}{\tau - t}\, d\tau \quad (t \in \Gamma), \tag{1.1}$$

which is understood in the sense of Cauchy's principal value. We denote the value of this integral by $(S_\Gamma \varphi)(t)$. Further we mean by a *simple closed curve* a closed oriented curve in the complex plane which bounds a simply connected domain and satisfies a Lyapunov condition. By a *simple curve* we mean a simple closed curve or a simple arc. Finally, a *composed curve* consists, by its definition, of a finite number of pairwise disjoint simple curves.

6.1.1. *Let* Γ *be a non-simple closed curve and* $r = r_+ + r_-$ *with* $r_\pm \in R^\pm(\Gamma)$ *and* $r_-(\infty) = 0$. *Then, for any non-singular point* $t \in \Gamma$,

$$S_\Gamma(r_+ + r_-) = r_+ - r_-, \quad S_\Gamma^2 r = r \tag{1.2}$$

(see [GK1, 1.1]).

6.2 The space $L_p(\Gamma, \rho)$

Let Γ be a non-simple curve. By $L_p(\Gamma,\rho)$ we denote the L_p-space on the curve Γ with weight ρ . The norm in $L_p(\Gamma,\rho)$ is defined by

$$\|\varphi\|_{L_p(\Gamma,\rho)} = \left(\int_\Gamma |\varphi(t)|^p \, \rho(t) \, |dt| \right)^{\frac{1}{p}} \tag{2.1}$$

where $|dt|$ refers to arc length differential. As is customary, we presume that

$$\rho(t) = \prod_{k=1}^{n} |t - t_k|^{\beta_k} \tag{2.2}$$

for certain pairwise distinct points $t_1, \ldots, t_n \in \Gamma$ and real numbers $\beta_1, \ldots, \beta_n$ satisfying

$$-1 < \beta_k < p - 1 \quad (k = 1, \ldots, n) . \tag{2.3}$$

These conditions guarantee that

$$\rho \in L_1(\Gamma) \quad \text{and} \quad \rho^{1-q} \in L_1(\Gamma) \quad (p^{-1} + q^{-1} = 1) \tag{2.4}$$

and that the set $R(\Gamma)$ is dense in the spaces $L_p(\Gamma,\rho)$ and $L_q(\Gamma,\rho^{1-q})$.

The Banach space $L_p(\Gamma,\rho)^*$ is isometrically isomorphic to $L_q(\Gamma,\rho^{1-q})$, and the isomorphism $\varphi \to f_\varphi$ is given by

$$f_\varphi(\psi) = \int_\Gamma \varphi(t)\overline{\psi}(t) \, |dt| . \tag{2.5}$$

6.3 Singular integral operators

Operators of the form

$$A = aI + bS_\Gamma \quad \text{and} \quad B = aI + S_\Gamma bI \tag{3.1}$$

with $a, b \in L_\infty(\Gamma)$, I the identity operator, and S_Γ the operator of singular integration defined by (1.1) will be called *singular integral operators.*

6.3.1. *Let* Γ *be a non–simple curve and* ρ *be the weight defined by (2.2). Then the operator* S_Γ *is bounded on the space* $L_p(\Gamma, \rho)$ *if and only if the numbers* β_k *satisfy the condition (2.3)*([GK1, Chapter 1]).

6.3.2. *If* $a \in C(\Gamma)$ *then the operator* $T = aS_\Gamma - S_\Gamma aI$ *is compact on* $L_p(\Gamma, \rho)$ ([GK1, 1.7]).

From now on if no restrictions for Γ, ρ , and the Banach space $\mathcal{B}$ on which the singular integral operators act are made, we always assume that Γ is a non–simple curve, the weight ρ is of the form (2.2) satisfying (2.3), and $\mathcal{B} = L_p(\Gamma, \rho)$.

We introduce the notation

$$P_\Gamma = \frac{1}{2}(I + S_\Gamma), \quad Q_\Gamma = \frac{1}{2}(I - S_\Gamma).$$

If Γ is a closed curve then the operators P_Γ and Q_Γ are projectors in $L_p(\Gamma, \rho)$. The singular integral operators defined by (3.1) can be rewritten as

$$A = a_1 P_\Gamma + b_1 Q_\Gamma, \quad B = P_\Gamma a_1 I + Q_\Gamma b_1 I \tag{3.2}$$

where $a_1 = a + b$ and $b_1 = a - b$.

Sometimes, if it is obvious which curve Γ is meant, we shall simply write S, P and Q instead of S_Γ, P_Γ and Q_Γ . The circle $\{z \in \mathbf{C} : |z| = 1\}$ will be denoted by $\mathbf{T}$. For any non–simple curve Γ we introduce a function $h(t)$ defined by

$$dt = h(t)\,|dt|. \tag{3.3}$$

6.3.3. *The operator* S_Γ^* *is connected with the operator* S_Γ *acting on* $L_q(\Gamma, \rho^{1-q})$ *via*

$$S_\Gamma^* = -H_\Gamma S_\Gamma H_\Gamma \tag{3.4}$$

where

$$(H_\Gamma \varphi)(t) = \overline{h_\Gamma(t)\varphi(t)}\,. \tag{3.5}$$

6.3.4. *If* Γ *is a composed curve then the operator* $S_\Gamma^* - S_\Gamma$ *is compact in* $L_q(\Gamma, \rho^{1-q})$ ([GK1, 1.7]).

Let Γ_1 and Γ_2 be two composed curves, $z = \beta(t)$ be a function which maps Γ_1 onto Γ_2 in a one–to–one manner,

$$\rho_1(t) = \prod_{k=1}^{n} |t - t_k|^{\beta_k}, \quad \rho_2(z) = \prod_{k=1}^{n} |z - z_k|^{\beta_k}$$

with $z_k = \beta(t_k)$, and let V stand for the linear bounded operator acting from $L_p(\Gamma_1, \rho_1)$ into $L_p(\Gamma_2, \rho_2)$ via $(V\varphi)(z) = \varphi(\alpha(z))$ where $\alpha(\beta(t)) = t$ and α is assumed to have a Hölder continuous derivative α' on Γ_2 with exponent μ $(0 < \mu < 1)$.

6.3.5. *The operator* $S_{\Gamma_2} - VS_{\Gamma_1}V^{-1}$ *is compact on the space* $L_p(\Gamma_2, \rho_2)$ ([GK1, 1.4]).

The following proposition describes the connection between the operator S_0 acting on $L_p(\mathbf{T}, \rho)$ and the operator $S_\mathbf{R}$ acting in $L_p(\mathbf{R}, \rho_1)$ where $\mathbf{R}$ stands for $(-\infty, \infty)$.

Let

$$\rho(z) = \prod_{k=0}^{n} |z - z_k|^{\beta_k}, \tag{3.6}$$

where $z_0 = 1$ and the numbers $\beta_0, \ldots, \beta_n$ satisfy the conditions (2.3). We choose

$$t_k = i(z_k + 1)(z_k - 1)^{-1} \quad (k = 1, \ldots, n)$$

and

$$\rho_1(t) = |t + i|^\beta \prod_{k=1}^{n} |t - t_k|^{\beta_k} \tag{3.7}$$

with

$$\beta = p - 2 - \beta_0 - \sum_{k=1}^{n} \beta_k .$$

6.3.6. *The operator* C *defined by*

$$(C\varphi)(z) = \frac{1}{z-1} \varphi\left(i\frac{z+1}{z-1}\right) \tag{3.8}$$

is linear and bounded from $L_p(\mathbf{R}, \rho_1)$ *into* $L_p(\mathbf{T}, \rho)$ *,*

$$(C^{-1}\psi)(t) = \frac{2i}{t-i} \psi\left(\frac{t+i}{t-i}\right) , \tag{3.9}$$

$\|C\varphi\| = \delta\|\varphi\|$ *for some constant* δ *independent of* φ *, and*

$$C^{-1} S_\mathbf{T} C = -S_\mathbf{R} . \tag{3.10}$$

6.4 The spaces $L_p^+(\Gamma,\rho)$, $L_p^-(\Gamma,\rho)$ and $\overset{\circ}{L_p^-}(\Gamma,\rho)$

If the curve Γ is closed and $1 < p < \infty$ then the operators P_Γ and Q_Γ are linear and bounded projectors in $L_p(\Gamma,\rho)$. We set

$$L_p^+(\Gamma,\rho) = P_\Gamma(L_p(\Gamma,\rho)), \quad \overset{\circ}{L_p^-}(\Gamma,\rho) = Q_\Gamma(L_p(\Gamma,\rho)) \tag{4.1}$$

and

$$L_p^-(\Gamma,\rho) = \overset{\circ}{L_p^-}(\Gamma,\rho) \dot{+} \{\text{const.}\}. \tag{4.2}$$

We next formulate a criterion to decide whether a function belongs to the spaces $L_p^\pm(\Gamma,\rho)$ (see [GK1, 2.4]). Denote by $F_1^+,\dots,F_m^+$ and $F_1^-,\dots,F_k^-$ the connected components of the sets F_Γ and F_Γ^-, respectively.

6.4.1. *Let $\varphi\in L_p(\Gamma,\rho)$. For the inclusion $\varphi\in L_p^+(\Gamma,\rho)$ to hold it is necessary and sufficient that the following conditions are satisfied:*

$$\int_\Gamma \frac{\varphi(\tau)\,d\tau}{(\tau-\alpha_j^-)^n} = 0 \quad (j=1,\dots,k;\ n=1,2,\dots) \tag{4.3}$$

for certain points $\alpha_j^-\in F_j^-$.

6.4.2. *Let $\varphi\in L_p(\Gamma,\rho)$. For the inclusion $\varphi\in \overset{\circ}{L_p^-}(\Gamma,\rho)$ to hold it is necessary and sufficient that the conditions*

$$\int_\Gamma \frac{\varphi(\tau)\,d\tau}{(\tau-\alpha_j^+)^n} = 0 \quad (j=1,\dots,m;\ n=1,2,\dots) \tag{4.4}$$

are satisfied for certain points $\alpha_j^+\in F_j^+$.

If, in particular, Γ is a simple closed curve and $0\in F_\Gamma^+$ then the conditions (4.3) and (4.4) can be rewritten as

$$\int_\Gamma \varphi(t)t^n\,dt = 0 \quad (n=0,1,\dots) \tag{4.3'}$$

and

$$\int_\Gamma \varphi(t)t^{-n}\,dt = 0 \quad (n=1,2,\dots) \tag{4.4'}$$

The operator S_Γ (and, hence, the operators P_Γ and Q_Γ) is not bounded in either $L_1(\Gamma)$ or $L_\infty(\Gamma)$. Therefore we introduce $L_1^+(\Gamma)$ and $\overset{\circ}{L_1^-}(\Gamma)$ the sets of all functions $\varphi\in L_1(\Gamma)$ which are subject to the conditions (4.3) and (4.4), or, in case of a simple closed curve, to the simpler conditions (4.3') and (4.4'), respectively. By $L^+_\infty(\Gamma)$ and $L^-_\infty(\Gamma)$ we denote the classes of all functions which are holomorphic and bounded in F^+_Γ and F^-_Γ, respectively. The functions $\varphi \in L^\pm_\infty(\Gamma)$ possess limits $\varphi(t) = \lim_{z\to t}\varphi(z)$ with $z \in F^\pm_\Gamma$ almost everywhere on Γ, and $\varphi \in L_\infty(\Gamma)$. Consequently, the classes $L^\pm_\infty(\Gamma)$ can be characterized as subspaces of $L_\infty(\Gamma)$ consisting of all functions φ which are limits of holomorphic and bounded in $F^\pm_\Gamma$ functions almost everywhere.

Let us further remark that if $\rho^{1-q} \in L_1$ then $L_p(\Gamma,\rho) \subset L_1(\Gamma)$, and for these weights we have the inclusions

$$L^\pm_\infty(\Gamma) \subset L^\pm_p(\Gamma,\rho) \subset L^\pm_1(\Gamma) \quad (p \in (1,\infty)). \tag{4.5}$$

Let $\varphi \in L_p(\Gamma,\rho)$ and

$$\Phi_\varphi(z) = \frac{1}{2\pi i}\int_\Gamma \frac{\varphi(\tau)\,d\tau}{\tau - z} \quad (z \in \mathbf{C}\backslash\Gamma).$$

6.4.3. *The function $\Phi_\varphi(z)$ is holomorphic on $\mathbf{C}\backslash\Gamma$, possesses the limits*

$$\Phi^+_\varphi(t) = \lim_{\substack{z\to t\\ z\in F^+_\Gamma}} \Phi_\varphi(z), \quad \Phi^-_\varphi(t) = \lim_{\substack{z\to t\\ z\in F^-_\Gamma}} \Phi_\varphi(z) \tag{4.6}$$

almost everywhere on Γ, and

$$\Phi^+_\varphi = P_\Gamma\varphi, \quad \Phi^-_\varphi = -Q_\Gamma\varphi. \tag{4.7}$$

6.5 Factorization [1]

Let Γ be an oriented closed curve, $a \in C(\Gamma)$, and $a(t) \neq 0$ on Γ. The *index* of the function a is, by definition, the integer

$$\operatorname{ind} a = \frac{1}{2\pi}[\arg a(t)]_\Gamma$$

where $[\arg a(t)]_\Gamma$ stands for the total increment of the argument of the function a as t runs through the curve Γ.

6.5.1. *If $a, b \in C(\Gamma)$, and $a(t) \neq 0$, $b(t) \neq 0$ $(t \in \Gamma)$, then*

$$\operatorname{ind} ab = \operatorname{ind} a + \operatorname{ind} b, \quad \operatorname{ind}\frac{a}{b} = \operatorname{ind} a - \operatorname{ind} b. \tag{5.1}$$

If $m \in C(\Gamma)$ and $|m(t)| < 1$ then

$$\operatorname{ind}(1+m) = 0. \tag{5.2}$$

[1] See [GK1, 3.3 and 3.9].

Let $C_\pm(\Gamma)$ refer to the classes of continuous functions on Γ which possess a holomorphic extension in $F_\Gamma^\pm$ and which are continuous on $F_\Gamma^\pm \cup \Gamma$, respectively.

A *factorization* of a continuous function a with respect to the closed curve Γ is a representation of the form $a(t) = a_-(t)t^\kappa a_+(t)$ where κ is an integer, $a_\pm \subset C_\pm(\Gamma)$, and the functions $a_\pm(z)$ do not vanish on $F_\Gamma^\pm$, respectively.

6.5.2. *If* $r \in R(\Gamma)$ *and* $r(t) \neq 0$ *on* Γ *then* r *admits a factorization,* $r = r_- t^\kappa r_+$, *with* $\kappa = \operatorname{ind} r$.

The same result holds for Hölder continuous functions, for functions in the Wiener algebra, and for other classes. On the other hand, there are examples of continuous and non–vanishing functions on Γ which are not factorizable with respect to Γ.

By a *generalized factorization* of a function $a \in L_\infty(\Gamma)$ with respect to the curve Γ we mean its representation in the form $a(t) = a_-(t)t^\kappa a_+(t)$, where $a_\pm \in L_1^\pm(\Gamma)$, $a_\pm^{-1} \in L_1^\pm(\Gamma)$, and the operator

$$a_+^{-1} P_\Gamma a_-^{-1} I \tag{5.3}$$

is bounded on $L_p(\Gamma)$ for all $p \in (1, \infty)$.

6.5.3. *Let* $a \in C(\Gamma)$. *If the function* a *admits a generalized factorization then the functions* $a_\pm$ *amd* $a_\pm^{-1}$ *belong to* $L_p^\pm(\Gamma, \rho)$ *and the operator* $a_+^{-1} P_\Gamma a_-^{-1} I$ *is bounded on every space* $L_p(\Gamma, \rho)$.

6.5.4. *Any continuous and non–vanishing function* a *on a closed curve* Γ *admits a generalized factorization* $a(t) = a_-(t)t^\kappa a_+(t)$ *with respect to* Γ. *Hence* $\kappa = \operatorname{ind} a$.

6.6 One–sided invertibility of singular integral operators [1]

Let Γ be a closed curve, $a, b \in C(\Gamma)$, $A = aP_\Gamma + bQ_\Gamma$ and $B = P_\Gamma aI + Q_\Gamma bI$.

6.6.1. *For the one-sided invertibility of the operator* A *(the operator* B *) on* $L_p(\Gamma, \rho)$ *it is necessary and sufficient that*

$$a(t) \neq 0\,, \quad b(t) \neq 0 \quad (t \in \Gamma)\,. \tag{6.1}$$

Let the conditions (6.1) be satisfied and $c(t) = c_-(t)t^\kappa c_+(t)$ *be the generalized factorization of the function* $c(t) = a(t)/b(t)$. *Then:*

---

[1] See [GK1, 3.7 and 3.10].

1.° For $\kappa = 0$, the operators A and B are invertible with

$$A^{-1} = x(a^{-1}P_\Gamma + b^{-1}Q_\Gamma)x^{-1}I , \tag{6.2}$$

where $x = c_- b$, and

$$B^{-1} = y(P_\Gamma a^{-1} + Q_\Gamma b^{-1})y^{-1}I , \tag{6.3}$$

where $y = c_+^{-1}b^{-1}$.

2.° For $\kappa > 0$, the operators A and B are invertible from the left with left inverses given by (6.2) and (6.3), respectively. The equation $A\varphi = f$ is solvable if and only if

$$\int_\Gamma f(t)b^{-1}(t)c_-^{-1}(t)t^{-j}\, dt = 0 \quad (j = 1, 2, \dots, \kappa) , \tag{6.4}$$

and the equation $B\varphi = f$ is solvable if and only if

$$\int_\Gamma f(t)(c_-^{-1}(t) - c_+(t))t^{-j}\, dt = 0 \quad (j = 1, 2, \dots, \kappa) . \tag{6.5}$$

3.° For $\kappa < 0$, the operators A and B are invertible from the right. The operators (6.2) and (6.3) are right inverses of A and B, respectively. Also

$$\ker A = \operatorname{span}\{g, gt, \dots, gt^{-\kappa-1}\} \tag{6.6}$$

with $g = c_+^{-1} - c_- t^\kappa$, and

$$\ker B = \operatorname{span}\{ut, ut^2, \dots, ut^{-\kappa}\} \tag{6.7}$$

with $u = c_+^{-1}b^{-1}$.

In the above, $\operatorname{span}\{.\}$ stands for the linear hull.

6.7 Fredholm operators[2].

Let $\mathcal{B}_1$ and $\mathcal{B}_2$ be Banach spaces, $L(\mathcal{B}_1, \mathcal{B}_2)$ the collection of all linear bounded operators acting from $\mathcal{B}_1$ into $\mathcal{B}_2$, and $\mathcal{T}(\mathcal{B}_1, \mathcal{B}_2)$ the subset of $L(\mathcal{B}_1, \mathcal{B}_2)$ consisting of all compact operators. In place of $L(\mathcal{B}, \mathcal{B})$ and $\mathcal{T}(\mathcal{B}, \mathcal{B})$ we shall write $L(\mathcal{B})$ and $\mathcal{T}(\mathcal{B})$. We denote by $\|\cdot\|$ the usual operator norm in $L(\mathcal{B}_1, \mathcal{B}_2)$ and by $|\cdot|$ the quotient norm; that is $|A| = \inf_{T \in \mathcal{T}(\mathcal{B}_1, \mathcal{B}_2)} \|A + T\|$.

[2]See [GK1, Ch. 4]

An operator $A(\in L(\mathcal{B}_1,\mathcal{B}_2))$ is said to be *normally solvable* in the situation that the equation $A\varphi = f$ has a solution if and only if the right hand side f is orthogonal to any solution ψ of the homogeneous adjoint equation $A^*\psi = 0$.

6.7.1. *The operator* A *is normally solvable if and only if its image is closed; that is* $\overline{\text{im } A} = \text{im } A$.

An operator A is called *Φ–operator* (or Fredholm operator) if it is normally solvable and if the numbers $\dim\ker A$ and $\dim \text{coker } A$ are finite. Remember that $\text{coker } A = \mathcal{B}_2/\overline{\text{im } A}$. The *index* of a Φ–operator A is defined as $\text{Ind } A = \dim\ker A - \dim \text{coker } A$.

6.7.2. *If the operator* A *is normally solvable then* $\dim(\mathcal{B}_2/\text{im A}) = \dim\ker A^*$.

The class of all Φ–operators acting from $\mathcal{B}_1$ into $\mathcal{B}_2$ will be denoted by $\Phi(\mathcal{B}_1,\mathcal{B}_2)$.

6.7.3. *If* $A \in \Phi(\mathcal{B}_1,\mathcal{B}_2)$ *and* $T \in \mathcal{T}(\mathcal{B}_1,\mathcal{B}_2)$ *then* $A+T \in \Phi(\mathcal{B}_1,\mathcal{B}_2)$ *and* $\text{Ind }(A+T) = \text{Ind } A$.

6.7.4. *If* $A \in \Phi(\mathcal{B}_1,\mathcal{B}_2)$ *then there is a number* $\delta > 0$ *such that, for all operators* $M \in L(\mathcal{B}_1,\mathcal{B}_2)$ *with* $\|M\| < \delta$, $A+M \in \Phi(\mathcal{B}_1,\mathcal{B}_2)$ *and* $\text{Ind }(A+M) = \text{Ind } A$.

One says that the operator $A \in L(\mathcal{B}_1,\mathcal{B}_2)$ *admits a regularization* if there exists an operator $R \in L(\mathcal{B}_2,\mathcal{B}_1)$ such that the operators $AR - I$ and $RA - I$ are compact.

6.7.5. *The operator* A *is a* Φ*–operator if and only if it admits a regularization.*

6.7.6. *If* $A \in \Phi(\mathcal{B}_1,\mathcal{B}_2)$ *and* $B \in \Phi(\mathcal{B}_2,\mathcal{B}_3)$ *then* $BA \in \Phi(\mathcal{B}_1,\mathcal{B}_3)$ *and* $\text{Ind } BA = \text{Ind } B + \text{Ind } A$.

An operator $A \in L(\mathcal{B}_1,\mathcal{B}_2)$ is called Φ_+*–operator* if it is normally solvable, $\dim\ker A < \infty$ and $\dim \text{coker } A = \infty$. An operator $A \in L(\mathcal{B}_1,\mathcal{B}_2)$ is a Φ_-*–operator* if it is normally solvable, $\dim\ker A = \infty$ and $\dim \text{coker } A < \infty$. The classes of all Φ_+– and Φ_-–operators acting from $\mathcal{B}_1$ into $\mathcal{B}_2$ will be designated by $\Phi_+(\mathcal{B}_1,\mathcal{B}_2)$ and $\Phi_-(\mathcal{B}_1,\mathcal{B}_2)$, respectively.

The propositions 6.7.3 and 6.7.4 remain valid for $\Phi_\pm$–operators, too. We note one other property of $\Phi_\pm$–operators.

6.7.7. *Let* $A \in L(\mathcal{B}_1,\mathcal{B}_2)$ *and* $B \in L(\mathcal{B}_2,\mathcal{B}_3)$. *If the product* BA *is a* Φ*– or a* Φ_-*–operator then* B *is a* Φ*– or a* Φ_-*–operator. If* BA *is a* Φ*– or a* Φ_+*–operator then* A *is a* Φ*– or* Φ_+*–operator.*

6.7.8. *Let* Γ *be a closed curve and* $a,b \in C(\Gamma)$. *The operator* $A = aP_\Gamma + bQ_\Gamma$ $(A = P_\Gamma aI + Q_\Gamma bI)$ *is a* Φ*–operator if and only if* $a(t) \neq 0$ *and* $b(t) \neq 0$ *on* Γ. *If this condition is fulfilled then* $\text{Ind } A = -\text{ind}(a/b)$.

6.7.9. *Let* Γ *be a closed curve and* $a,b \in C(\Gamma)$. *If at least one of the functions* a

and b *has a zero on* Γ *then the operator* $A = aP_\Gamma + bQ_\Gamma$ $(A = P_\Gamma aI + Q_\Gamma bI)$ *is neither a* Φ_+*– nor a* Φ_-*–operator.*

6.8 The local principle for singular integral operators

We call the singular integral operators $A = aP_\Gamma + bQ_\Gamma$ and $A_1 = a_1P_\Gamma + b_1Q_\Gamma$ (or the operators $B = P_\Gamma aI + Q_\Gamma bI$ and $B_1 = P_\Gamma a_1I + Q_\Gamma b_1I$) locally equivalent at the point $t_0 \in \Gamma$ if, for any $\varepsilon > 0$, there is a neighbourhood $U(t_0)$ such that

$$\operatorname*{ess\ sup}_{t\in U(t_0)} |a(t) - a_1(t)| < \varepsilon, \quad \operatorname*{ess\ sup}_{t\in U(t_0)} |b(t) - b_1(t)| < \varepsilon\,. \tag{8.1}$$

6.8.1. *Assume the operator* $A = aP_\Gamma + bQ_\Gamma$ $(A = P_\Gamma aI + Q_\Gamma bI)$ *to be locally equivalent to a* Φ*–operator* $A_\tau = a_\tau P_\Gamma + b_\tau Q_\Gamma$ $(A_\tau = P_\Gamma a_\tau I + Q_\Gamma b_\tau I)$ *at each point* $\tau \in \Gamma$. *Then the operator* A *is a* Φ*–operator.*

If, in particular, $a, b \in C(\Gamma)$ *and* $a_\tau = a(\tau)$ *and* $b_\tau = b(\tau)$ *then*

$$|A|_{L_p(\Gamma)} = \max_{\tau\in\Gamma} |A_\tau|_{L_p(\Gamma)} \tag{8.2}$$

and

$$|A|_{L_p(\Gamma,\rho)} = \max_{\tau\in\Gamma} |A_\tau|_{L_p(\Gamma,\rho_\tau)}\,, \text{ [1]} \tag{8.3}$$

where $\rho(t) = \prod_{k=1}^{n} |t - t_k|^{\beta_k}$ *and*

$$\rho_\tau(t) = \begin{cases} 1 & \text{if} \quad \tau \notin \{t_1, \ldots, t_n\} \\ |t - t_k|^{\beta_k} & \text{if} \quad \tau = t_k. \end{cases}$$

Proof. Consider the quotient $\widehat{L} = L(\mathcal{B})/\mathcal{T}(\mathcal{B})$ of the algebra $L(\mathcal{B})$ of all linear bounded operators acting on the Banach space $\mathcal{B} = L_p(\Gamma, \rho)$ and the algebra $\mathcal{T}(\mathcal{B})$ of all compact operators. The cosets $\widehat{A}_\tau$ containing the operators A_τ are invertible in the algebra $\widehat{L}$. As in Example 1.1 from [GK 1, 5.1] we introduce localizing classes $M_\tau (\tau \in \Gamma)$ in the algebra $\widehat{L}$ in the following way. Let F_τ refer to the set of all non-negative and continuous functions on Γ which are equal to one in a certain neighbourhood of the point τ . Denote by N_τ the class of all operators of multiplication by functions $f \in F_\tau$ and by M_τ the class of all cosets $\widehat{R} (\in \widehat{L})$ with $R \in N_\tau$.

Due to Proposition 6.3.2, the cosets $\widehat{A}_\tau$ and $\widehat{N}_t$ commute with each other for all $\tau, t \in \Gamma$, and from condition (8.1) one easily derives that $\widehat{A}$ is M_τ–equivalent to the

[1] $|\cdot|$ denotes the quotient norm in the corresponding spaces (cf. p. 22).

(invertible) coset $\widehat{A}_\tau$ $(\tau \in \Gamma)$. Now [GK1, 5.1, Theorem 1.1] implies that $\widehat{A}$ is invertible in $\widehat{L}$ and, hence, A is a Φ-operator in $L_p(\Gamma, \rho)$. Analogously one shows that B is a Φ-operator.

The equalities (8.2) and (8.3) are verified in [GK1, 5.4]. ∎

6.9 The interpolation theorem

Let $A \in L(L_{p_1}(\Gamma, \rho_1)) \cap L(L_{p_2}(\Gamma, \rho_2))$.[1]

If

$$\frac{1}{p} = \frac{1-\theta}{p_1} + \frac{\theta}{p_2} \quad (0 \le \theta \le 1) \tag{9.1}$$

and

$$\rho^{\frac{1}{p}} = \rho_1^{\frac{1-\theta}{p_1}} \cdot \rho_2^{\frac{\theta}{p_2}} \tag{9.2}$$

then $A \in L(L_p(\Gamma, \rho))$ *and*

$$\|A\|_{L_p(\Gamma,\rho)} \le \|A\|^{1-\theta}_{L_{p_1}(\Gamma,\rho_1)} \cdot \|A\|^{\theta}_{L_{p_2}(\Gamma,\rho_2)} . \tag{9.3}$$

The proof of this theorem was obtained by E. M. Stein [1].

[1] Notice that all spaces under consideration are linear spaces over the field of complex numbers.

Chapter 7

General theorems on singular integral operators

In this chapter we formulate and prove theorems for operators of the form $A = aP_\Gamma + bQ_\Gamma$ or $B = P_\Gamma aI + Q_\Gamma bI$ under the simple assumption that the coefficients a and b belong to $L_\infty(\Gamma)$. It is demonstrated how these operators change if the curve is changed in a special way. This yields necessary conditions for the operators A and B to be Φ–, Φ_+– or Φ_-–operators. Further we show that, under certain natural conditions, at least one of the numbers $\dim \ker A$ or $\dim \operatorname{coker} A$ ($\dim \ker B$ or $\dim \operatorname{coker} B$) is equal to zero. In Section 7.2 the computation of the quotient norm of the operators A and B is reduced to the case where Γ is a circle. In the third section we consider the principle of the separation of singularities in the coefficients. In the concluding sections we generalize some theorems obtained in Chapter 3 of [GK 1] for the case of continuous coefficients.

Throughout this chapter we consider singular integral operators on the space $L_p(\Gamma, \rho)$ under the usual restrictions

$$\rho(t) := \prod_{k=1}^{n} |t - t_k|^{\beta_k} ; \quad t_1, t_2, \dots, t_n \in \Gamma ;$$

$$1 < p < \infty; \ -1 < \beta_k < p - 1 \quad (k = 1, 2, \dots, n) .$$

7.1 Change of the curve

In this section we state two theorems on the connections between singular integral operators on different curves.

Let Γ and $\widetilde{\Gamma}$ be nonsimple curves with $\Gamma \subset \widetilde{\Gamma}$. The space $L_p(\widetilde{\Gamma}, \rho)$ can be identified with the direct sum of the spaces $L_p(\Gamma, \rho)$ and $L_p(\widetilde{\Gamma} \backslash \Gamma, \rho)$. By R_1 we denote the projection operator from the space $L_p(\widetilde{\Gamma},\rho)$ onto $L_p(\Gamma,\rho)$ parallel to $L_p(\widetilde{\Gamma}\backslash\Gamma,\rho)$ [1], and by R_2 the complementary projection.

Theorem 7.1. *Let* $\widetilde{a},\widetilde{b}\in L_\infty(\widetilde{\Gamma})$ *and* $a := \widetilde{a}|\Gamma$, $b := \widetilde{b}|\Gamma$. *If*

$$\widetilde{a}(t) = \widetilde{b}(t) = 1 \quad (t\in\widetilde{\Gamma}\backslash\Gamma) \tag{1.1}$$

then the operator $A = aP_\Gamma + bQ_\Gamma$ *is the restriction of the operator* $\widetilde{A} = \widetilde{a}P_{\widetilde{\Gamma}} + \widetilde{b}Q_{\widetilde{\Gamma}}$ *to* $L_p(\Gamma,\rho)$,

$$A = \widetilde{A}|L_p(\Gamma,\rho)\,. \tag{1.2}$$

Also the equality

$$\widetilde{A} = (I + R_1\widetilde{A}R_2)(AR_1 + R_2) \tag{1.3}$$

holds, in which the operator $I + R_1\widetilde{A}R_2$ *is invertible.*

Proof. Because

$$R_2\widetilde{A} = R_2\widetilde{a}P_{\widetilde{\Gamma}} + R_2\widetilde{b}Q_{\widetilde{\Gamma}} = R_2P_{\widetilde{\Gamma}} + R_2Q_{\widetilde{\Gamma}} = R_2$$

one has

$$\widetilde{A} = (R_1 + R_2)\widetilde{A} = R_1\widetilde{A} + R_2\,. \tag{1.4}$$

This in particular implies that $R_1\widetilde{A}R_1 = \widetilde{A}R_1 = AR_1$ and, thus, the space $L_p(\Gamma,\rho)$ is invariant for A , and (1.2) holds. Further, (1.4) shows that

$$\widetilde{A} = R_1\widetilde{A}R_1 + R_1\widetilde{A}R_2 + R_2 = (I + R_1\widetilde{A}R_2)(R_1\widetilde{A}R_1 + R_2)\,,$$

and so (1.3) is verified. Obviously,

$$(I + R_1\widetilde{A}R_2)^{-1} = I - R_1\widetilde{A}R_2\,.$$

■

Let us mention three important consequences of this theorem.

Corollary 1.1. *Let the condition (1.1) be satisfied. Then the operator* A *is normally solvable in* $L_p(\Gamma,\rho)$ *if and only if the operator* $\widetilde{A}$ *is normally solvable in* $L_p(\widetilde{\Gamma},\rho)$. *Moreover*

$$\ker A = \ker \widetilde{A}|\Gamma \quad \text{and} \quad \operatorname{coker} A = \operatorname{coker} \widetilde{A}|\Gamma\,.^2 \tag{1.5}$$

This corollary follows at once from (1.2) and (1.4).

---

[1]More precisely, ρ has to be replaced by $\rho|\Gamma$ or $\rho|\widetilde{\Gamma}\backslash\Gamma$,where $a|\Gamma$ denotes the restriction of the function a to the curve Γ .

[2]For a subspace X of functions defined on $\widetilde{\Gamma}$, $X|\Gamma$ denotes the subspace at all restrictions to Γ .

Corollary 1.2. *Assume that (1.1) is satisfied. Then the operator A admits a regularization from one side if and only if the operator $\widetilde{A}$ is regularizable from the same side. If the operator $\widetilde{M}$ is a one-sided regularization of $\widetilde{A}$ then the operator $M := R_1\widetilde{M}|L_p(\Gamma,\rho)$ is a regularization of A from the same side.*

The first assertion is immediate from equation (1.3). If the operator $\widetilde{M}$ is a left-sided regularizator of $\widetilde{A}$ then the second assertion is evident. On the other hand, if $\widetilde{M}$ is a right-sided regularization of $\widetilde{A}$ then the operator $\widetilde{A}\widetilde{M}-\widetilde{I}$ is compact. But then the operator $R_2\widetilde{A}\widetilde{M}R_1 = R_2\widetilde{M}R_1$ is compact and, consequently, the operator

$$\widetilde{M_1} := R_1\widetilde{M}R_1 + R_1\widetilde{M}R_2 + R_2\widetilde{M}R_2$$

is also a right-sided regularization of $\widetilde{A}$. This gives the second assertion of the corollary in the right-sided case. ■

Corollary 1.3. *For the invertibility from one side of the operator A it is necessary and sufficient that the operator $\widetilde{A}$ is invertible from the same side. If $\widetilde{A}^{-1}$ is a one-sided inverse of $\widetilde{A}$ then the operator $R_1\widetilde{A}^{-1}|L_p(\Gamma,\rho)$ is an inverse of A from the same side.*

This corollary can be derived in the same way as the preceding one.

A particular case of Theorem 1.1 and of its corollaries had already been used in the proof of Theorem 11.1 in Chapter 3 of [GK 1]. There this method was employed to reduce the problem of inversion of a singular integral operator along a composed closed curve to the problem of inversion of operators along a simple closed curve.

Let us examine one more application of Theorem 1.1 and its corollaries. Let the curve Γ be an interval , say $[\alpha,\beta]$, of the real axis; $a\in C(\alpha,\beta)$, $a(\alpha)=a(\beta)=1$ and $A = aP_\Gamma + Q_\Gamma$. Denote by $\widetilde{\Gamma}$ a simple closed curve containing the interval $[\alpha,\beta]$. Then the function $\widetilde{a}$ defined by $\widetilde{a}|\Gamma := a$ and $\widetilde{a}|\widetilde{\Gamma}\backslash\Gamma := 1$ is continuous on the curve $\widetilde{\Gamma}$. Theorem 1.1 and Theorem 7.1 of Chapter 3, [GK 1] imply that the operator A is one-sided invertible in the space $L_p(\Gamma,\rho)$ if and only if $a\in GC(\alpha,\beta)$.[1] Therefore, if $a\in GC(\alpha,\beta)$ and $\kappa = (1/2\pi)[\arg a(t)]_\Gamma$, then A is invertible, invertible only from the right, or invertible only from the left depending on whether κ is equal to zero, negative or positive, respectively. Moreover, $\dim\ker A = -\kappa$ if $\kappa<0$ and $\dim\operatorname{coker} A = \kappa$ if $\kappa>0$. If $a\in GC(\alpha,\beta)$ then the function $\widetilde{a}$ admits a generalized factorization [2] $\widetilde{a} = \widetilde{a}_- t^\kappa \widetilde{a}_+$ with respect to $\widetilde{\Gamma}$. From Corollary 1.3 one concludes further that a one-sided inverse of the operator A is given by $A^{-1} = (a_+^{-1}(t)t^{-\kappa}P_\Gamma + a_-(t)Q_\Gamma)\cdot a_-^{-1}(t)I$ where $a_\pm$ stand for the restrictions of the functions $\widetilde{a}_\pm$ to the interval $[\alpha,\beta]$.

[1]For an algebra $\mathcal{A}$, $G\mathcal{A}$ denotes the class of all invertible elements of $\mathcal{A}$.

[2]Here we assume that $0\notin[\alpha,\beta]$. Thus, $\widetilde{\Gamma}$ can be chosen in such a way that $0\in F^+_{\widetilde{\Gamma}}$.

Theorem 1.2. *Let* $\Gamma_1, \ldots, \Gamma_n$ *be non-simple curves with pairwise empty intersections, and let* $a, b \in L_\infty(\Gamma)$ *with* $\Gamma := \Gamma_1 \cup \ldots \cup \Gamma_n$. *Then the operator* $A = aP_\Gamma + bQ_\Gamma$ *is a* Φ*-operator if and only if each of the operators* $A_j = a_j P_{\Gamma_j} + b_j Q_{\Gamma_j}$ *with* $a_j = a|\Gamma_j$ *and* $b_j = b|\Gamma_j$ *is a* Φ*-operator.*

The operator A *is a* Φ_+*- (*Φ_-*-) operator if and only if all operators* A_j *are* Φ_+*- (*Φ_-*-) or* Φ*-operators with at least one of them a* Φ_+*- (*Φ_-*-) operator. In every case,*

$$\operatorname{Ind}(aP_\Gamma + bQ_\Gamma) = \sum_{j=1}^{n} \operatorname{Ind}(a_j P_\Gamma + b_j Q_\Gamma) .$$

Proof. The space $L_p(\Gamma, \rho)$ can be naturally identified with the direct sum $L_p(\Gamma_1, \rho) \dot{+} L_p(\Gamma_2, \rho) \dot{+} \ldots \dot{+} L_p(\Gamma_n, \rho)$. We let R_k $(k = 1, 2, \ldots, n)$ refer to the projection operator projecting the space $L_p(\Gamma, \rho)$ onto $L_p(\Gamma_k, \rho)$ parallel to the direct sum of the other spaces $L_p(\Gamma_j, \rho)$ $(j = 1, \ldots, n,\ j \neq k)$. Because of $R_1 \dot{+} R_2 \dot{+} \ldots \dot{+} R_n = I$ we can write the operator A in the form

$$A = \sum_{j,k=1}^{n} R_j A R_k .$$

Clearly,

$$R_j A R_j = A_j R_j \quad \text{and} \quad R_j A R_k = \frac{1}{2}(a_j - b_j) R_j S_{\Gamma_k} \quad (j \neq k) .$$

Since the curves Γ_j $(j = 1, \ldots, n)$ are assumed to be pairwise disjoint, the operators $R_j A R_k$ are compact for all $j \neq k$. Hence,

$$A = \sum_{j=1}^{n} A_j R_j + T \tag{1.6}$$

where T is compact. But obviously the operator $A_1 R_1 \ldots + A_j R_j$ is, in principle, the direct sum of the operators $A_j R_j$ acting on $L_p(\Gamma_j, \rho)$. This fact yields the assertion of the theorem. ■

Corollary 1.4. *The operator* A *admits a left-sided (right-sided) regularization if and only if all operators* A_j *are left-sided (right-sided) regularizable.*

This corollary is a result of identity (1.6) which was obtained in the course of the proof of Theorem 1.2.

7.2 The quotient norm of singular integral operators

Let Γ be a composed curve and S stand for the operator of singular integration along Γ . The norm $\|S\|_{L_p(\Gamma)}$ of the operator S thought of as acting on the space $L_p(\Gamma)$ depends on p and Γ. Indeed, it will be shown in Section 9.9 [1] that $\|S\|_{p,\mathbf{T}} = \cot(\pi/2p)$ if $p = 2^n$. Hence, the norm depends on p. Its dependence on the curve Γ will now be explained. Consider Γ to be the ellipse defined by $t = \alpha \sin\theta + i\cos\theta$ $(\alpha \neq 1,\ 0 \leq \theta < 2\pi)$ and set $f(t) := t + t^{-1} (t \in \Gamma)$. Then $(Sf)(t) = t - t^{-1}$ and

$$\|Sf\|^2_{L_2(\Gamma)} - \|f\|^2_{L_2(\Gamma)} = 4\int_0^{2\pi} \frac{\alpha^2 \sin^2\theta - \cos^2\theta}{\alpha^2\sin^2\theta + \cos^2\theta}\, d\theta = 8\pi\frac{\alpha - 1}{\alpha + 1}.$$

But if $\alpha > 1$ then $\|Sf\|_{L_2(\Gamma)} > \|f\|_{L_2(\Gamma)}$ and, hence, $\|S\|_{L_2(\Gamma)} > 1$. Since $\|S\|_{L_2(\mathbf{T})} = 1$, the norm $\|S\|_{L_2(\Gamma)}$ depends on the curve Γ.

In contrast to this, the general theorems established in this section will yield that the quotient norm of S is independent of Γ.

Let us recall that the quotient norm is defined by

$$|A|_{L_p(\Gamma,\rho)} = \inf_{T \in \mathcal{T}(L_p(\Gamma,\rho))} \|A + T\|_{L_p(\Gamma,\rho)} \tag{2.1}$$

for $A \in L(L_p(\Gamma,\rho))$ where $\mathcal{T}(L_p(\Gamma,\rho))$ denotes the set of all compact operators in $L(L_p(\Gamma,\rho))$.

Lemma 2.1. *Let Γ be a simple closed curve, $\mathbf{T}$ the unit circle, and let $\alpha : \Gamma \to \mathbf{T}$ be a bijective mapping belonging to $H_\mu(\Gamma)$ $(0 < \mu < 1)$ [2] whose derivative $\alpha'(z)$ exists [3], and which does not vanish on Γ. Further let $\rho(t) := \prod_{k=1}^n |t - t_k|^{\beta_k}$ $(t, t_k \in \Gamma)$, $\rho_0(z) := \prod_{k=1}^n |z - z_k|^{\beta_k}$ $(z, z_k \in \mathbf{T};\ t_k = \alpha(z_k))$, $A = aI + bS_\Gamma$ and $A_0 = a_0 I + b_0 S_{\mathbf{T}}$ where $a_0(z) := a(\alpha(z)), b_0(z) = b(\alpha(z))$. Then*

$$|A|_{L_p(\Gamma,\rho)} = |A_0|_{L_p(\mathbf{T},\rho_0)}. \tag{2.2}$$

Proof. Denote by B the linear bounded operator acting from $L_p(\Gamma,\rho)$ into $L_p(\mathbf{T},\rho_0)$ by

$$(B\varphi)(z) := f(z)\varphi(\alpha(z))$$

where $f(z) := |\alpha'(z)|^{1/p}\rho^{1/p}(\alpha(z))\rho_0^{-1/p}(z)$. One easily gets that

$$\|B\varphi\|_{L_p(\mathbf{T},\rho_0)} = \|\varphi\|_{L_p(\Gamma,\rho)}$$

and hence that B is an invertible operator.

---

[1]In Lemma 2.1 of [GK 1, 1.2] we have already seen that $\|S\|_{p,\mathbf{T}} \leq \cot(\pi/2p)$.

[2] $\mathbf{H}_\mu(\Gamma)$ denotes the class of all functions in Γ satisfying a Hölder condition with exponent μ.

[3]See [GK 1, 1.2.].

Now let $\varphi \in L_p(\mathbf{T},\rho_0)$. Then, clearly, $BS_\Gamma B^{-1} - S_\mathbf{T} = T_1 + T_2$ with

$$(T_1\varphi)(z) := \frac{1}{\pi i} f(z) \int\limits_{\mathbf{T}} \left(\frac{\alpha'(\zeta)}{\alpha(\zeta)-\alpha(z)} - \frac{1}{\zeta - z} \right) f^{-1}(\zeta)\varphi(\zeta)\, d\zeta$$

and

$$(T_2\varphi)(z) := \frac{1}{\pi i} \int\limits_{\mathbf{T}} \frac{f(z)-f(\zeta)}{f(\zeta)(\zeta - z)} \varphi(\zeta)\, d\zeta\ .$$

Lemma 2.2 of Chapter 1 [GK 1] guarantees that the kernel function of the operator T_1 is weakly singular. So, by Theorem 4.2 in Chapter 1, the operator T_1 is compact on $L_p(\mathbf{T},\rho_0)$. In order to see that T_2 is also compact we represent it in the form $T_2 = (fS_\mathbf{T} - S_\mathbf{T} f)f^{-1}I$. Since $\alpha'(z)$ is continuous on $\mathbf{T}$ and $\alpha'(z) \neq 0$, the function

$$f(z) = |\alpha'(z)|^p \prod_{k=1}^{n} \left| \frac{\alpha(z)-\alpha(z_k)}{z - z_k} \right|^{\beta_k/p}$$

is continuous on $\mathbf{T}$ as well, and Theorem 4.3, Chapter 1 [GK 1] ensures the compactness of T_2 . Hence,

$$B(aI + bS_\Gamma)B^{-1} = a_0 I + b_0 S_\mathbf{T} + T \tag{2.3}$$

where $T \in \mathcal{T}(L_p(\mathbf{T},\rho_0))$. Taking into account that $\|B\| = \|B^{-1}\| = 1$ the desired equality (2.2) follows immediately from (2.3). ■

Now let Γ stand for a simple non–closed arc and $\alpha : [0,1] \to \Gamma$ for a mapping possessing a derivate $\alpha'(z)$ which belongs to $H_\mu(0,1)$ $(0 < \mu < 1)$ and does not vanish on [0,1]. Such a mapping exists since Γ satisfies a Lyapunov condition.

If the weight functions $\rho(t)$ and $\rho_0(t)$ are connected via

$$\rho(t) := \prod_{k=1}^{n} |t - t_k|^{\beta_k} \quad (t, t_k \in \Gamma)$$

$$\rho_0(z) := \prod_{k=1}^{n} |z - z_k|^{\beta_k} \quad (z, z_k \in [0,1];\ t_k = \alpha(z_k))$$

then one can prove in analogy to Lemma 2.1 the following ■

Lemma 2.2. *Let the curve* Γ *be as above,* $a, b \in L_\infty(\Gamma)$, $a_0(z) := a(\alpha(z))$, $b_0(z) := b(\alpha(z))$, $\Gamma_0 := [0,1]$, $A = aI + bS_\Gamma$ *and* $A_0 := a_0 I + b_0 S_{\Gamma_0}$. *Then the equality*

$$|A|_{L_p(\Gamma,\rho)} = |A|_{L_p(\Gamma_0,\rho_0)} \tag{2.4}$$

holds.

Let us now consider the general case where Γ is a curve consisting of simple closed curves $\Gamma_1, \Gamma_2, \ldots, \Gamma_n$ and of simple non–closed arcs $\Gamma_{n+1}, \ldots, \Gamma_m$. Put $\rho_j(t) :=$

$\prod_{k_j=1}^{n_j}|t-t_{k_j}|^{\beta_{k_j}}$ $(t_{k_j}\in\Gamma_j)$; $\alpha_j:\Gamma'\to\Gamma_j$ $(j=1,2,\ldots,m)$ where Γ' stands for $\mathbf{T}$ if $j=1,\ldots,n$ and for $[0,1]$ if $j=n+1,\ldots,m$; $\rho_{j0}(z)=\prod_{k_j-1}^{n_j}|z-z_{k_j}|^{\beta_{k_j}}$ $(z_{k_j}\in\Gamma'$, $t_{k_j}=\alpha(z_{k_j}))$, and let $\rho(t)$ $(t\in\Gamma)$ be the weight function defined by $\rho(t):=\rho_j(t))$ if $t\in\Gamma_j$.

Theorem 2.1. *Let* $a,b\in L_\infty(\Gamma)$, $A=aI+bS_\Gamma$, $A_j=a_jI+b_jS_{\Gamma_j}$ *with* $a_j(z)=a_{j0}(\alpha_j(z)), b_j(z)=b_{j0}(\alpha_j(z))$ *and* $a_{j0}=a|\Gamma_j, b_{j0}=b|\Gamma_j$. *Then the equality*

$$|A|_{L_p(\Gamma,\rho)}=\max_{j=1,\ldots,m}|A_j|_{L_p(\Gamma',\rho_{j0})} \tag{2.5}$$

holds.

Proof. Write R_k for the projection operator projecting the space $L_p(\Gamma,\rho)$ onto $L_p(\Gamma_k,\rho_k)$ parallel to the direct sum of the remaining spaces $L_p(\Gamma_j,\rho_j)$ $(j\neq k, j=1,\ldots,n)$ (see Sect. 7.1). If T is an arbitrary operator in $\mathcal{T}(L_p(\Gamma,\rho))$ and $C:=A+T$, then the estimation

$$\|C\|_{L_p(\Gamma,\rho)}\geq\|R_jCR_j\|_{L_p(\Gamma,\rho)}=\|A_j+R_jTR_j\|_{L_p(\Gamma_j,\rho_j)}\geq|A_j|_{L_p(\Gamma_j,\rho_j)}$$

holds. Consequently

$$|A|_{L_p(\Gamma,\rho)}\geq\max_{j=1,\ldots,m}|A_j|_{L_p(\Gamma',\rho_{j0})}\,. \tag{2.6}$$

For the reverse direction notice that, by Lemma 2.1 and 2.2, given $j=1,\ldots,m$ and $\varepsilon>0$ there is an operator $T_j\in L_p(\Gamma_j,\rho_j)$ such that

$$|A_j|_{L_p(\Gamma',\rho_{j0})}+\varepsilon>\|A_j+T_j\|_{L_p(\Gamma_j,\rho_j)}\,.$$

Set $T=\sum_{j=1}^{n}R_jT_jR_j$. Without much ado it can be seen that

$$\|\sum_{j=1}^{n}R_j(A+T)R_j\|_{L_p(\Gamma,\rho)}=\max_{j=1,\ldots,m}\|A_j+T_j\|_{L_p(\Gamma_j,\rho_j)}$$

and, thus,

$$\max_{j=1,\ldots,m}|A_j|_{L_p(\Gamma',\rho_{j0})}+\varepsilon>\|\sum_{j=1}^{n}R_j(A+T)R_j\|_{L_p(\Gamma,\rho)}\,. \tag{2.7}$$

Due to the compactness of the operator R_jAR_k $(j\neq k)$, we have

$$\sum_{j=1}^{n}R_j(A+T)R_j=A+T'$$

for some $T'\in\mathcal{T}(L_p(\Gamma,\rho))$. This in combination with (2.7) yields

$$\max_{j=1,\ldots,m}|A|_{L_p(\Gamma',\rho)}\geq|A|_{L_p(\Gamma,\rho_{j0})}\,. \tag{2.8}$$

The equality (2.5) follows from (2.6) and (2.8). ∎

7.3 The principle of the separation of singularities

Throughout this section we assume Γ is a non–simple closed curve. Remember that this implies that $P_\Gamma^2 = P_\Gamma$.

Let a be a function in $L_\infty(\Gamma)$. The smallest closed subset of the curve Γ on the complement of which the function a is continuous will be called the *singularity support* of the function a. We denote this set by $\Delta(a)$.

The main result in this section is

Theorem 3.1. *Suppose the singularity support $\Delta(a)$ of the function $a \in GL_\infty(\Gamma)$ to consist of n pairwise disjoint closed arcs $\gamma_1, \dots, \gamma_n$, and let $a_j \in GL_\infty(\Gamma)$ $(j = 1, \dots, n)$ be functions possessing the following properties:*

α) the function a_j is continuous on the curve $\Gamma \backslash \gamma_j$ $(j = 1, \dots, n)$;

β) $a_j(t) = a(t)$ if $t \in \gamma_j$.

Then the operator $aP_\Gamma + Q_\Gamma$ is a Φ–operator if and only if each of the operators $a_jP_\Gamma + Q_\Gamma$ $(j = 1, \dots, n)$ is a Φ–operator, and the operator $aP_\Gamma + Q_\Gamma$ is a Φ_+- $(\Phi_-$–) operator if and only if all operators $a_jP_\Gamma + Q_\Gamma$ $(j = 1, \dots, n)$ are Φ_+- $(\Phi_-$-) or Φ–operators with at least one of them a Φ_+- $(\Phi_-$–) operator. In each case,

$$\mathrm{Ind}\,(aP_\Gamma + Q_\Gamma) = \sum_{j=1}^{n} \mathrm{Ind}\,(a_jP_\Gamma + Q_\Gamma) - \mathrm{ind}\, a_0 ,$$

where a_0 is the function in $GC(\Gamma)$ defined by

$$a_0 = aa_1^{-1}a_2^{-1} \dots a_n^{-1} .$$

Before proving this theorem we formulate and prove two technical lemmas.

Lemma 3.1. *Let $a_1, a_2, \dots, a_n$ be functions in $L_\infty(\Gamma)$ with pairwise disjoint singularity supports. Then the operator*

$$P_\Gamma \widetilde{a} P_\Gamma - P_\Gamma a_1 P_\Gamma P_\Gamma a_2 P_\Gamma \dots P_\Gamma a_n P_\Gamma \quad (\widetilde{a} := a_1a_2 \dots a_n)$$

is compact on $L_p(\Gamma, \rho)$. ■

Proof. Clearly, one can assume $n = 2$ without loss of generality. In case one of the functions a_1, a_2 is continuous, the assertion of the lemma is a consequence of Proposition 3.2 of Chapter 6 (see also Theorem 4.3 of Chapter 1, [GK 1]), which gives the compactness of the operator $bP_\Gamma - P_\Gamma bI$ for arbitrary functions $b \in C(\Gamma)$.

Now consider the general case. In this situation the set $\triangle(a_1)$ lies in the complement of the closed set $\triangle(a_2)$. This complement consists of finitely or countably many pairwise disjoint open arcs covering the closed set $\triangle(a_1)$. Pick a finite subcovering $\{\delta_j\}_1^r$ of this covering, and choose open arcs γ_j and $\triangle_j$ such that

$$\triangle(a_1) \cap \delta_j \subset \gamma_j, \quad \overline{\gamma_j} \subset \triangle_j, \quad \overline{\triangle}_j \subset \delta_j \quad (j = 1, 2, \ldots, r),$$

which is always possible.

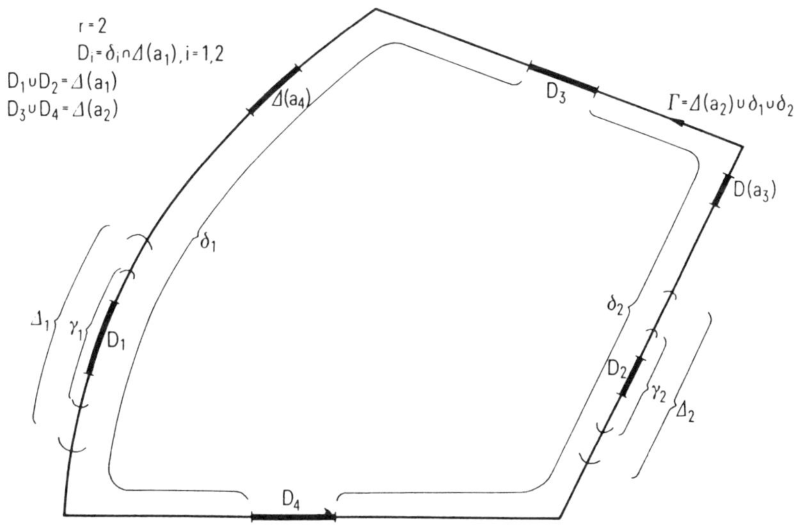

Figure 7.1

Put

$$\gamma := \bigcup_{j=1}^{r} \gamma_j \; (\supset \triangle(a_1)) \quad \text{and} \quad \triangle := \bigcup_{j=1}^{r} \triangle_j \, .$$

Then, obviously, there exists a function $b \in C(\Gamma)$ which coincides with the function a_2 on the set γ and which differs from a_2 at each point of the boundary of $\overline{\triangle}$.

Consider two functions $c_1(t)$ and $c_2(t)$ $(t \in \Gamma)$, the first of which satisfies the conditions

$$c_1 \in C(\Gamma), \; c_1(t) = a_2(t) - b(t) \quad (t \in \triangle), \; c_1(t) \neq 0 \quad (t \in \Gamma \backslash \triangle),$$

while the second one is defined by

$$c_2(t) := \begin{cases} 1 & \text{if} \quad t \in \triangle \\ [a_2(t) - b(t)]/c_1(t) & \text{if} \quad t \in \Gamma \backslash \triangle \, . \end{cases}$$

Due to the special choice of the functions c_1 and c_2 we have $a_2 - b = c_1c_2$ and, hence, $a_1a_2 = a_1c_1c_2 + a_1b$. Since the function c_1 vanishes on $\Delta(a_1)$, the function a_1c_1 is continuous on Γ and so

$$P_\Gamma a_1c_1c_2P_\Gamma = P_\Gamma a_1c_1P_\Gamma c_2P_\Gamma + T_1 \tag{3.1}$$

for some compact operator T_1. Taking into account the continuity of c_1 we further obtain

$$P_\Gamma a_1c_1P_\Gamma = P_\Gamma a_1P_\Gamma c_1P_\Gamma + T_2 \quad \text{and} \quad P_\Gamma c_1P_\Gamma c_2P_\Gamma = P_\Gamma c_1c_2P_\Gamma + T_3 \tag{3.2}$$

with compact operators T_2 and T_3. The identities (3.1), (3.2) and

$$P_\Gamma a_1bP_\Gamma = P_\Gamma a_1P_\Gamma bP_\Gamma + T_4$$

imply that

$$\begin{aligned} P_\Gamma\tilde{a}P_\Gamma &= P_\Gamma a_1c_1c_2P_\Gamma + P_\Gamma a_1bP_\Gamma \\ &= P_\Gamma a_1P_\Gamma(P_\Gamma c_1c_2P_\Gamma + P_\Gamma bP_\Gamma) + T_5 \end{aligned}$$

where T_4 and T_5 stand for certain compact operators. Thus

$$P_\Gamma\tilde{a}P_\Gamma = P_\Gamma a_1P_\Gamma P_\Gamma a_2P_\Gamma + T_5 \,.$$

■

Lemma 3.2. *If the singularity supports of the functions* a_1 *and* $a_2 \in L_\infty(\Gamma)$ *are disjoint then the operator* $P_\Gamma a_1P_\Gamma a_2P_\Gamma - P_\Gamma a_2P_\Gamma a_1P_\Gamma$ *is compact.*

This is an immediate consequence of the preceding proposition.

Proof of Theorem 3.1 Define the function a_0 by

$$a_0 := aa_1^{-1}a_2^{-1}\dots a_n^{-1}\,.$$

Obviously, $a_0 \in C(\Gamma)$. Setting

$$A := P_\Gamma aP_\Gamma + Q_\Gamma \quad \text{and} \quad A_j := P_\Gamma a_jP_\Gamma + Q_\Gamma \quad (j = 0, 1, \dots, n)$$

we have, by Proposition 3.1,

$$A = A_0A_1\dots A_n + T \tag{3.3}$$

where T is compact. This equality in combination with the Lemmas 3.1 and 3.2 and Proposition 7.6 of Chapter 6 (see also Theorem 6.1 of Chapter 4 [GK 1]) yields that Theorem 3.1 is valid if the operator $aP_\Gamma + Q_\Gamma$ is replaced by A and the operators $a_jP_\Gamma + Q_\Gamma$ are replaced by A_j. Taking into account the identities

$$aP_\Gamma + Q_\Gamma = A(I + Q_\Gamma aP_\Gamma) \quad \text{and} \quad a_jP_\Gamma + Q_\Gamma = A_j(I + Q_\Gamma a_jP_\Gamma)\,, \tag{3.4}$$

and the fact that the operators $I + Q_\Gamma a P_\Gamma$ and $I + Q_\Gamma a_j P_\Gamma$ are invertible with inverses given by

$$(I + Q_\Gamma a P_\Gamma)^{-1} = I - Q_\Gamma a P_\Gamma, \quad (I + Q_\Gamma a_j P_\Gamma)^{-1} = I - Q_\Gamma a_j P_\Gamma \; ; \tag{3.5}$$

the theorem follows. ■

Corollary 3.1. *Under the assumptions of Theorem 3.1, the operator* $aP_\Gamma + Q_\Gamma$ *admits a left–, right– or two–sided regularization, respectively.*

This corollary is a consequence of Proposition 3.2 and the equality obtained in the course of the proof of Theorem 3.1, namely,

$$\begin{aligned} aP_\Gamma + Q_\Gamma &= (a_0 P_\Gamma + Q_\Gamma)(I - Q_\Gamma a_0 P_\Gamma)(a_1 P_\Gamma + Q_\Gamma)(I - Q_\Gamma a_1 P_\Gamma)\ldots \\ &\quad \ldots (a_n P_\Gamma + Q_\Gamma)(I - Q_\Gamma a_n P_\Gamma)(I + Q_\Gamma a P_\Gamma) + T \, , \end{aligned} \tag{3.6}$$

where the operator T is compact.

7.4 A necessary condition

In the present section we prove the following

Theorem 4.1. *Let* Γ *be a non–simple curve and* $a \in L_\infty(\Gamma)$. *If the operator* $A = aP_\Gamma + bQ_\Gamma$ *is a* Φ–, Φ_+– *or* Φ_-*–operator in* $L_p(\Gamma, \rho)$ *then*

$$\operatorname*{ess\,inf}_{t \in \Gamma} |a(t)| > 0 \, , \quad \operatorname*{ess\,inf}_{t \in \Gamma} |b(t)| > 0 \, . \tag{4.1}$$

An analogous assertion holds for the operator $B = P_\Gamma aI + Q_\Gamma bI$ *in place of* A .

In preparation of the proof we need

Lemma 4.1. *Let* Γ *be a non–simple closed curve which bounds the open set* F_Γ^+ , *and let* $c \in L_\infty(\Gamma)$ *be a function such that, for certain subsets* γ_1 *and* γ_2 *of* Γ *of positive measure,*

$$c(t) = 0 \quad if \quad t \in \gamma_1 \quad and \quad c(t) \neq 0 \quad if \quad t \in \gamma_2 \, .$$

Assume further the measure of the intersection of γ_2 *with the boundary of each connected component of* F_Γ^+ *to be positive. Then for* A *equal to either* $cP_\Gamma + Q_\Gamma$ *or* $P_\Gamma + cQ_\Gamma$,

$$\dim \ker A = \dim \ker A^* = 0$$

Proof. Let $F_1, F_2, \ldots, F_m$ be the set of all (not necessarily different) connected components of F_Γ^+ and $\Gamma_1, \Gamma_2, \ldots, \Gamma_m$ their boundaries. They are assumed to be numerated in such a way that the following two properties are satisfied: mes $(\gamma_1 \cap \Gamma_1) > 0$, and the points in F_j and F_{j+1} can be joined by a straight line which intersects the curve Γ only at the boundaries Γ_j and Γ_{j+1} . Obviously, for each vector $\varphi_0 \in \ker(cP_\Gamma + Q_\Gamma)$ and for all $t \in \gamma_1 \cap \Gamma_1$ one has $(Q_\Gamma \varphi_0)(t) = 0$. So, the Lusin–Privalov theorem (see PRIVALOV [1], p. 232) implies that $(Q_\Gamma \varphi_0)(t) = 0$ for $t \in \Gamma_1$. Since Γ_2 is contained in the boundary of the region in which the function $Q_\Gamma \varphi_0$ vanishes, we obtain $\varphi_0(t) = 0$ for $t \in \Gamma_2$ by repeating the previous arguments. Iterating this procedure, we arrive at $\varphi_0 = 0$.

Analogously, one can show that $\dim \ker(P_\Gamma + cQ_\Gamma) = 0$. The operator $A^* = (cP_\Gamma + Q_\Gamma)^*$ can be written in the form $A^* = H_\Gamma(P_\Gamma + \overline{c}Q_\Gamma)H_\Gamma$ where $(H_\Gamma\varphi)(t) := \overline{[\exp(i\alpha(t))]\varphi(t)}$ with $\alpha(t)$ denoting the angle between the tangent of the curve Γ at the point t and the positive direction of the x–axis (see 6.3.3 in Chapter 6). Further we have $P_\Gamma + Q_\Gamma cI = (I + Q_\Gamma cP_\Gamma)(P_\Gamma + cQ_\Gamma)(I - P_\Gamma cQ_\Gamma)$ where the outer factors on the right-hand side are invertible: $(I + Q_\Gamma cP_\Gamma)^{-1} = (I - Q_\Gamma cP_\Gamma)$. Thus, by the above, $\dim \ker(cP_\Gamma + Q_\Gamma)^* = 0$. Analogously, $\dim \ker(P_\Gamma + cQ_\Gamma)^* = 0$. ■

Proof of Theorem 4.1 To start with we consider the case when Γ is a closed non–simple curve. Assume the operator to be a Φ– or $\Phi_\pm$–operator and that one of the conditions in (4.1) is violated, say $\operatorname{ess\,inf} |a(t)| = 0$ without loss of generality. Now define functions a_1 and b_1 by

$$a_1(t) := \begin{cases} a(t) & \text{if} \quad |a(t)| \geq \varepsilon \\ 0 & \text{if} \quad |a(t)| < \varepsilon \end{cases} \quad \text{and} \quad b_1(t) := \begin{cases} b(t) & \text{if} \quad |b(t)| \geq \varepsilon \\ \varepsilon & \text{if} \quad |b(t)| < \varepsilon \end{cases}$$

where ε is a sufficiently small number. Evidently, $|a(t) - a_1(t)| < \varepsilon$ and $|b(t) - b_1(t)| < 2\varepsilon$ for each $t \in \Gamma$.

Put $A_1 = a_1 P_\Gamma + b_1 Q_\Gamma$. The estimation

$$\|A - A_1\| < 2\varepsilon(\|P_\Gamma\| + \|Q_\Gamma\|)$$

shows that A_1 is a Φ– or $\Phi_\pm$–operator whenever ε is small enough.

Next set $c := a_1/b_1$ and $D := cP_\Gamma + Q_\Gamma$. Due to our assumption, the set $\gamma_1 = \{t \in \Gamma : c(t) = 0\}$ has a positive measure, and since D is a Φ– or $\Phi_\pm$–operator, the intersection of the set $\gamma_2 = \{t \in \Gamma : c(t) \neq 0\}$ and each of the curves Γ_j (see the proof of the preceding proposition for their definition) is of positive measure, too. So Lemma 4.1 implies that $\dim \ker D = \dim \operatorname{coker} D = 0$, from which the invertibility of the operator D follows.

Let φ_0 be the solution of the equation $D\varphi_0 = 1$. Because of $c(t) = 0$ if $t \in \gamma_1$, the equation $cP_\Gamma\varphi_0 = 1 - Q_\Gamma\varphi_0$ implies that $1 - (Q_\Gamma\varphi_0)(t) = 0$ on γ_1 , and the Lusin–Privalov theorem gives that $1 - (Q_\Gamma\varphi_0)(t) = 0$ on Γ_1 . Repeating the arguments from the proof of Lemma 4.1 we finally get $1 - (Q_\Gamma\varphi_0)(z) = 0$ for all $z \in F_\Gamma^-$. But this contradicts $(Q_\Gamma\varphi_0)(\infty) = 0$, which finishes the proof for the case of a closed curve.

If Γ is an arbitrary non-simple curve we complete Γ to a closed non-simple curve $\widetilde{\Gamma}$ and consider functions $\widetilde{a}(t)$ and $\widetilde{b}(t)$ $(t \in \widetilde{\Gamma})$ which coincide with a and b on Γ, respectively, and which are equal to one on $\widetilde{\Gamma}\backslash\Gamma$. From Theorem 1.1 we infer that $\widetilde{a}P_{\widetilde{\Gamma}} + \widetilde{b}Q_{\widetilde{\Gamma}}$ is a $\Phi-$ or $\Phi_{\pm}$-operator on $L_p(\widetilde{\Gamma}, \rho)$ [1], and the above considerations show that

$$\operatorname{ess}\inf_{t\in\widetilde{\Gamma}} |\widetilde{a}(t)| > 0 , \quad \operatorname{ess}\inf_{t\in\widetilde{\Gamma}} |\widetilde{b}(t)| > 0 ,$$

which implies (4.1). ■

7.5 Theorems on kernel and cokernel of singular integral operators

Let Γ be a non-simple curve, $a, b \in L_\infty(\Gamma)$, and let A be defined by $A := aP_\Gamma + bQ_\Gamma$. In this section we shall state some necessary conditions which guarantee that one of the numbers $\dim\ker A$ or $\dim\ker A^*$ is equal to zero.

Theorem 5.1. *Let $a, b \in L_\infty(\Gamma)$ and $A := aP_\Gamma + bQ_\Gamma$ where Γ is a simple closed curve. If neither of the funtions a or b vanishes identically and if*

$$\operatorname{mes}\{t \in \Gamma : a(t) = b(t) = 0\} = 0 \tag{5.1}$$

then at least one of the numbers $\dim\ker A$ and $\dim\ker A^*$ is equal to zero. An analogous assertion holds for the operator $P_\Gamma aI + Q_\Gamma bI$.

Proof. Let $\varphi_0 \in \ker A$ and $\widetilde{\psi} \in \ker A^*$, and write the operator A^* as $A^* = H_\Gamma(P_\Gamma b + Q_\Gamma a)H_\Gamma$ (see Proposition 3.3 of Chapter 6 or Section 1.7, [GK 1]). Then $(P_\Gamma b + Q_\Gamma a)\psi_0 = 0$ where ψ_0 abbreviates the function $H_\Gamma\widetilde{\psi}$. Applying the operators P_Γ and Q_Γ to both sides of the equation $(P_\Gamma b + Q_\Gamma a)\psi_0 = 0$ we obtain $P_\Gamma b\psi_0 = 0$ and $Q_\Gamma a\psi_0 = 0$, respectively. This implies $h_+ := a\psi_0 \in L_q^+(\Gamma, \rho^{q-1})$ and $h_- := b\psi_0 \in L_q^-(\Gamma, \rho^{q-1})$ where $p^{-1} + q^{-1} = 1$. Set $\varphi_+ := P_\Gamma\varphi_0$ and $\varphi_- := Q_\Gamma\varphi_0$. Multiplying both sides of the equation $a\varphi_+ = -b\varphi_-$ by ψ_0 we find $h_+\varphi_+ = -h_-\varphi_-$, from which it follows $h_+\varphi_+ = 0$ and $h_-\varphi_- = 0$, since $h_+\varphi_+ \in L_1^+(\Gamma), h_-\varphi_- \in \overset{\circ}{L_1^-}$ and $L_1^+(\Gamma) \cap \overset{\circ}{L_1^-}(\Gamma) = \{0\}$.

If the set of all points at which one of the functions φ_+ and φ_- vanishes has measure zero then $h_+ = h_- = 0$. But in this case we have $a\psi_0 = b\psi_0 = 0$, meaning that $(|a| + |b|)|\psi_0| = 0$, and so, due to condition (5.1), $\psi_0 = 0$. We conclude that $\dim\ker A^* = 0$.

On the other hand, if one of the functions φ_+ and φ_- vanishes on a set $\gamma(\subset\Gamma)$ of positive measure then, by the Lusin–Privalov theorem, this function is identically zero.

[1]The weight function is defined on $\widetilde{\Gamma}$ by the same expression as on Γ.

Now the equality $b\varphi_- = -a\varphi_+$ the other function must likewise vanish on a set of positive measure. Invoking the Lusin-Privalov theorem once more we see that this function is also identically zero. Thus, $\varphi_0 = \varphi_+ + \varphi_- = 0$, that is, $\dim \ker A = 0$. ■

The following examples illustrate that the conditions in Theorem 5.1 are essential.

1. If $A = P_\Gamma$ then $b = 0$ and $\dim \ker A = \dim \ker A^* = \infty$.

2. If $A = \chi I$ where χ denotes the characteristic function of a certain subarc of the curve Γ then $\dim \ker A = \dim \ker A^* = \infty$.

Notice further that, whenever condition (5.1) for the operator $A = aP_\Gamma + bQ_\Gamma$ is violated or one of the functions a and b is identically zero, $\dim \ker A = \infty$ or $\dim \ker A^* = \infty$. Indeed, if $a = 0$ then $L_p^+(\Gamma,\rho) \subset \ker A$, whereas $b = 0$ implies $\overset{\circ}{L_p^-}(\Gamma,\rho) \subset \ker A$. If, finally, condition (5.1) is violated then there exists a set γ of positive measure such that $a|\gamma = b|\gamma = 0$. Thus, any function the support of which is contained in γ belongs to the kernel of the operator A^* , that is $\dim \ker A^* = \infty$.

As a consequence of this remark and of Theorem 5.1 we obtain

Theorem 5.2. *Let Γ be a simple closed curve, $a, b \in L_\infty(\Gamma)$, and $A := aP_\Gamma + bQ_\Gamma$. If both numbers $\dim \ker A$ and $\dim \ker A^*$ are finite then one of them is equal to zero.*

If, in particular, Γ is a simple closed curve and A is a Φ–operator then Theorem 5.2 states that A is one-sided invertible. The more general theorem that if A is a Φ– or $\Phi_\pm$–operator then one of the numbers $\dim \ker A$ and $\dim \ker A^*$ must be zero, can be proved by means of Theorems 4.1 and 5.1. This result remains true even for arbitrary non–simple curves. For its derivation we need the following

Theorem 5.3. *Let Γ be an arbitrary non–simple curve, $a, b \in GL_\infty(\Gamma)$ [1], and $A := aP_\Gamma + bQ_\Gamma$. Then one of the numbers $\dim \ker A$ or $\dim \ker A^*$ is equal to zero.*

Proof. First of all we consider the case when Γ is a closed curve. Let $\varphi_0 \in \ker A$ and $\widetilde{\psi} \in \ker A^*$. Repeating the first part of the proof of Theorem 5.1 yields the equality $h_+\varphi_+ = 0$ where $h_+ := a\psi_0$, $\psi_0 := H_\Gamma\widetilde{\psi}$, and $\varphi_+ = P_\Gamma\varphi_0$. If $h_+ = 0$ then $\psi_0 = 0$ and thus, $\dim \ker A^* = 0$.

On the other hand, if $h_+ \neq 0$ then $\varphi_+(t)$ vanishes on a set $\gamma \subset \Gamma$ of positive measure. Obviously, the set F_Γ^+ splits into a finite number of connecting components, and we can pick one of them, F_1 , the boundary Γ_1 where the condition $\mathrm{mes}\,(\gamma \cap \Gamma_1) > 0$ is satisfied. By the Lusin–Privalov theorem we find $\varphi_+|\Gamma_1 = 0$. Set $\varphi_- := Q_\Gamma\varphi_0$. Since $a\varphi_+ = -b\varphi_-$ and $b \in GL_\infty(\Gamma)$, it follows that $\varphi_-|\Gamma_1 = 0$, $\varphi_+|\Gamma_1 = \varphi_-|\Gamma_1 = 0$. Now let $\tau_0 \in \Gamma_1$, and let τ be an arbitrary, that is fixed point of Γ . Further we choose a curve l joining the points τ_0 and τ and in turn running through the regions $F_2, \ldots, F_m$ each of

[1]See first footnote at p. 29.

which is a connecting component of one of the open sets F_Γ^+ or F_Γ^- . Now we can argue as follows: one of the functions φ_+ and φ_- admits a holomorphic continuation into F_2 and so, by the Lusin–Privalov theorem, it is identically zero in F_2 . Our assumption that $a, b \in GL_\infty(\Gamma)$ in combination with the equality $a\varphi_+ = -b\varphi_-$ yields $\varphi_+|\Gamma_2 = \varphi_-|\Gamma_2 = 0$, Γ_k the boundary of F_k . Applying the same arguments to the sets $F_3, \ldots . F_m$ we arrive at $\varphi_+|\Gamma_m = \varphi_-|\Gamma_m = 0$ and, thus, $\varphi_+(\tau) = \varphi_-(\tau) = 0$. Hence, $\varphi_+ = \varphi_- = 0$, and so $\varphi_0 = 0$ and $\dim \ker A = 0$.

To finish the proof let Γ be an arbitrary non-simple curve. We complete Γ to the closed curve $\widetilde{\Gamma}$ bounding a set $F_{\widetilde{\Gamma}}^+$ which consists of a finite number of connecting components. Further, let $\widetilde{a}$ and $\widetilde{b}$ be functions which coincide with a and b on Γ , respectively, and which are equal to one on $\widetilde{\Gamma}\backslash\Gamma$. The above arguments transferred to the operator $\widetilde{A} = \widetilde{a}P_{\widetilde{\Gamma}} + \widetilde{b}Q_{\widetilde{\Gamma}}$ give that one of the numbers $\dim \ker \widetilde{A}$ or $\dim \ker \widetilde{A}^*$ is equal to zero. Now it remains to invoke Corollary 1.1 which states that either $\dim \ker A = \dim \ker \widetilde{A}$ or $\dim \ker A^* = \dim \ker \widetilde{A}^*$. ∎

Combining Theorems 4.1 and 5.3 we get

Theorem 5.4. *Let a and b be bounded measurable functions on the non-simple curve Γ . If $A := aP_\Gamma + bQ_\Gamma$ is a $\Phi-$ or $\Phi_\pm$-operator then one of the numbers* $\dim \ker A$ *or* $\dim \ker A^*$ *is equal to zero.*

An analogous result holds for the operator $P_\Gamma aI + Q_\Gamma bI$ (see Theorem 6.1 and Corollary 6.1). If, in particular, the operator A admits a one-sided regularization then it is in fact one-sided invertible.

7.6 Two theorems on connections between singular integral operators

Besides the operator $aP_\Gamma + bQ_\Gamma$ one usually considers the operator $P_\Gamma aI + Q_\Gamma bI$. In the sequel we verify a simple theorem on connections between these two operators.

Theorem 6.1. *Let Γ be a non-simple closed curve and $a, b \in GL_\infty(\Gamma)$. Then the operators $A := aP_\Gamma + bQ_\Gamma$ and $B := P_\Gamma aI + Q_\Gamma bI$ are connected via the equality*

$$D_1(aP_\Gamma + bQ_\Gamma)D_2 = P_\Gamma aI + Q_\Gamma bI \tag{6.1}$$

with the invertible operators D_1 and D_2 being defined by

$$D_1 := (I + P_\Gamma ab^{-1}Q_\Gamma)b^{-1}I \quad \text{and} \quad D_2 := (I - Q_\Gamma ab^{-1}P_\Gamma)bI . \tag{6.2}$$

Proof. Evidently, the operator A can be represented in the form

$$A = b(ab^{-1}P_\Gamma + Q_\Gamma) = b(P_\Gamma ab^{-1}P_\Gamma + Q_\Gamma)(I + Q_\Gamma ab^{-1}P_\Gamma) , \tag{6.3}$$

and for the operator B we find analogously

$$B = (P_\Gamma ab^{-1}I + Q_\Gamma)bI = (P_\Gamma ab^{-1}Q_\Gamma + I)(P_\Gamma ab^{-1}P_\Gamma + Q_\Gamma)bI . \tag{6.4}$$

The operators bI, $I + Q_\Gamma ab^{-1}P_\Gamma$ and $I + P_\Gamma ab^{-1}Q_\Gamma$ are invertible with

$$(I + Q_\Gamma ab^{-1}P_\Gamma)^{-1} = I - Q_\Gamma ab^{-1}P_\Gamma$$

and

$$(I + P_\Gamma ab^{-1}Q_\Gamma)^{-1} = I - P_\Gamma ab^{-1}Q_\Gamma .$$

Equalities (6.3) and (6.4) imply that

$$\begin{aligned} P_\Gamma ab^{-1}P_\Gamma + Q_\Gamma &= b^{-1}(aP_\Gamma + bQ_\Gamma)(I - Q_\Gamma ab^{-1}P_\Gamma) \\ &= (I - P_\Gamma ab^{-1}Q_\Gamma)(P_\Gamma aI + Q_\Gamma bI)b^{-1}I \end{aligned}$$

whence (6.1) follows. This proves the theorem. ■

Corollary 6.1. *Let Γ be an arbitrary non-simple curve. For the operator $aP_\Gamma + bQ_\Gamma$ $(a, b \in L_\infty(\Gamma))$ to be a Φ-, Φ_+- or Φ_--operator on the space $L_p(\Gamma, \rho)$ it is necessary and sufficient that the operator $P_\Gamma aI + Q_\Gamma bI$ be so on $L_p(\Gamma, \rho)$ is so.*

The operator $aP_\Gamma + bQ_\Gamma$ admits a left (right) regularization on $L_p(\Gamma, \rho)$ if and only if $P_\Gamma aI + Q_\Gamma bI$ admits a left (right) regularization on $L_p(\Gamma, \rho)$. Finally,

$$\dim\ker(aP_\Gamma + bQ_\Gamma) = \dim\ker(P_\Gamma aI + Q_\Gamma bI)$$

and

$$\dim\operatorname{coker}(aP_\Gamma + bQ_\Gamma) = \dim\operatorname{coker}(P_\Gamma aI + Q_\Gamma bI) .$$

This is a simple conclusion of Theorems 6.1 and 1.1.

Theorem 6.2. *Let Γ be a non-simple closed curve and $a, b \in GL_\infty(\Gamma)$. Then the operators $A := aP_\Gamma + bQ_\Gamma$ and $B := a^{-1}P_\Gamma + b^{-1}Q_\Gamma$ acting on $L_p(\Gamma, \rho)$ and on its dual space $L_q(\Gamma, \rho^{1-q})(p^{-1} + q^{-1} = 1)$, respectively, are connected via the equality*

$$A^* = D_3BD_4 , \tag{6.5}$$

where

$$\begin{aligned} D_3 &:= H_\Gamma(I + P_\Gamma a^{-1}bQ_\Gamma)bI , \\ D_4 &:= (I - Q_\Gamma a^{-1}bP_\Gamma)aH_\Gamma , \end{aligned}$$

and H_Γ *refers to the invertible operator defined by* $(H_\Gamma\varphi)(t) := \overline{h_\Gamma(t)\varphi(t)}$ *(see Section 6.3 or 1.7, [GK 1]).*

Indeed, by Proportion 3.3 of Chapter 6 (Theorem 7.1 of Chapter 1, [GK 1]),

$$A^* = H_\Gamma(P_\Gamma bI + Q_\Gamma aI)H_\Gamma ,$$

which immediately implies equality (6.5).

As a consequence of (6.5) we remark

Corollary 6.2. *Let* Γ *be an arbitrary non-simple curve. Then the operator* $aP_\Gamma + bQ_\Gamma$ $(a, b \in GL_\infty(\Gamma))$ *is a* Φ-, Φ_+- *or* Φ_-*-operator on* $L_p(\Gamma, \rho)$ *if and only if the operator* $a^{-1}P_\Gamma + b^{-1}Q_\Gamma$ *is a* Φ-, Φ_+- *or* Φ_-*-operator on the dual space* $L_q(\Gamma, \rho^{1-q})$ $(p^{-1} + q^{-1} = 1)$. *If the operator* $aP_\Gamma + bQ_\Gamma$ *admits a left (right) regularization on* $L_p(\Gamma, \rho)$ *then the operator* $a^{-1}P_\Gamma + b^{-1}Q_\Gamma$ *admits a right (left) regularization on* $L_q(\Gamma, \rho^{1-q})$. *Furthermore*

$$\dim\ker (aP_\Gamma + bQ_\Gamma)|L_p(\Gamma, \rho) = \dim\operatorname{coker} (a^{-1}P_\Gamma + b^{-1}Q_\Gamma)|L_q(\Gamma, \rho)^{1-q} .$$

7.7 Index cancellation and approximative inversion of singular integral operators

Suppose the operator $A := aP_\Gamma + bQ_\Gamma$ with coefficients in $L_\infty(\Gamma)$ to be a Fredholm operator, and set $\kappa := \operatorname{Ind} (aP_\Gamma + bQ_\Gamma)$. By Theorem 5.4, the operator A is right–, left– or two–sided invertible depending on whether the number κ is positive, negative, or equal to zero.

Consider further an operator B_κ defined by

$$B_\kappa := at^\kappa P_\Gamma + bQ_\Gamma .$$

From Proposition 3.2 of Chapter 6 (Theorem 4.3 of Chapter 1, [GK 1]) we infer that the difference $B_\kappa - A(t^\kappa P_\Gamma + Q_\Gamma)$ is a compact operator. Hence $\operatorname{Ind} B_\kappa = 0$. We have $a, b \in GL_\infty(\Gamma)$ by Theorem 4.1, and so the operator B_κ is invertible by Theorem 5.4.

In case $\kappa < 0$ the operator A can be written as

$$A = B_\kappa(t^{-\kappa}P_\Gamma + Q_\Gamma) .$$

Thus, a left inverse to A is given by

$$(aP_\Gamma + bQ_\Gamma)^{-1} = (t^\kappa P_\Gamma + Q_\Gamma)(at^\kappa P_\Gamma + bQ_\Gamma)^{-1} . \tag{7.1}$$

A straightforward computation shows that, in case $\kappa > 0$, (7.1) describes a right inverse operator to A .

Theorem 7.1. *Let the operator* $A = aP_\Gamma + bQ_\Gamma$ *be Fredholm. Then* A *is one-sided invertible, the operator* $at^\kappa P_\Gamma + bQ_\Gamma$ *with* $\kappa := \text{Ind}\, A$ *is two-sided invertible, and a one-sided inverse of* A *is given by (7.1). Furthermore, in case* $\kappa > 0$ *we have*

$$\ker\,(aP_\Gamma + bQ_\Gamma) = \text{span}\,\{g, gt, \ldots, gt^{\kappa-1}\}\,, \tag{7.2}$$

where $g := P_\Gamma\varphi - t^{-\kappa}(1 - Q_\Gamma\varphi)$ *and* φ *is the solution to the equation*

$$at^\kappa P_\Gamma\varphi + bQ_\Gamma\varphi = b\,. \tag{7.3}$$

Proof. The first assertion has already been verified. For the other ones notice that equation (7.3) yields

$$(aP_\Gamma + bQ_\Gamma)(gt^j) = t^{j-\kappa}(at^\kappa P_\Gamma\varphi + bQ_\Gamma\varphi - b) = 0 \quad (j = 0, 1, \ldots, \kappa - 1)\,,$$

from which it follows $\text{span}\,\{g, gt, \ldots, gt^{\kappa-1}\} \subseteq \ker\,(aP_\Gamma + bQ_\Gamma)$. Taking into account the linear independence of the functions $g, gt, \ldots, gt^{\kappa-1}$ as well as the equality $\dim\ker\,(aP_\Gamma + bQ_\Gamma) = \kappa$ one arrives at (7.2). ∎

Let us remark that the first assertion of Theorem 7.1 can be restated as follows: If $\kappa < 0$ then the equation

$$aP_\Gamma\varphi + bQ_\Gamma\varphi = f \tag{7.4}$$

is solvable if and only if the solution ψ to the equation

$$at^\kappa P_\Gamma\psi + bQ_\Gamma\psi = f \tag{7.5}$$

satisfies the condition $t^\kappa P_\Gamma\psi \in L_p^+(\Gamma, \rho)$. If this condition is fulfilled then the function $\varphi := t^\kappa P_\Gamma\psi + Q_\Gamma\psi$ solves equation (7.4).

If $\kappa > 0$ then the vector

$$\varphi = t^\kappa P_\Gamma\psi + Q_\Gamma\psi$$

is always a solution to our original equation (7.4) whenever ψ is a solution to (7.5).

Theorem 7.1 admits a generalization of the Theorems 8.1 and 8.2 of Chapter 3, [GK 1] concerning the approximative solution of singular integral operators.

Theorem 7.2. *Let* $a, b \in GL_\infty(\Gamma)$ *and assume the operator* $A = aP_\Gamma + bQ_\Gamma$ *to be Fredholm. Further, let* a_n *and* b_n *be functions in* $L_\infty(\Gamma)$ *satisfying*

$$\sup_{t\in\Gamma} |a_n(t) - a(t)| \le \varepsilon_n,\; \sup_{t\in\Gamma} |b_n(t) - b(t)| \le \varepsilon_n$$

with some sequence $\{\varepsilon_n\}$ *tending to zero as* $n \to \infty$. *Then, for sufficiently large* n *, the operators* $a_n t^\kappa P_\Gamma + b_n Q_\Gamma$ *are invertible and the operators*

$$A_n^{-1} := (t^\kappa P_\Gamma + Q_\Gamma)(a_n t^\kappa P_\Gamma + b_n Q_\Gamma)^{-1} \quad (n = 1, 2, \ldots)$$

with $\kappa = \mathrm{Ind}\, A_n$ *are one-sided inverses of the operators* $A_n = a_n P_\Gamma + b_n Q_\Gamma$. *Moreover the sequence* $\{A_n^{-1}\}$ *converges to the operator*

$$A^{-1} = (t^\kappa P_\Gamma + Q_\Gamma)(a t^\kappa P_\Gamma + b Q_\Gamma)^{-1} ,$$

which is a one-sided inverse of A *, and*

$$\|A_n^{-1} - A^{-1}\| = O(\varepsilon_n) \quad (n \to \infty) .$$

In case $\kappa > 0$ *the solutions* φ_n *to the equations*

$$a_n t^\kappa P_\Gamma \varphi_n + b_n Q_\Gamma \varphi_n = b_n$$

converge in the $L_p(\Gamma, \rho)$ *norm to the function* φ *, and the equality*

$$\ker\,(a P_\Gamma + b Q_\Gamma) = \mathrm{span}\,\{g, gt, \ldots, gt^{\kappa-1}\}$$

holds, where $g = P_\Gamma \varphi - t^{-\kappa}(1 - Q_\Gamma \varphi)$.

7.8 Exercises

7.1. Show that, for $\Gamma = [0, 1]$, the operator $(t^2 - t + 1) P_\Gamma + Q_\Gamma$ is invertible in the space $L_p(\Gamma)$ $(1 < p < \infty)$.

7.2. Let Γ be a semi-circle. Show that the operator $t^{2n} P_\Gamma + Q_\Gamma$ is a Φ-operator on $L_p(\Gamma)$, and compute its index.

7.3. Is the operator $t^{2n} P_\Gamma + Q_\Gamma$ a Φ-operator on $L_p(\Gamma)$ if Γ denotes the interval [-1,1]?

7.4. Let Γ be the boundary of the annulus $1 \le |z| \le R$. Determine both the norm and the essential norm of the operators P_Γ, Q_Γ and S_Γ on $L_2(\Gamma)$. Show that

$$\begin{aligned} \|P_\Gamma\| = \|Q_\Gamma\| &= \frac{\|S_\Gamma\| + \|S_\Gamma\|^{-1}}{2} , \\ |P_\Gamma| = |Q_\Gamma| &= \frac{|S_\Gamma| + |S_\Gamma|^{-1}}{2} , \\ \|S_\Gamma\| &\ne |S_\Gamma| . \end{aligned}$$

7.5. Let Γ_1 and Γ_2 refer to the boundaries of a semi–disk and of a quadrate, respectively. Verify

$$|\alpha I + \beta S_{\Gamma_1}|_{L_p(\Gamma_1)} = |\alpha I + \beta S_{\Gamma_2}|_{L_p(\Gamma_2)}$$

for any arbitrary pair (α, β) of complex numbers.

7.6. Assume the singularity supports of the functions a and $b\ (\in L_\infty(\mathbf{T}))$ are located on disjoint open arcs. Show under this assumption that if $A = aP_{\mathbf{T}} + Q_{\mathbf{T}} \in \Phi_+(L_p(\mathbf{T}))$ and $B = bP_{\mathbf{T}} + Q_{\mathbf{T}} \in \Phi_-(L_p(\mathbf{T}))$, then the operator $C = abP_{\mathbf{T}} + Q_{\mathbf{T}}$ fails to be normally solvable.

7.7. Let $a, b \in L_\infty(\Gamma)$. Prove that if the operator

$$(A\varphi)(t) = \varphi(t) + \frac{a(t)}{\pi i} \int_\Gamma \frac{b(\tau)\varphi(\tau)\, d\tau}{\tau - t}$$

is a Φ–operator on $L_p(\Gamma)$ $(1 < p < \infty)$, then

$$\operatorname{ess}\inf_{t\in\Gamma} |1 - a^2(t)b^2(t)| > 0 \, .$$

7.8. Show that, under the assumption $h \in H_\mu[0,1]$ $(0 < \mu < 1)$, the equation

$$((t^2 - t + 1)P_{[0,1]} + Q_{[0,1]})\varphi = h$$

possesses a unique solution in $H_\mu[0,1]$.

7.9. Let $a \in C(\mathbf{T})$ and $a(t) \not\equiv 0$. Prove that then the operator T_a on l_2 generated by the Toeplitz matrix $(a_{i-j})_0^\infty$, where a_i denotes the i–th Fourier coefficient of a (see Section 2.6) is normally solvable on l_2 if and only if $a(t) \neq 0$ for every $t \in \mathbf{T}$.

7.10. Show that, if $a \in L_\infty(\mathbf{T})$ and $\operatorname{ess}\sup_{t\in\mathbf{T}} |a(t)| > 0$, then the normal solvability of the operator T_a on l_2 implies that

$$\operatorname{ess}\inf_{t\in\mathbf{T}} |a(t)| > 0 \, .$$

7.11. Let $a \in L_\infty(\Gamma)$ and assume the operator T_a is invertible on l_2 . Verify that the operator $T_{a^{-1}}$ is also invertible on l_2 .

7.12. Let $a, b \in C(\mathbf{T})$. Prove or disprove the following proposition: The operator $aP_{\mathbf{T}} + bQ_{\mathbf{T}}$ is normally solvable on $L_p(\mathbf{T})$ if and only if the operators $a(\tau)P_{\mathbf{T}} + b(\tau)Q_{\mathbf{T}}$ are normally solvable on $L_p(\mathbf{T})$ for every $\tau \in \mathbf{T}$.

Comments and references

7.1. Under more special assumptions a result similar to 1.1 was established by Gakhov, Muskhelishvili and others (see, e.g. GAKHOV [1], §42, Sect. VI).

7.2. This section is an extended version of GOHBERG/KRUPNIK [5].

7.3. The material of this section has been taken from GOHBERG/SEMENCUL [1].

7.4. Under some additional restrictions, Theorem 4.1 goes back to SIMONENKO [6].

7.5. For $b = 1$, $\Gamma = \mathbf{T}$, $p = 2$ and $\rho(t) \equiv 1$, Theorem 5.1 is Coburn's (see [1]). Assertions similar to Theorem 5.4 can be found in GOHBERG [1], SIMOǸENKO [6], and GOHBERG/KRUPNIK [8].

In this chapter we restricted ourselves to the treatment of non–simple curves composed by Lyapunov arcs and of weights of power form. But as it turns out the basic results of this section remain true for an essentially wider classes of curves and weights.

For denote by R the collection of all Carleson curves (see Section 1.3 [GK 1] for their definition) and by $W_p(\Gamma)$ the class of all weights ρ satisfying on Γ the Hunt–Muckenhaupt–Wheeden condition (of the form (4.6)). To maintain Theorem 1.2 it suffices to suppose the pairwise disjoint curves Γ_k to be Carleson and the weight ρ to possess restrictions $\rho|\Gamma_k$ belonging to $W_p(\Gamma_k)$. For Theorem 1.1 one has only to require the curve Γ and the weight ρ be extendible to a closed curve $\widetilde{\Gamma} \in R$ and a weight $\widetilde{\rho} \in W_p(\widetilde{\Gamma})$.

The results of Section 7.3 carry over to the case of arbitrary closed curves $\Gamma \in R$ and weights $\rho \in W_p(\Gamma)$ without changes in the proofs.

In the remaining theorems of this chapter we make no use of the specific nature of the curve Γ and the weight ρ. The only thing we had always to suppose is the boundedness of the operator S_Γ on $L_p(\Gamma, \rho)$.

Chapter 8

The generalized factorization of bounded measurable functions and its application

In Chapter 3, [GK1], it was shown that any continuous function $a \in GC(\Gamma)$ admits a generalized factorization with respect to the closed curve Γ, and we exhibited the role of the factorization for the inversion of singular integral operators with continuous coefficients. But it turns out that these results are not immediately transferable to the case of functions in $GL_\infty(\Gamma)$.

In the present chapter we introduce a new concept of factorization: *the generalized factorization of functions with respect to a curve* Γ *in the space* $L_p(\Gamma, \rho)$. This concept will allow us to generalize all essential results of Chapter 3 to the case of functions in $GL_\infty(\Gamma)$ which admit such a factorization.

The central assertion of this chapter is Theorem 3.1 which provides us with a criterion for the generalized factorizability in $L_p(\Gamma, \rho)$. In the concluding section we present some applications of this factorization to the inversion of singular integral operators.

8.1 Sketch of the problem

As we have already mentioned above there exist functions $a \in GL_\infty(\Gamma)$ which do not admit a generalized factorization with respect to the closed curve Γ.

For the example of Γ being the unit circle $\mathbf{T}$, set $a(t) = t^{1/2} (= e^{i(\theta/2)}$, $0 < \theta \le 2\pi)$. Suppose the function a possesses the generalized factorization $a = a_- t^\kappa a_+$ (see Section 3.9, [GK1] or 6.5.3). Besides this factorization we can represent a in the form $a(t) = (t-1)^{1/2} (1 - 1/t)^{-1/2}$. Setting $b_-(t) := (1 - 1/t)^{-1/2}$ and $b_+(t) := (t-1)^{1/2}$ we obtain $a_- t^\kappa a_+ = b_- b_+$. By Theorem 4.8 of Chapter 2, [GK1], $b_+^{\pm 1} \in L_p^+(\Gamma)$ and $b_-^{\pm 1} \in L_p^-(\Gamma)$ if

$1 < p < 2$. Suppose that $\kappa < 0$. Then the left hand side of the equality $t^{\kappa}a_-b_-^{-1} = b_+a_+^{-1}$ belongs to $\overset{\circ}{L_1^-}(\Gamma)$ whereas its right-hand side is contained in $L_1^+(\Gamma)$. Due to the fact that $L_1^+(\Gamma) \cap \overset{\circ}{L_1^-}(\Gamma) = \{0\}$ we conclude that $b_+a_+^{-1} = 0$, which is impossible.

If, on the other hand, $\kappa > 0$, then the equality $t^{-1}b_-a_-^{-1} = t^{\kappa-1}a_+b_+^{-1}$ analogously implies that $b_-a_-^{-1} = 0$, which is also impossible. Finally, in case $\kappa = 0$, we consider the equality $b_-a_-^{-1} = a_+b_+^{-1}$. Since $b_-a_-^{-1} \in L_1^-(\Gamma)$ and $a_+b_+^{-1} \in L_1^+(\Gamma)$ we derive that $a_- = \lambda b_-$ $(\lambda \in \mathbf{C})$. But this is again impossible because of $b_- \notin L_q(\Gamma)$ $(q > 2)$. Thus, the function $a(t) = t^{1/2}$ does not permit a generalized factorization with respect to the circle $\mathbf{T}$.

As we explained in the third chapter, the problem of inversion of singular operators is closely related to the factorization of functions. In case the coefficients of the singular operator are not continuous it may happen that the singular operator is (two-sided) invertible in some spaces but not invertible from any side on other spaces (see Chapter 9). So it is natural to expect the factorizability of functions to depend on the space, too. These reflections lead us to the following definition.

Let Γ be a closed non-simple curve bounding the set F_Γ^+ , and assume that $0 \in F_\Gamma^+$. A *generalized factorization of the function* $a \in L_\infty(\Gamma)$ *with respect to the curve* Γ *in the space* $L_p(\Gamma, \rho)$ is a representation in the form

$$a(t) = a_-(t)t^{\kappa}a_+(t) \tag{1.1}$$

where κ is an integer and the factors $a_\pm$ are subject to the following conditions:

1.

$$a_- \in L_p^-(\Gamma, \rho),\ a_+ \in L_q^+(\Gamma, \rho^{1-q}),\ a_-^{-1} \in L_q^-(\Gamma, \rho^{1-q}) \quad \text{and} \quad a_+^{-1} \in L_p^+(\Gamma, \rho)\,. \tag{1.2}$$

2. The operator $a_+^{-1}P_\Gamma a_-^{-1}I$ is bounded on $L_p(\Gamma, \rho)$.

Remember that $L_q(\Gamma, \rho^{1-q}) = L_p(\Gamma, \rho)^*$ (see Section 1.7, [GK1]).

If a function admits a generalized factorization with respect to the curve Γ in the sense of Chapter 3, [GK1] then, obviously, it also admits a generalized factorization with respect to Γ in every space $L_p(\Gamma, \rho)$ $(1 < p < \infty,\ \rho(t) = \prod_{k=1}^n |t - t_k|^{\beta_k},\ -1 < \beta_k < p - 1)$.

Precisely as in Section 3.9 one can state that if the function a admits a factorization with respect to the curve Γ in the space $L_p(\Gamma, \rho)$, then the number κ is uniquely determined. This number will be called the *index of the function* a *in* $L_p(\Gamma, \rho)$, and we denote it by $\operatorname{ind} a|L_p(\Gamma, \rho)$. The factors $a_\pm$ are uniquely determined up to a constant factor. If we, for example, assume that $a_-(\infty) = 1$ then the factors $a_\pm$ are unique.

Notice, however, that the same function can possess different generalized factorizations with respect to Γ in different spaces. For example, let $\mathbf{T}$ be the unit circle and $a(t) =$

$t^{1/2}$ ($= e^{i\theta/2}$, $0 < \theta \leq 2\pi$) again. The function a admits the generalized factorization $a = a_- a_+$ (with $a_+(t) = (t-1)^{1/2}, a_-(t) = (1-t^{-1})^{-1/2}$) in $L_p(\mathbf{T})$ if $1 < p < 2$ and the generalized factorization $a = b_- t b_+$ (with $b_+(t) = (t-1)^{-1/2}$, $b_-(t) = (1-t^{-1})^{1/2}$) in $L_p(\mathbf{T})$ if $2 < p < \infty$) . In Section 1.3, [GK1] it had been shown the function $a_\pm$ and $b_\pm$ satisfy the conditions (1.2) with the corresponding p . The boundedness of the operator $a_+^{-1} P_{\mathbf{T}} a_-^{-1} I$ on $L_p(\mathbf{T})$ $(1 < p < 2)$ as well of the boundedness of $b_+^{-1} P_{\mathbf{T}} b_-^{-1} I$ on $L_p(\mathbf{T})$ $(2 < p < \infty)$ is a consequence of Theorem 4.1 of Chapter 1, [GK1]. It is worth mentioning that the function $a(t) = t^{1/2}$ does not permit a generalized factorization with respect to $\mathbf{T}$ in $L_2(\mathbf{T})$ (see Exercise 8.6 at the end of this chapter).

For brevity, we shall simply write in what follows *"factorization of a function in $L_p(\Gamma,\rho)$"* in place of *"generalized factorization of a function with respect to the curve Γ in $L_p(\Gamma,\rho)$"* or simply *"p,ρ-factorization"*.

8.2 Functions admitting a generalized factorization with respect to a curve in $L_p(\Gamma, \rho)$

In this section we give some classes of functions which are factorizable with respect to a closed curve Γ in certain spaces.

Theorem 2.1. *Let a be a real-valued measurable function defined on a closed curve Γ satisfying the conditions*

$$0 < \operatorname{ess}\inf_{t\in\Gamma} a(t) \quad \text{and} \quad \operatorname{ess}\sup_{t\in\Gamma} a(t) < \infty .$$

Then the function a permits a factorization $a = a_- a_+$ with the factors

$$a_+ := \exp\left(P_\Gamma \ln a\right) \quad \text{and} \quad a_- := \exp\left(Q_\Gamma \ln a\right) .$$

Thus, $a_+^{\pm 1} \in L_\infty^+(\Gamma)$ and $a_-^{\pm 1} \in L_\infty^-(\Gamma)$.

Proof. It is well known (cf. PRIVALOV [1], pp. 137-139) that the functions $a_+^{\pm 1}$ and $a_-^{\pm 1}$ are almost everywhere the boundary values of the function

$$F(z) := \exp\left(\pm\frac{1}{2\pi i}\int_\Gamma \frac{\ln a(\tau)}{\tau - z}\, d\tau\right)$$

in F_Γ^+ and F_Γ^- , respectively.

Let $\Gamma_k(\subset \Gamma)$ be a simple closed curve dividing the plane into the regions F_k^+ and F_k^- , and define

$$u_k(z) := \operatorname{Re}\left(\frac{1}{2\pi i}\int_{\Gamma_k} \frac{\ln a(\tau)}{\tau - z}\, d\tau\right) .$$

Since $\ln a(\tau)$ is an essentially bounded and real-valued function on Γ_k, the function $u_k(z)$ proves to be harmonic in each of the regions F_k^+ and F_k^- and, moreover, $u_k(\infty) = 0$ and $|u_k(z)| \leq \text{const}$ $(z \in \mathbf{C})$ (see PRIVALOV [1], pp. 82 and 188). Hence, the function

$$u(z) = \operatorname{Re}\left(\frac{1}{2\pi i}\int_\Gamma \frac{\ln a(\tau)}{\tau - z}\, d\tau\right)$$

is bounded in both F_Γ^+ and F_Γ^-. This, in turn, implies the holomorphy and boundedness of the functions $a_+^{\pm 1}$ $(a_-^{\pm 1})$ in F_Γ^+ (F_Γ^-), and hence finally follows

$$a_+^{\pm 1} \in L_\infty^+(\Gamma) \quad \text{and} \quad a_-^{\pm 1} \in L_\infty^-(\Gamma)\,.$$

■

As an easy consequence of Theorem 2.1 one obtains:

Theorem 2.2. *Let a be an arbitrary function in $L_\infty(\Gamma)$ and $b \in L_\infty(\Gamma)$ be a real-valued function satisfying*

$$\operatorname{ess}\inf_{t\in\Gamma} b(t) > 0\,.$$

Then the factorizability of the function ab and the function a with respect to Γ in $L_p(\Gamma, \rho)$ are equivalent. If the function a admits a factorization then

$$\operatorname{ind} ab|L_p(\Gamma, \rho) = \operatorname{ind} a|L_p(\Gamma, \rho)\,.$$

Proof. By Theorem 2.1, the function b is factorizable into $b = b_- b_+$ with factors $b_\pm$ distinguished by $b_+^{\pm 1} \in L_\infty^+(\Gamma)$ and $b_-^{\pm 1} \in L_\infty^-(\Gamma)$. Let $a = a_- t^\kappa a_+$ be a factorization of a with respect to Γ in $L_p(\Gamma, \rho)$. Then, evidently, the equality $ab = g_- t^\kappa g_+$ with $g_- := a_- b_-$ and $g_+ := a_+ b_+$ provides us with a factorization of ab with respect to Γ in $L_p(\Gamma, \rho)$. ■

Let us formulate some consequences of Theorem 2.1.

Corollary 2.1. *Assume the function a is continuous on Γ with the exception of finitely many points $t_1, \ldots, t_n$ where a possesses finite limits $a(t_k \pm 0)$ $(k = 1, \ldots, n)$, and let $\inf |a(t)| > 0$. If every pair of limits $a(t_k + 0)$ and $a(t_k - 0)$ $(k = 1, \ldots, n)$ is located on some ray starting from zero then the function a admits a generalized factorization with respect to Γ in all spaces $L_p(\Gamma, \rho)$.* [1]

Indeed, let f be a real-valued function which is continuous at any point of $\Gamma \setminus \{t_1, \ldots, t_n\}$, possessing finite limits $f(t_k \pm 0)$ and subject to the conditions

$$f(t) > 0 \quad \text{and} \quad \frac{f(t_k + 0)}{f(t_k - 0)} = \frac{a(t_k + 0)}{a(t_k - 0)} \quad (k = 1, \ldots, n)\,.$$

[1] In Section 9.2 we shall establish a certain converse to Corollary 2.1.

Clearly, the function $b := af^{-1}$ is continuous and so, by Theorem 2.2 and Theorem 9.1 of Chapter 3, the function $a = bf$ admits a factorization with respect to Γ in all spaces $L_p(\Gamma,\rho)$.

Another immediate consequence of Theorem 2.2 is

Corollary 2.2. *Let $a \in GL_\infty(\Gamma)$. In order that the function a to admit a generalized factorization with respect to Γ in $L_p(\Gamma,\rho)$ it is necessary and sufficient that the function $a/|a|$ possesses this property. Hence,* $\operatorname{ind} a|L_p(\Gamma,\rho) = \operatorname{ind}(a/|a|)|L_p(\Gamma,\rho)$.

As a third consequence of Theorem 2.2 we remark

Corollary 2.3. *Assume the hypotheses of Theorem 2.2. Then the operators $A_1 = abP_\Gamma + Q_\Gamma$ and $A_2 = aP_\Gamma + Q_\Gamma$ are connected via the equality*

$$A_1 = B_1 A_2 B_2$$

where B_1 and B_2 are invertible operators on every space $L_p(\Gamma,\rho)$.

Indeed, if the equality $b = b_- b_+$ ($b_-^{\pm 1} \in L_\infty^-(\Gamma)$ and $b_+^{\pm 1} \in L_\infty^+(\Gamma)$) represents a factorization of b with respect to the curve Γ then

$$A_1 = b_-(ab_+P_\Gamma + b_-^{-1}Q_\Gamma) = b_-(aP_\Gamma + Q_\Gamma)(b_+P_\Gamma + b_-^{-1}Q_\Gamma).$$

The operators $B_1 := b_- I$ and $B_2 := b_+P_\Gamma + b_-^{-1}Q_\Gamma$ are invertible with

$$B_1^{-1} = b_-^{-1}I\,, \quad B_2^{-1} = b_+^{-1}P_\Gamma + b_-Q_\Gamma\,.$$

8.3 Factorization in the spaces $L_p(\Gamma,\rho)$

In this section we present criteria for the factorizability of functions in $L_\infty(\Gamma)$ with respect to a curve Γ and mention some of the consequences.

Theorem 3.1. *The function $a \in L_\infty(\Gamma)$ admits a factorization in $L_p(\Gamma,\rho)$ if and only if the operator $A = aP_\Gamma + Q_\Gamma$ is Fredholm on $L_p(\Gamma,\rho)$. If A is a Fredholm operator then* $\operatorname{ind} a|L_p(\Gamma,\rho) = -\operatorname{Ind} A|L_p(\Gamma,\rho)$.

Proof. First we assume the function a possesses a factorization $a = a_+a_-$ in $L_p(\Gamma,\rho)$ with $\operatorname{ind} a|L_p(\Gamma,\rho) = 0$ and consider the operator $B := (a_+^{-1}P_\Gamma + a_-Q_\Gamma)a_-^{-1}I$. We claim that B is the inverse of A.

For let r be an arbitrary rational function in $R(\Gamma)$. By the definition of the factorization we have $a_-^{-1}r \in L_q(\Gamma,\rho^{1-q})$, $a_+^{-1} \in L_p^+(\Gamma,\rho)$ and $a_- \in \overset{\circ}{L_p^-}(\Gamma,\rho)$. Thus, $a_+^{-1}P_\Gamma a_-^{-1}r \in L_1^+(\Gamma)$ and $a_-Q_\Gamma a_-^{-1}r \in \overset{\circ}{L_1^-}(\Gamma)$. Since, moreover, the operators $a_+^{-1}P_\Gamma a_-^{-1}I$

and $a_-Q_\Gamma a_-^{-1}I = I - aa_+^{-1}P_\Gamma a_-^{-1}I$ are bounded on $L_p(\Gamma,\rho)$ we conclude that $a_+^{-1}P_\Gamma a_-^{-1}r \in L_p^+(\Gamma,\rho)$ and $a_-Q_\Gamma a_-^{-1}r \in \overset{\circ}{L_1^-}(\Gamma,\rho)$. Hence the identity

$$\begin{aligned} ABr &= (aP_\Gamma + Q_\Gamma)(a_+^{-1}P_\Gamma a_-^{-1} + a_-Q_\Gamma a_-^{-1})r \\ &= aa_+^{-1}P_\Gamma a_-^{-1}r + a_-Q_\Gamma a_-^{-1}r = r \end{aligned}$$

follows. Analogously, the inclusions $a_+P_\Gamma r \in L_q^+(\Gamma,\rho^{1-q})$ and $a_-^{-1}Q_\Gamma r \in \overset{\circ}{L_q^-}(\Gamma,\rho^{1-q})$ imply $BAr = (a_+^{-1}P_\Gamma + a_-Q_\Gamma)(a_+P_\Gamma + a_-^{-1}Q_\Gamma)r = r$.

Taking into account the boundedness of the operator $B = I + (1-a)a_+^{-1}P_\Gamma a_-^{-1}I$ on $L_p(\Gamma,\rho)$ the claim now follows, that is, the operator A is invertible, and B is its inverse.

Now we consider the general case when κ is an arbitrary integer. In this case, the function $at^{-\kappa}$ admits the factorization $at^{-\kappa} = a_-a_+$, and by the above, the operator $at^{-\kappa}P_\Gamma + Q_\Gamma$ is invertible on the space $L_p(\Gamma,\rho)$. If $\kappa > 0$ then the operator A can be rewritten in the form $A = (at^{-\kappa}P_\Gamma + Q_\Gamma)(t^\kappa P_\Gamma + Q_\Gamma)$. Since $t^\kappa P_\Gamma + Q_\Gamma$ is a Fredholm operator with index $-\kappa$ the operator A is Fredholm with index $-\kappa$, too. In case $\kappa < 0$ we consider the identity $at^{-\kappa}P_\Gamma + Q_\Gamma = A(t^{-\kappa}P_\Gamma + Q_\Gamma)$. The operator $at^{-\kappa}P_\Gamma + Q_\Gamma$ is invertible in $L_p(\Gamma,\rho)$, and the operator $t^{-\kappa}P_\Gamma + Q_\Gamma$ is Fredholm with index κ (see Section 2.7). So the operator A must be a Fredholm operator with index $-\kappa$ which proves the necessity of the condition in the theorem.

For its sufficiency assume the operator $A = aP_\Gamma + Q_\Gamma$ to be a Fredholm operator with index $-\kappa$. Then, by Theorem 7.1 of Chapter 7, the operator $C := at^{-\kappa}P_\Gamma + Q_\Gamma$ is invertible on $L_p(\Gamma,\rho)$ and, by Theorem 4.1 of Chapter 7, $a \in GL_\infty(\Gamma)$. Now an application of Corollary 6.2 of Chapter 7 shows that the operator $a^{-1}t^\kappa P_\Gamma + Q_\Gamma$ is invertible on $L_q(\Gamma,\rho^{1-q})$.

Let $\varphi_0 \in L_p(\Gamma,\rho)$ and $\psi_0 \in L_q(\Gamma,\rho^{1-q})$ be the solutions to the equations $(cP_\Gamma + Q_\Gamma)\varphi = 1$ and $(c^{-1}P_\Gamma + Q_\Gamma)\psi = 1$ with $c := at^{-\kappa}$. Then, $cP_\Gamma\varphi_0 = 1 - Q_\Gamma\varphi_0$ and $c^{-1}P_\Gamma\psi_0 = 1 - Q_\Gamma\psi_0$ and, consequently,

$$(P_\Gamma\varphi_0)(P_\Gamma\psi_0) = (1 - Q_\Gamma\varphi_0)(1 - Q_\Gamma\psi_0) . \tag{3.1}$$

The function $(1 - Q_\Gamma\varphi_0)(1 - Q_\Gamma\psi_0) - 1$ belongs to the subspace $\overset{\circ}{L_1^-}(\Gamma)$ whereas $(P_\Gamma\varphi_0)(P_\Gamma\psi_0) - 1$ is contained in $L_1^+(\Gamma)$. Since $L_1^+(\Gamma) \cap \overset{\circ}{L_1^-}(\Gamma) = \{0\}$ we conclude that $(P_\Gamma\varphi_0)(P_\Gamma\psi_0) = (1 - Q_\Gamma\varphi_0)(1 - Q_\Gamma\psi_0) = 1$. Now setting $a_+ := P_\Gamma\psi_0$ and $a_- := 1 - Q_\Gamma\varphi_0$ gives $c = a_+a_-$ and, thus, $a = a_+t^\kappa a_-$. Evidently, the functions $a_+ = P_\Gamma\psi_0$, $a_- = 1 - Q_\Gamma\varphi_0$, $a_+^{-1} = P_\Gamma\varphi_0$ and $a_-^{-1} = 1 - Q_\Gamma\psi_0$ satisfy the conditions (1.2).

To finish the proof it remains to check the boundedness of the operator $a_+^{-1}P_\Gamma a_-^{-1}I$ on $L_p(\Gamma,\rho)$. Assume without loss of generality that $|c(t)| \le m < 1$ and consider the operators $C = cP_\Gamma + Q_\Gamma$ and $B = (a_+^{-1}P_\Gamma + a_-Q_\Gamma)a_-^{-1}I$.

As above, it is not hard to show that the equality $BC\varphi = \varphi$ holds for every function $\varphi \in L_p(\Gamma,\rho)$. Hence, the operator $B = C^{-1}$ is bounded on $L_p(\Gamma,\rho)$, and since this operator can be rewritten as $B = I + (1-c)a_+^{-1}P_\Gamma a_-^{-1}I$, the boundedness of the operator $a_+^{-1}P_\Gamma a_-^{-1}I$ on $L_p(\Gamma,\rho)$ follows. ∎

Theorem 3.2. *Let Γ be a non-simple closed curve [1] which is the union of the pairwise disjoint non-simple closed curves $\Gamma_1,\Gamma_2,\dots,\Gamma_n$, and let $a \in L_\infty(\Gamma)$. Then the function a admits a generalized factorization with respect to Γ on $L_p(\Gamma,\rho)$ if and only if each of the functions $a|\Gamma_j$ $(j=1,\dots,n)$ has a generalized factorization with respect to Γ_j on the space $L_p(\Gamma_j,\rho)$. If the function a admits a generalized factorization then*

$$\operatorname{ind} a|L_p(\Gamma,\rho) = \sum_{i=1}^{n} \operatorname{ind}\,(a|\Gamma_j)|L_p(\Gamma_j,\rho)\,.$$

This theorem is an easy consequence of Theorem 3.1 and Theorem 1.2 of Chapter 7.

Theorem 3.3. *Assume the singularity support $\Delta(a)$ of the function $a \in GL_\infty(\Gamma)$ to decompose into n pairwise disjoint arcs $\gamma_1,\dots,\gamma_n$ and the functions $a_j \in GL_\infty(\Gamma)$ $(j = 1,2,\dots,n)$ to be subject to the conditions*

$$a_j|(\Gamma\backslash\gamma_j) \in C(\Gamma\backslash\gamma_j) \quad \textit{and} \quad a_j|\gamma_j = a|\gamma_j \quad (j=1,\dots,n)\,.$$

Then the function a is factorizable in $L_p(\Gamma,\rho)$ if and only if every function a_j $(j = 1,\dots,n)$ is factorizable in $L_p(\Gamma,\rho)$. If a is factorizable then

$$\operatorname{ind} a|L_p(\Gamma,\rho) = \sum_{j=1}^{n} \operatorname{ind} a_j|L_p(\Gamma,\rho) + \operatorname{ind} a_0\,,$$

where a_0 defined by

$$a_0 := a a_1^{-1} a_2^{-1} \dots a_n^{-1}$$

is a function in $GC(\Gamma)$.

This theorem follows immediately from Theorem 3.1 and Theorem 3.1 of Chapter 7.

Theorem 3.4. *Let $a \in L_\infty(\Gamma)$ and $b \in GC(\Gamma)$. The factorizability of a in $L_p(\Gamma,\rho)$ is equivalent to the factorizability of the function ab in $L_p(\Gamma,\rho)$.*

If the function a admits a factorization in $L_p(\Gamma,\rho)$ and if

$$a = a_- t^{\kappa_1} a_+\,, \quad b = b_- t^{\kappa_2} b_+$$

are the factorization of a in $L_p(\Gamma,\rho)$ and the generalized factorization of the function a, respectively, then the equality

$$ab = a_- b_- t^{\kappa_1+\kappa_2} a_+ b_+$$

---

[1] See Section 1.1 or 6.1 for the definition of a non-simple closed curve.

is a factorization of the function ab *in* $L_p(\Gamma,\rho)$.

Proof. Due to the continuity of the function b on Γ , the operator $abP_\Gamma + Q_\Gamma - (aP_\Gamma + Q_\Gamma)(bP_\Gamma + Q_\Gamma)$ is compact on $L_p(\Gamma,\rho)$. Since the operator $bP_\Gamma + Q_\Gamma$ is Fredholm we conclude that the operator $abP_\Gamma + Q_\Gamma$ is Fredholm if and only if the operator $aP_\Gamma + Q_\Gamma$ is as well. Taking into account Theorem 3.1 this proves the first assertion of the theorem.

Now let $ab = c_-t^m c_+$ be a generalized factorization of the function ab in $L_p(\Gamma,\rho)$. Then one has $a_-b_-t^{\kappa_1+\kappa_2}a_+b_+ = c_-t^m c_+$, and the definition of the generalized factorization entails that $(a_+b_+)^{\pm 1},\ c_+^{\pm 1} \in L_r^+(\Gamma)$ and $(a_-b_-)^{\pm 1},\ c_-^{\pm 1} \in L_r^-(\Gamma)$ for some $r > 1$. From this we conclude by standard arguments (see Section 3.9) that $\kappa_1 + \kappa_2 = m,\quad a_-b_- = \lambda c_-$ and $a_+b_+ = \lambda^{-1}c_+$ with a certain constant λ . ■

8.4 Application of the factorization to the inversion of singular integral operators

The following theorem summarizes results from previous sections.

Theorem 4.1 *Let* Γ *be an arbitrary closed curve and* $a, b \in L_\infty(\Gamma)$. *If the operator* $A = aP_\Gamma + bQ_\Gamma$ *is a* Φ*-operator or a* $\Phi_\pm$*-operator in* $L_p(\Gamma,\rho)$ *then the conditions*

$$a \in GL_\infty(\Gamma) \quad \text{and} \quad b \in GL_\infty(\Gamma) \tag{4.1}$$

are fulfilled.

Let the condition (4.1) be fulfilled. Then the operator A *is a* Φ*-operator if and only if the function* ab^{-1} *admits a generalized factorization with respect to* Γ *in the space* $L_p(\Gamma,\rho)$. *If* A *is a* Φ*-operator then* $\operatorname{Ind} A = -\operatorname{ind} ab^{-1}|L_p(\Gamma,\rho)$. *Further, if the function* ab^{-1} *is generalized factorizable with respect to* Γ *in* $L_p(\Gamma,\rho)$, $ab^{-1} = c_-t^\kappa c_+$ *with* $\kappa := \operatorname{ind} ab^{-1}|L_p(\Gamma,\rho)$, *then* A *is left–, right– or two–sided invertible on* $L_p(\Gamma,\rho)$ *depending on whether the number* κ *is positive, negative or equal to zero, respectively. In any case, a (one–sided) inverse operator to* A *is given by*

$$(aP_\Gamma + bQ_\Gamma)^{-1} = (t^{-\kappa}P_\Gamma + Q_\Gamma)(c_+^{-1}P_\Gamma c_-^{-1} + ab^{-1}t^{-\kappa}c_+^{-1}Q_\Gamma c_-^{-1})b^{-1}I\ . \tag{4.2}$$

This theorem is a consequence of Theorem 3.1 of the present chapter and of Theorems 4.1, 5.3 and 7.1 of Chapter 7.

Let us remark that Theorem 4.1 remains true if the operator $aP_\Gamma + bQ_\Gamma$ is replaced by the operator $P_\Gamma aI + Q_\Gamma bI$ and the equality (4.2) is replaced by

$$(P_\Gamma aI + Q_\Gamma bI)^{-1} = b^{-1}(c_+^{-1}P_\Gamma c_-^{-1}I + c_+^{-1}Q_\Gamma c_-^{-1}t^{-\kappa}ab^{-1}I)(P_\Gamma t^{-\kappa}I + Q_\Gamma)\ .$$

As a completion to Theorem 4.1 we mention the following

Theorem 4.2. *Let* $a, b \in GL_\infty(\Gamma)$ *and assume the function* $c = ab^{-1}$ *possesses the factorization* $c = c_- t^\kappa c_+$ *in* $L_p(\Gamma, \rho)$. *Then, if* $\kappa = \text{ind}\, c|L_p(\Gamma, \rho) < 0$,

$$\ker\, (aP_\Gamma + bQ_\Gamma) = \text{span}\, \{g, gt, \ldots, gt^{\kappa-1}\}\,, \tag{4.3}$$

with $g := c_+^{-1} - c_- t^\kappa$.

In case $\kappa > 0$ *one has*

$$\text{coker}\, (aP_\Gamma + bQ_\Gamma) = \text{span}\, \{bc_-, bc_- t, \ldots, bc_- t^{\kappa-1}\}\,, \tag{4.4}$$

and the equation $aP_\Gamma\varphi + bQ_\Gamma\varphi = f$ *is solvable if and only if*

$$\int_\Gamma f(t) b^{-1}(t) c_-^{-1}(t) t^{-j}\, dt = 0 \quad (j = 1, \ldots, \kappa)\,. \tag{4.5}$$

Proof. To start with we verify the first assertion of the theorem. Define

$$\varphi_k := gt^k \; (= c_+^{-1} t^k - c_- t^{\kappa+k})\,.$$

Then, $P_\Gamma \varphi_k = c_+^{-1} t^k$ and $Q_\Gamma \varphi_k = -c_- t^{\kappa+k}$ from which it follows that $cP_\Gamma\varphi_k + Q_\Gamma\varphi_k = 0$ and, thus, span $\{g, gt, \ldots, gt^{\kappa-1}\} \subseteq \ker A$. Taking into account the equality $\dim \ker A = |\kappa|$ one arrives at (4.3).

Now let $\kappa > 0$. Our goal is the construction of a basis of the subspace $\ker A^*$. For remember that $A^* = H_\Gamma(P_\Gamma bI + Q_\Gamma aI)H_\Gamma$ where the operator H_Γ is defined by $(H_\Gamma\varphi)(t) := \overline{H_\Gamma(t)\varphi(t)}$ (see Section 1.7, [GK1] or 6.3.2).

Consider the functions $y_j := H_\Gamma(c_-^{-1} b^{-1} t^{-j})$ $(j = 1, \ldots, \kappa)$. They satisfy the equalities $H_\Gamma A^* y_j = (P_\Gamma b + Q_\Gamma a) c_-^{-1} b^{-1} t^{-j} = P_\Gamma c_-^{-1} t^{-j} + Q_\Gamma c_+ t^{\kappa-j} = 0$ and so we find span $\{y_1, \ldots, y_\kappa\} \subseteq \ker A^*$. By means of the equality $\dim \ker A^* = \kappa$ one finds

$$\text{span}\, \{y_1, \ldots, y_\kappa\} = \ker A^*\,.$$

The operator A is normally solvable. So the inclusion $f \in \text{im}\, A$ holds if and only if $f \in L_p(\Gamma, \rho)$ and

$$\int_\Gamma f(t)\overline{y_i(t)}\, |dt| = 0 \quad (j = 1, 2, \ldots, \kappa)\,. \tag{4.6}$$

Since $\overline{y_i(t)}\, |dt| = hb^{-1} c_-^{-1} t^{-j}\, |dt| = b^{-1} c_-^{-1} t^{-j}\, dt$, the conditions (4.6) and (4.5) coincide.

Now define $x_j := bc_- t^j$ $(j = 0, 1 \ldots, \kappa - 1)$. Then

$$\int_\Gamma x_j(t) b^{-1}(t) c_-^{-1}(t) t^{-1-j}\, dt = \int_\Gamma \frac{dt}{t} \neq 0$$

and, consequently, $x_j \notin \operatorname{im} A$. Since $\dim \operatorname{coker} A = \kappa$ we obtain (4.4). ■

An analogous theorem is valid for the operator $P_\Gamma aI + Q_\Gamma bI$. Simply replace the equalities (4.3) and (4.4) by

$$\ker (P_\Gamma aI + Q_\Gamma bI) = \operatorname{span} \{b^{-1}c_+^{-1}, b^{-1}c_+^{-1}t, \ldots, b^{-1}c_+^{-1}t^{-\kappa-1}\} , \tag{4.7}$$

$$\operatorname{coker} (P_\Gamma aI + Q_\Gamma bI) = \operatorname{span} \{g, gt, \ldots, gt^{\kappa-1}\} \tag{4.8}$$

with $g := c_-^{-1} - c_+t^\kappa$, and condition (4.5) by

$$\int_\Gamma f(t)(c_-^{-1}(t) - c_+(t)t^\kappa)t^{-j}\, dt = 0 \quad (j = 1, \ldots, \kappa) . \tag{4.9}$$

The proof proceeds analogously to the proof of Theorem 4.2.

8.5 Exercises

8.1. Let Γ be a closed curve and suppose the function $a \in L_\infty(\Gamma)$ to admit a factorization of the form $a = a_-t^\kappa a_+$ where $\kappa \in \mathbf{Z}$,

$$a_-^{\pm 1} \in L_1^-(\Gamma) , \quad a_+^{\pm 1}(\Gamma) \in L_1^+(\Gamma) , \tag{8.1}$$

and the operator $a_+^{-1}P_\Gamma a_+$ is bounded on $L_p(\Gamma, \rho)$. Show that $a_-, a_+^{-1} \in L_p(\Gamma, \rho)$ and $a_-^{-1}, a_+ \in L_q(\Gamma, \rho^{1-q})$ $(p^{-1} + q^{-1} = 1)$.

8.2 Show that the conditions 1) and 2) in the definition of the generalized factorization of a function a with respect to the curve Γ in $L_p(\Gamma, \rho)$ are equivalent to the conditions (8.1) and 2).

8.3. Assume the function $a \in L_\infty(\Gamma)$ permits two factorizations in the space $L_p(\Gamma, \rho)$:

$$a = a_-t^m a_+ \quad \text{and} \quad a = b_-t^n b_+ .$$

Show then that $m = n$, $b_- = ca_-$ and $b_+ = c^{-1}a_+$ for some constant c .

8.4. Let the function $a(t) = t^\gamma$, which is defined on $\mathbf{T}$ by $a(e^{i\theta}) = e^{i\gamma\theta}$ $(0 \le \theta < 2\pi)$, admit the generalized factorization $a = a_-t^\kappa a_+$ with respect to $\mathbf{T}$ in a certain space $L_p(\mathbf{T})$. Show that

$$a_+(t) = c(t-1)^\alpha \quad \text{and} \quad a_-(t) = c^{-1}(1 - t^{-1})^{-\alpha}$$

with $\alpha = \gamma - \kappa$ and a certain constant c .

8.5. Show that, under the assumptions of the previous exercise,

$$\mathrm{Re}\gamma - 1/p < \kappa < \mathrm{Re}\gamma - 1/p + 1 .$$

8.6. Verify that the function $a(t) = t^\gamma$ defined as in Exercise 8.4 admits a factorization in $L_p(\mathbf{T})$ if and only if

$$\mathrm{Re}\gamma - 1/p \notin \mathbf{Z} .$$

In particular, the function $a(t) = t^{1/2}$ does not admit a generalized factorization with respect to $\mathbf{T}$ in $L_2(\mathbf{T})$.

8.7. Let the function $a(t) = t^\gamma$ defined in Exercise 8.4 be factorizable in the space $L_{p_0}(\mathbf{T}, |t-1|^{\beta_0})$ $(1 < p_0 < \infty, -1 < \beta_0 < p_0 - 1)$. Show that there exist neighbourhoods $U(p_0)$ and $V(\beta_0)$ such that the function $a(t)$ is factorizable in $L_p(\mathbf{T}, |t-1|^\beta)$ for all $p \in U(p_0)$ and $\beta \in V(\beta_0)$.

8.8. Prove that for any pair of numbers p_1, p_2 $(p_1 \neq p_2,\ 1 < p_1, p_2 < \infty)$ one can find a function $a \in L_\infty(\mathbf{T})$ which admits a factorization in $L_{p_1}(\mathbf{T})$ but not in $L_{p_2}(\mathbf{T})$.

8.9. Let (m, n) be an arbitrary pair of integers, $p_1 \in (1, \infty)$, $p_2 \in (1, \infty)$ and $p_1 \neq p_2$. Find a function a possessing the factorizations

$$a = a_- t^m a_+ \quad \text{in} \quad L_{p_1}(\mathbf{T})$$

and

$$a = b_- t^n b_+ \quad \text{in} \quad L_{p_2}(\mathbf{T}) .$$

8.10. Let the function $a \in L_\infty(T)$ admit the factorizations $a = a_- t^\kappa a_+$ in $L_p(\mathbf{T})$ and $a = b_- t^\kappa b_+$ in $L_q(\mathbf{T})$ $(p^{-1} + q^{-1} = 1)$. Show that then the function a is factorizable in $L_r(\mathbf{T})$ for all $r = \lambda p + (1-\lambda) q$ $(0 \leq \lambda \leq 1)$.

8.11. Let $1 < p_1 < p_2 < \infty$ and suppose the function $a \in L_\infty(T)$ possesses the factorizations $a = a_- t^\kappa a_+$ in $L_{p_1}(\mathbf{T})$ and $a = b_- t^\kappa b_+$ in $L_{p_2}(\mathbf{T})$. Verify that then the function a is factorizable in $L_p(\mathbf{T})$ for all p belonging to the interval $[p_1, p_2]$.

8.12. Prove that the function $a(t)$ being defined on $\mathbf{T}$ by

$$a(t) = \begin{cases} 1 & \text{if} \quad \mathrm{Im}\, t > 0 \\ -1 & \text{if} \quad \mathrm{Im}\, t < 0 \end{cases} \tag{8.2}$$

can be represented as the product

$$a(t) = a_-(t) t a_+(t) \tag{8.3}$$

where $a_-(t)$ and $a_+(t)$ are certain analytic branches of the functions $(1 - t^{-2})^{1/2}$ and $(t^2 - 1)^{-1/2}$ on the regions $\mathbf{C} \setminus [-1, 1]$ and $\mathbf{C} \setminus ((-\infty, -1] \cup [1, \infty))$, respectively.

8.13. Show that the equality (8.3) exhibits a factorization of the function a defined by (8.2) in the space $L_p(\mathbf{T})$ whenever $p > 2$, whereas the identity

$$a(t) = a_-^{-1}(t)t^{-1}a_+^{-1}(t) \tag{8.4}$$

yields a factorization of the same function in $L_p(\mathbf{T})$ if $1 < p < 2$.

Comments and references

The factorization of functions in spaces $L_p(\Gamma)$ (the so-called p-factorization) goes back to SIMONENKO [6]. He also established the basic result of this chapter, namely Theorem 3.1 (in case $\rho(t) \equiv 1$).

The results of this chapter carry over to spaces $L_p(\Gamma, \rho)$ with more general curves and weight functions. For details, see LITVINCHUK/SPITKOVSKI [1].

Chapter 9

Singular integral operators with piecewise continuous coefficients and their applications

In the present chapter we study singular integral operators the coefficients of which possess finitely (or countably) many points of discontinuity where all these discontinuities are assumed to be of the first kind. The curves are throughout assumed to be without intersections. The consideration of general composed curves will be postponed until the next chapter.

The chapter is divided into 16 sections the first two of which are of supporting character. The basic theorems are presented in the third and fourth sections. In the fifth section we consider in detail singular integral operators with continuous coefficients on non–closed curves, and in Sections 6 and 7 we present different methods for the inversion of singular integral operators on the real axis as well as on the half–axis. The eighth section is devoted to the case of coefficients possessing countably many discontinuities. Furthermore, based on the material of the previous sections we establish estimates for the norms of the operators P_Γ, Q_Γ and S_Γ in Section 9.

In the following two sections we deal with singular operators with piecewise continuous coefficients on weighted Hölder spaces and on symmetric spaces, respectively, and Section 12 is devoted to singular operators on L_p–spaces with arbitrary weights not necessarily of power form.

In Sections 13 and 14 we investigate operators defined by Toeplitz matrices in l_p as well as paired operators with piecewise continuous generating functions. Some applications of the results of this chapter will be given in Section 15, and the concluding section consists of exercises.

9.1 Non–singular functions and their index

Let Γ be a closed curve without intersections. By $PC(\Gamma)$ we denote the class of all functions a in $L_\infty(\Gamma)$ with the following properties:

1. The function a is continuous on Γ with the possible exception of finitely many points;

2. the limits

$$a(t_0+0) = \lim_{\substack{t\to t_0 \\ t\succ t_0}} a(t) \quad \text{and} \quad a(t_0-0) = \lim_{\substack{t\to t_0 \\ t\prec t_0}} a(t)$$

exist at each point t_0 of discontinuity of a and are finite. Note that by $t \prec t_0$ we mean that the point t is located before t_0 with respect to the orientation of Γ;

3. at any point of discontinuity one has $a(t_0-0) = a(t_0)$.

Now, given a pair (z_1, z_2) of points in the complex plane $\mathbf{C}$ and a number δ lying in the interval $(0,\pi)$, we designate by $l(z_1, z_2; \delta)$ the circular arc joining the point z_1 to z_2 and being distinguished by the following property:

From any point z $(z \neq z_1, z_2)$ of the arc $l(z_1, z_2; \delta)$ one sees the straight line between z_1 and z_2 under the angle δ, and running through the arc $l(z_1, z_2; \delta)$ from z_1 to z_2 this straight line is located at the left hand side.

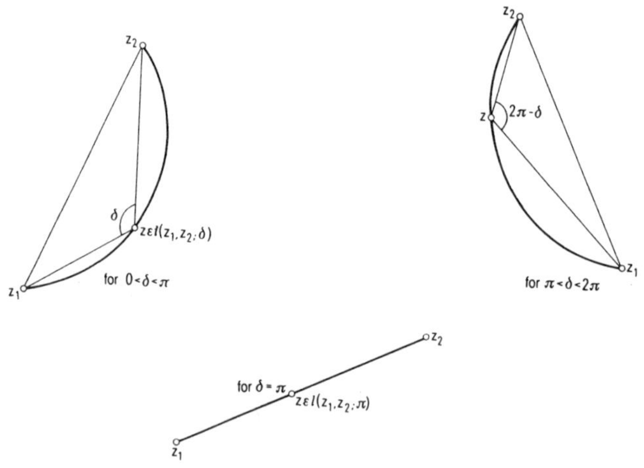

Figure 9.1

Further, for numbers δ in the interval $\pi < \delta < 2\pi$ we define $l(z_1, z_2; \delta) = l(z_2, z_1; 2\pi - \delta)$, and we denote by $l(z_1, z_2; \pi)$ the straight line between z_1 and z_2. The arc $l(0, 1; \delta)$ $(0 < \delta < \pi)$ can analytically be represented in the parameter form

$$z = \frac{\sin \theta\mu}{\sin \theta} e^{i\theta(\mu-1)} \quad (0 \leq \mu \leq 1), \tag{1.1}$$

where $\theta = \pi - \delta$, and the parameter representation of the arc $l(z_1, z_2; \delta)$ $(0 < \delta < \pi)$ is

$$z = z_1 + (z_2 - z_1) f_\delta(\mu) ,$$

where $f_\delta(\mu)$ refers to the function on the right-hand side of equality (1.1).

As μ increases from zero to one then the value of the function $1 - f_\delta(\mu)$ runs through the arc $l(1, 0; 2\pi - \delta)$. Hence, if $\pi < \delta < 2\pi$ then the parameter representation of the arc $l(z_1, z_2; \delta)$ is given by

$$z = z_2 + (z_1 - z_2)(1 - f_\delta(\mu)) .$$

Summarizing we have the equation of the arc $l(z_1, z_2; \delta)$ as $z = z_2 f_\delta(\mu) + z_1(1 - f_\delta(\mu))$ with

$$f_\delta(\mu) := \begin{cases} \frac{\sin\theta\mu}{\sin\theta} e^{i\theta(\mu-1)} \quad (\theta = \pi - \delta) & \text{if} \quad 0 < \delta < 2\pi, \delta \neq \pi \\ \mu & \text{if} \quad \delta = \pi . \end{cases} \tag{1.2}$$

Let p and $\beta_1 \ldots, \beta_r$ be real numbers satisfying $1 < p < \infty$ and $-1 < \beta_k < p - 1$ $(k = 1, \ldots, r)$. Consider the function

$$\rho(t) := \prod_{j=1}^{r} |t - t_j|^{\beta_j} ,$$

where the t_j are certain pairwise distinct points of the curve Γ .

To each function $a \in PC(\Gamma)$ we associate the function $a^{p,\rho} : \Gamma \times [0, 1] \to \mathbf{C}$ which is defined by

$$a^{p,\rho}(t, \mu) := a(t + 0) f(t, \mu) + a(t)(1 - f(t, \mu))$$

$(t \in \Gamma,\ 0 \leq \mu \leq 1)$, where we set $f(t, \mu) := f_{\delta(t)}(\mu)$ and

$$\delta(t) := \begin{cases} \frac{2\pi}{p} & \text{if} \quad t \in \Gamma \backslash \{t_1, \ldots, t_r\} \\ \frac{2\pi(1+\beta_k)}{p} & \text{if} \quad t = t_k \quad (k = 1, \ldots, r) . \end{cases} \tag{1.3}$$

If the function a is continuous at t_0 then $a^{p,\rho}(t_0, \mu) = a(t_0)$, but if t_0 is a point of discontinuity of a then the range of the function $a^{p,\rho}(t_0, \mu)$ $(0 \leq \mu \leq 1)$ coincides with the arc (or possibly straight line) $l(a(t_0), a(t_0 + 0); \delta(t_0))$.

Let us denote by $W_{p,\rho}(a)$ the plane curve which results from the range of the function a by adding the arcs $l(a(\tau_k),\ a(\tau_k + 0); \delta(\tau_k))$ for all points τ_k $(k = 1, \ldots, m)$ of discontinuity of a . Obviously, the curve $W_{p,\rho}(a)$ coincides with the range of the function $a^{p,\rho}$. We orient the curve $W_{p,\rho}(a)$ in the natural manner. On intervals of continuity of a , the motion along $W_{p,\rho}(a)$ agrees with the motion of t along Γ in positive direction, and along the supplementary arcs the curve $W_{p,\rho}(a)$ is oriented from $a(\tau_k)$ to $a(\tau_k + 0)$.

A function a $(\in PC(\Gamma))$ will be called *$\{p, \rho\}$-non-singular* if the curve $W_{p,\rho}(a)$ does not contain the origin, i.e. if $a^{p,\rho}(t, \mu) \neq 0$ $(t \in \Gamma, \mu \in [0, 1])$. If the function a

is $\{p,\rho\}$–non–singular then the winding number of the curve $W_{p,\rho}(a)$ around the point $z=0$ is called its *$\{p,\rho\}$–index.* This index will be abbreviated by $\operatorname{ind} a^{p,\rho}$. If a is a continuous function on Γ and $a(t)\neq 0\ (t\in\Gamma)$ then $\operatorname{ind} a^{p,\rho}=\operatorname{ind} a$, but if the function a is discontinuous then its $\{p,\rho\}$–index depends both on p and ρ .

Let us consider some examples. Denote by $\mathbf{T}$ the counter–clockwise oriented unit circle $|\zeta|=1$, and set $\rho(t):=|t-1|^{\beta}$ and $a(t)=t^{1/2}\ (=e^{i(\theta/2)},\ 0<\theta\leq 2\pi)$. The range of the function a is the upper semi–circle, and $a(1)=-1,\ a(1+0)=1$. If $(1+\beta)/p<1/2$ then the circular arc $l(-1,1;2\pi(1+\beta)/p)$ is located in the lower half plane, and if $(1+\beta)/p>1/2$, in the upper half plane. For $2(1+\beta)=p$ one has $l(-1,1;\pi)=[-1,1]$. In the latter case, the function a possesses a zero (at the point $t=1,\mu=1/2$) . Otherwise, the function a is $\{p,\rho\}$–nonsingular if $2(1+\beta)\neq p$, and it is not hard to see that

$$\operatorname{ind} a^{a,\rho}=\begin{cases}1 & \text{if}\quad 2(1+\beta)<p\\ 0 & \text{if}\quad 2(1+\beta)>p\end{cases}.$$

Next we explain that the product of two $\{p,\rho\}$–non–singular functions needs not itself be $\{p,\rho\}$–non–singular. Consider, for example, $a(t)=t^{1/4}\ (=e^{i(\theta/4)},\ 0<\theta\leq 2\pi)$ and $2(1+\beta)=p$. As one easily checks, the function a is $\{p,\rho\}$–non–singular whereas the function a^2 is not.

Further we mention that, even in case that the functions a,b and ab are $\{p,\rho\}$–non–singular, the $\{p,\rho\}$–index of the product ab does not need to coincide with the sum of the $\{p,\rho\}$–indices of the functions a and b . For example, if $a(t)=b(t)=t^{1/3}$ and $2(1+\beta)=p$ then $\operatorname{ind} a^{p,\rho}=\operatorname{ind} b^{p,\rho}=0$ but $\operatorname{ind}(ab)^{p,\rho}=1$. However one has the following

Theorem 1.1. *If the two $\{p,\rho\}$–non–singular functions a and b do not have common points of discontinuity then their product $c=ab$ is also $\{p,\rho\}$–non–singular, and*

$$\operatorname{ind} c^{p,\rho}=\operatorname{ind} a^{p,\rho}+\operatorname{ind} b^{p,\rho}\ . \tag{1.4}$$

Proof. One easily checks the validity of the equality

$$c^{p,\rho}(t,\mu)-a^{p,\rho}(t,\mu)b^{p,\rho}(t,\mu)=(a(t+0)-a(t))(b(t+0)-b(t))f(t,\mu)(1-f(t,\mu))\ .$$

Thus, if the $\{p,\rho\}$–non–singular functions a and b do not possess common discontinuities then $c^{p,\rho}=a^{p,\rho}b^{p,\rho}$ from which it follows that $c^{p,\rho}(t,\mu)\neq 0\quad(t\in\Gamma,\ 0\leq\mu\leq 1)$ and (1.4) holds. ∎

We will write *"p–non–singular function"* instead of "$\{p,1\}$–non–singular function" in case $\rho(t)\equiv 1$. Now we establish a criterion for the $\{p,\rho\}$–non–singularity of a function a is established.

Theorem 1.2. *The following two conditions are necessary and sufficient for the* $\{p,\rho\}$*-non-singularity of the function* a *:*

1. $a(t \pm 0) \neq 0$ *for all points* $t \in \Gamma$;
2. *at any point* t_k *of discontinuity of* a *, the quotient* $a(t_k)/a(t_k+0)$ *can be written as* $\exp(i\gamma_k)$ *where* $\delta(t_k) - 2\pi < \operatorname{Re}\gamma_k < \delta(t_k)$ *and the function* $\delta(t)$ *is defined by (1.3).*

Proof. Clearly, the first condition is necessary. So assume this condition is satisfied. Then the quotient $a(t_k)/a(t_k+0)$ can be expressed in the form $\exp(i\gamma_k)$ with $\delta(t_k)-2\pi < \operatorname{Re}\gamma_k \leq \delta(t_k)$. It is easy to verify that the arc $l(a(t_k), a(t_k+0); \delta(t_k))$ contains the point $z = 0$ if and only if $\operatorname{Re}\gamma_k = \delta(t_k)$. Hence, a function a being subject to the condition $a(t \pm 0) \neq 0$ $(t \in \Gamma)$ is $\{p,\rho\}$-non-singular if and only if $\operatorname{Re}\gamma_k \neq \delta(t_k)$. ■

9.2 Criteria for the generalized factorizability of power functions

In this section we formulate necessary and sufficient conditions for the factorizability in $L_p(\Gamma,\rho)$ of functions of the form $(t-z)^\gamma$ and of their products. It will be seen that these functions admit a factorization in $L_p(\Gamma,\rho)$ if and only if they are $\{p,\rho\}$-non-singular.

First suppose Γ to be a simple closed curve, z_0 a point in F_Γ^+, γ a complex number, and $\Psi_0(z)$ a fixed branch of the function $(z-z_0)^\gamma$ in the complex plane $\mathbf{C}$ with a cut from z_0 to ∞ intersect the curve Γ at only one point t_0. The function $\Psi_0(z)$ is continuous at every point $t \in \Gamma$ except, possibly, at $t_0; \Psi_0(t \pm 0) \neq 0$ $(t \in \Gamma)$ and

$$\frac{\Psi_0(t_0)}{\Psi_0(t_0+0)} = \exp(2\pi i\gamma). \tag{2.1}$$

Theorem 2.1. *The function* Ψ_0 *is* $\{p,\rho\}$*-non-singular if and only if the difference* $\operatorname{Re}\gamma - \delta(t_0)/2\pi$ *is not an integer. If this condition is fulfilled and if* κ *refers to the integer satisfying* [1]

$$0 < \kappa + \delta(t_0)/2\pi - \operatorname{Re}\gamma < 1$$

then $\operatorname{ind}\Psi_0^{p,\rho} = \kappa$.

Proof. The first assertion of the theorem is a consequence of equality (2.1) and Theorem 1.2.

Define $\gamma' = \gamma - \kappa = \alpha + \beta i$; then $\Psi_0(t) = (t-z_0)^{\gamma'}(t-z_0)^\kappa$. The function $(t-z_0)^\kappa$ is continuous on Γ, and its index equals κ. So, by Theorem 1.2, it suffices to verify that

[1] Remember that the function $\delta(t)$ is defined by (1.3).

ind $a^{p,\rho} = 0$, where $a(t) := (t - z_0)^{\gamma'}$. If the point t runs along the curve Γ starting at the point t_0 then the increase in the argument of the function a amounts to $2\pi\alpha$, and hence

$$\text{ind } a^{p,\rho} = \alpha + \frac{1}{2\pi}[\arg a^{p,\rho}(t_0, \mu)]_{0 \le \mu \le 1}.$$

Because of $a^{p,\rho}(t_0, \mu) = a(t_0 + 0)\{f(t_0, \mu) + \exp\,(2\pi i\gamma')(1 - f(t_0, \mu))\}$ and $f(t_0, 0) = 0$, $f(t_0, 1) = 1$ we find $[\arg a^{p,\rho}(t_0, \mu)]_{0 \le \mu \le 1} = -2\pi\alpha$ from which it follows that ind $a^{p,\rho} = 0$. ■

Now let Γ be a closed curve. To each point $\tau \in \Gamma$ and each complex number γ we associate a certain function $\Psi_{\tau,\gamma}$ which we are now going to define. If τ_k is a point on the simple closed curve $\Gamma_k (\subset \Gamma)$ then we denote by $\Psi_{\tau_k,\gamma}$ the continuous function at each point $t \neq \tau_k$ defined by

$$\Psi_{\tau_k,\gamma}(t) := \begin{cases} (t - z_k)^{\varepsilon\gamma} & \text{if } \ t \in \Gamma_k \\ 1 & \text{if } \ t \in \Gamma \backslash \Gamma_k, \end{cases} \tag{2.2}$$

where $\varepsilon = 1$ if Γ_k is oriented in the positive direction, and $\varepsilon = -1$ otherwise. The point z_k is assumed to belong to F_Γ^+ if $\varepsilon = 1$ and to F_Γ^- if $\varepsilon = -1$, and it is chosen in such a way that the straight line between the points τ_k and z_k meets the curve Γ at only one point, namely τ_k .

Theorem 2.2. *Let* $\rho(t) = \prod_{k=1}^{r} |t - t_k|^{\beta_k}$ *and* $\Psi = \Psi_{t_1,\gamma_1} \cdots \Psi_{t_n,\gamma_n}$ $(n \le r)$. *For the* $\{p, \rho\}$*–non–singularity of the function* Ψ *it is necessary and sufficient that none of the numbers*

$$\nu_k := (1 + \beta_k)/p - \operatorname{Re} \gamma_k \quad (k = 1, \ldots, n)$$

be an integer. If this condition is satisfied and if $\kappa_1, \ldots, \kappa_n$ *are integers fulfilling the inequalities* $0 < \kappa_k + \nu_k < 1$ *then* ind $\Psi^{p,\rho} = \kappa_1 + \ldots + \kappa_n$.

Proof. Repeating the arguments from the proof of Theorem 2.1 one obtains that the function Ψ_{t_k,γ_k} is $\{p, \rho\}$–non–singular if and only if the number ν_k is not an integer and that ind $\Psi_{t_k,\gamma_k}^{p,\rho} = \kappa_k$ in this case. Since the functions Ψ_{t_k,γ_k} do not possess common discontinuities, is to state the theorem is a consequence of Theorem 1.1. ■

Our next goal is to state conditions which guarantee the factorizability of the function Ψ in the space $L_p(\Gamma, \rho)$. To start with we consider the case when Γ is a simple closed curve which implies that $\Psi_{\tau,\gamma} = (t - z_0)^\gamma$. Define

$$\Psi_{\tau,\gamma}^+ := (t - \tau)^\gamma \quad \text{and} \quad \Psi_{\tau,\gamma}^- := \left(\frac{t - \tau}{t - z_0}\right)^{-\gamma}, \tag{2.3}$$

where $\{(t - \tau)/(t - z_0)\}^{-\gamma}$ denotes a branch of this function which is holomorphic in the complex plane except for the cut along the straight line joining z_0 to τ and equal to one at infinity, and $(t-\tau)^{\tau}$ refers to a branch of this function which is holomorphic in the complex plane except for a cut along a ray starting from τ running to infinity and not meeting the region F_{Γ}^{+}. Further suppose this branch is chosen in such a way that the equality $\Psi_{\tau,\gamma}=\Psi_{\tau,\gamma}^{+}\Psi_{\tau,\gamma}^{-}$ holds.

Theorem 2.3. *Let* $t_1\in\Gamma$ *and*

$$\frac{1+\beta_1}{p}-1<\operatorname{Re}\gamma<\frac{1+\beta_1}{p}\,. \tag{2.4}$$

Then the function $\Psi_{t_1,\gamma}$ *admits the factorization* $\Psi_{t_1,\gamma}=\Psi_{t_1,\gamma}^{+}\Psi_{t_1,\gamma}^{-}$ *in the space* $L_p(\Gamma,\rho)$ *with*

$$\rho(t):=\prod_{k=1}^{r}|t-t_k|^{\beta_k}\quad(1<p<\infty,-1<\beta_k<p-1,t_k\in\Gamma)\,.$$

Proof. The functions $(\Psi_{t_1,\gamma}^{+})^{\pm1}$ and $(\Psi_{t_1,\gamma}^{-})^{\pm1}$ are holomorphic in F_{Γ}^{+} and F_{Γ}^{-}, respectively. They are continuous at each point $t\neq t_1$ of the curve Γ, and in a neighbourhood of the point t_1 satisfy the inequality

$$|\,(\Psi_{t_1,\gamma}^{\pm}(z))^{\pm1}\,|\leq\frac{C}{|z-t_1|^{|\operatorname{Re}\gamma|}}\quad(C\text{ a constant })\,.$$

It is not hard to see that $|\operatorname{Re}\gamma|<1$, and furthermore one has

$$(\Psi_{t_1,\gamma}^{+})^{-1},\ \Psi_{t_1,\gamma}^{-}\in L_p(\Gamma,\rho);\qquad\Psi_{t_1,\gamma}^{+},\ (\Psi_{t_1,\gamma}^{-})^{-1}\in L_q(\Gamma,\rho^{1-q})\,.$$

So, Theorem 4.8 of Chapter 2, [GK1] implies that

$$(\Psi_{t_1,\gamma}^{+})^{-1}\in L_p^{+}(\Gamma,\rho);\ \Psi_{t_1,\gamma}^{-}\in L_p^{-}(\Gamma,\rho)\,;$$

$$\Psi_{t_1,\gamma}^{+}\in L_q^{+}(\Gamma,\rho^{1-q});\ (\Psi_{t_1,\gamma}^{-})^{-1}\in L_q^{-}(\Gamma,\rho^{1-q})\,.$$

The boundedness of the operator $(\Psi_{t_1,\gamma}^{+})^{-1}P_{\Gamma}(\Psi_{t_1,\gamma}^{-})^{-1}I$ follows from Theorem 4.1 of Chapter 1, [GK1]. ■

Now we consider the more general case when Γ is a closed curve consisting of finitely many pairwise disjoint simple closed curves. Let $\Gamma_k(\subset\Gamma)$ be a simple closed curve containing the point t_k, and denote by F_k the bounded region in the complex plane which has Γ_k as its boundary. Furthermore, we set

$$\Psi_{t_k,\gamma}^{+}(t):=\begin{cases}(t-t_k)^{\gamma} & \text{if}\quad t\in\Gamma_k'\,,\\ \left(\frac{t-t_k}{t-z_k}\right)^{\gamma} & \text{if}\quad t\in\Gamma\backslash\Gamma_k'\,,\end{cases}$$

$$\Psi_{t_k,\gamma}^{-}(t):=\begin{cases}(t-t_k)^{-\gamma} & \text{if}\quad t\in\Gamma_k''\,,\\ \left(\frac{t-t_k}{t-z_k}\right)^{-\gamma} & \text{if}\quad t\in\Gamma\backslash\Gamma_k''\,,\end{cases}$$

where $\Gamma_k' := \Gamma_k \cup (\Gamma \cap F_k)$, $\Gamma_k'' := \Gamma \cap F_k$ in case Γ_k is oriented counterclockwise, and $\Gamma_k' := \Gamma \cap F_k$, $\Gamma_k'' := \Gamma_k \cup (\Gamma \cap F_k)$ otherwise. The branches of these functions are chosen as above.

Theorem 2.4. *The function* $\Psi = \Psi_{t_1,\gamma_1} \ldots \Psi_{t_n,\gamma_n}$ *is factorizable in* $L_p(\Gamma, \rho)$ *if and only if it is* $\{p, \rho\}$*–non–singular. If the function* Ψ *is* $\{p, \rho\}$ *–non–singular then its factorization in* $L_p(\Gamma, \rho)$ *is of the form*

$$\Psi = \Psi_- t^\kappa \Psi_+$$

with

$$\kappa := \operatorname{ind} \Psi^{p,\rho} ,$$

$$\Psi_+ := \Psi^+_{t_1,\gamma_1'} \ldots \Psi^+_{t_n,\gamma_n'} f_1^+ \ldots f_n^+ , \quad \Psi_- := \Psi^-_{t_1,\gamma_1'} \ldots \Psi^-_{t_n,\gamma_n'} f_1^- \ldots f_n^- , \tag{2.5}$$

$$\gamma_k' := \gamma_k - \kappa_k , \quad \kappa_k := \operatorname{ind} \Psi^{p,\rho}_{t_k,\gamma_k} \quad (k = 1, \ldots, n) ,$$

and

$$f_k^+(t) := \begin{cases} 1 & \text{if } t \in \Gamma_k' , \\ (t - z_k)^{-\kappa_k} & \text{if } t \in \Gamma \backslash \Gamma_k' , \end{cases}$$

$$f_k^-(t) := \begin{cases} t^{-\kappa_k} & \text{if } t \in \Gamma_k'' , \\ (\frac{t-z_k}{t})^{+\kappa_k} & \text{if } t \in \Gamma \backslash \Gamma_k'' . \end{cases}$$

Proof. Suppose the function Ψ to be $\{p, \rho\}$–non–singular. Since the functions Ψ_{t_k,γ_k} do not possess common discontinuities we find via Theorem 1.1 that each of the functions Ψ_{t_k,γ_k} is $\{p, \rho\}$–non–singular and that $\kappa = \kappa_1 + \ldots + \kappa_n$.

By Theorem 2.2, the numbers $\gamma_k' := \gamma_k - \kappa_k$ satisfy the conditions

$$\frac{1 + \beta_k}{p} - 1 < \operatorname{Re} \gamma_k' < \frac{1 + \beta_n}{p} .$$

Repeating the proof of Theorem 2.3 one can show that the function $\Psi_0 = \Psi_{t_1,\gamma_1} \ldots \Psi_{t_n,\gamma_n}$ admits the factorization $\Psi_0 = \Psi_0^- \Psi_0^+$ in $L_p(\Gamma, \rho)$ with the factors

$$\Psi_0^- = \Psi^-_{t_1,\gamma_1'} \ldots \Psi^-_{t_n,\gamma_n'}, \quad \Psi_0^+ = \Psi^+_{t_1,\gamma_1'} \ldots \Psi^+_{t_n,\gamma_n'} .$$

The function $f = \Psi_{t_1,\kappa_1} \ldots \Psi_{t_n,\kappa_n}$ is continuous on Γ . A little thought shows that the equality $f = f_- t^\kappa f_+$ with $f_- := f_1^- \ldots f_n^-$ and $f_+ := f_1^+ \ldots f_n^+$ yields a factorization of the function f (in the sense of Section 3.3). Since the factors $f_\pm$ and $f_\pm^{-1}$ are bounded on Γ , the equality $\Psi = \Psi_- t^\kappa \Psi_+$ with $\Psi_- = \Psi_0^- f_-$ and $\Psi_+ = \Psi_0^+ f_+$ describes a factorization of the function Ψ in $L_p(\Gamma, \rho)$. This proves the sufficiency of the conditions of the theorem. Their necessity will be a consequence of the following theorem.

Theorem 2.5. *The operator* $A = \Psi P_\Gamma + Q_\Gamma$ *is normally solvable in the space* $L_p(\Gamma,\rho)$ *if and only if the function* Ψ *is* $\{p,\rho\}$*-non-singular. If the function* Ψ *is* $\{p,\rho\}$ *non singular then* A *is a Fredholm operator with the index* $\operatorname{Ind} A = -\operatorname{ind} \Psi^{p,\rho}$.

Proof. The sufficiency of the condition of the theorem follows from Theorem 3.1 of Chapter 8 and from the factorizability of the $\{p,\rho\}$-non-singular function Ψ in $L_p(\Gamma,\rho)$ which was shown in the proof of the previous theorem.

The necessity of the conditions will be proved indirectly. Let the operator $A = \Psi P_\Gamma + Q_\Gamma$ be normally solvable on $L_p(\Gamma,\rho)$, and assume that there is a pair of numbers $(\beta,\gamma) \in \{(\beta_1,\gamma_1),\dots,(\beta_n,\gamma_n)\}$ such that the difference $\delta := (1+\beta)/p - \operatorname{Re}\gamma$ is an integer. For definiteness, let $\beta = \beta_1$ and $\gamma = \gamma_1$. First of all we verify that the operator function $A(\varepsilon_1,\dots,\varepsilon_n) := \Psi_{t_1,\gamma_1+\varepsilon_1}\dots\Psi_{t_n,\gamma_n+\varepsilon_n} P_\Gamma + Q_\Gamma$ is continuous at the point $(0,\dots,0)$.

Indeed, let $t_0 \in \Gamma$ and let Γ° be a curve which results from Γ by "splitting" the point t_0 into two points, t_0^+ and t_0^- . For each $\gamma_0 \in \mathbf{C}$ one can regard the function Ψ_{t_0,γ_0} as a continuous function on Γ° by defining $\Psi_{t_0,\gamma_0}(t_0^\pm) := \Psi_{t_0,\gamma_0}(t_0 \pm 0)$. Analogously, we think of the function $\Psi_{t_0,\gamma_0+\varepsilon}(t)$ as a continuous function depending on the two parameters t and ε $(t \in \Gamma^\circ, 0 \le \varepsilon \le \varepsilon_0)$ for some positive constant ε_0 . Due to the uniform continuity of the function $\Psi_{t_0,\gamma_0+\varepsilon}(t)$ on $\Gamma^\circ \times [0,\varepsilon_0]$ the equality

$$\lim_{\varepsilon\to 0}\sup_{t\in\Gamma} |\Psi_{t_0,\gamma_0+\varepsilon}(t) - \Psi_{t_0,\gamma_0}(t)| = 0$$

holds, and it follows that

$$\lim_{\substack{\varepsilon_j\to 0\\ j=1,\dots,n}} \sup_{t\in\Gamma} |\Psi_{t_1,\gamma_1+\varepsilon_1}(t)\dots\Psi_{t_n,\gamma_n+\varepsilon_n}(t) - \Psi_{t_1,\gamma_1}(t)\dots\Psi_{t_n,\gamma_n}(t)| = 0 \,.$$

In view of

$$\|A(\varepsilon_1,\dots,\varepsilon_n) - A(0,\dots,0)\| \le \sup_{t\in\Gamma} |\Psi_{t_1,\gamma_1+\varepsilon_1}(t)\dots\Psi_{t_n,\gamma_n+\varepsilon_n}(t) - \Psi_{t_1,\gamma_1}(t)\dots\Psi_{t_n,\gamma_n}(t)|\cdot\|P_\Gamma\| \,,$$

the operator function $A(\varepsilon_1,\dots,\varepsilon_n)$ is continuous at the point $(0,\dots,0)$.

By our assumption, the operator $A(= A(0,\dots,0))$ is normally solvable. Furthermore we have $\Psi(t\pm 0) \neq 0$ $(t \in \Gamma)$. Thus, one of the numbers $\dim\ker A$ or $\dim\operatorname{coker} A$ must be equal to zero (see Theorem 5.3 of Chapter 7), and consequently the operator A is a Φ- or $\Phi_\pm$-operator. Now the theorem on the stability of the index (see Chapter 4, [GK1]) implies the existence of sufficiently small positive numbers $\varepsilon_1,\varepsilon_2,\dots,\varepsilon_n$ satisfying the following conditions:

1. The operators $A_1 := A(\varepsilon_1,\varepsilon_2,\dots,\varepsilon_n)$ and $A_2 := A(-\varepsilon_1,\varepsilon_2,\dots,\varepsilon_n)$ are Φ- or $\Phi_\pm$-operators.
2. $\operatorname{Ind} A_1 = \operatorname{Ind} A_2 = \operatorname{Ind} A$.
3. The numbers $(1+\beta_k)/p - \operatorname{Re}\gamma_k - \varepsilon_k$ are not integer.

Let $\delta_k := (1+\beta_k)/p - \operatorname{Re}\gamma_k$ $(k = 2,\dots,n)$ and $\delta^{\pm} := (1+\beta_1)/p - \operatorname{Re}(\gamma_1 \pm \varepsilon_1)$, and let $\kappa^{\pm}, \kappa_2, \dots, \kappa_n$ refer to fixed integers satisfying $0 < \kappa^{\pm} + \delta^{\pm} < 1$ and $0 < \kappa_k + \delta_k < 1$ $(k = 2,\dots,n)$. Since $\delta^+ < \delta < \delta^-$ and δ is an integer, one has $\kappa^+ \neq \kappa^-$. This implies that $\kappa' \neq \kappa''$ where $\kappa' := \kappa^+ + \kappa_2 + \dots + \kappa_n$ and $\kappa'' := \kappa^- + \kappa_2 + \dots + \kappa_n$. But we have shown that $\operatorname{Ind} A_1 = -\kappa'$ and $\operatorname{Ind} A_2 = -\kappa''$, contradicting the equality $\operatorname{Ind} A_1 = \operatorname{Ind} A_2$. ■

From Theorem 3.1 of Chapter 8 and Theorem 2.5 we obtain the necessity of the condition in Theorem 2.4.

Let us remark that as a consequence of Theorem 2.4, the function $\varphi_{t_1,1/2}$ is factorizable in $L_p(\Gamma)$ $(1 < p < \infty)$ if and only if $p \neq 2$. We have already considered this example in Section 8.1.

9.3 The inversion of singular integral operators on a closed curve

In this section we lay out necessary and sufficient conditions for the one-sided invertibility of singular integral operators with piecewise continuous coefficients along a closed curve Γ in the space $L_p(\Gamma, \rho)$.

Theorem 3.1. *Let Γ be a closed curve, $\rho(t) := \prod_{k=1}^{n} |t-t_k|^{\beta_k}$ $(-1 < \beta_k < p-1,\ k = 1,\dots,n)$, and $a, b \in PC(\Gamma)$. Then the operator $A = aP_\Gamma + bQ_\Gamma$ is at least one-sided invertible on $L_p(\Gamma,\rho)$ if and only if the condition* [1]

$$a(t+0)b(t)f(t,\mu) + a(t)b(t+0)(1 - f(t,\mu)) \neq 0 \quad (t \in \Gamma, 0 \leq \mu \leq 1) \tag{3.1}$$

is fulfilled. If this condition is met and $c := a/b$ then the operator A is invertible, invertible only from the left or invertible only from the right depending on whether the number $\kappa = \operatorname{ind} c^{p,\rho}$ is equal to zero, positive or negative, respectively. If $\kappa > 0$ then $\dim \operatorname{coker} A = \kappa$, *and if $\kappa < 0$ then* $\dim \ker A = -\kappa$.

Proof. Suppose condition (3.1) is satisfied. Because $f(t,0) = 0$ and $f(t,1) = 1$, this condition immediately yields $a(t \pm 0)b(t \pm 0) \neq 0$ $(t \in \Gamma)$. Set $c := ab^{-1}$, and let $t_1,\dots,t_n$ denote all points of discontinuity of the function c. Then condition (3.1) implies additionally that $c^{p,\rho}(t,\mu) \neq 0$ $(t \in \Gamma,\ 0 \leq \mu \leq 1)$ and, hence, the function c is $\{p,\rho\}$-non-singular. From Theorem 1.2 we know that the quotient $c(t_k)/c(t_k+0)$ can be written in the form $\exp(2\pi i \gamma_k)$ with

$$\delta(t_k)/2\pi - 1 < \operatorname{Re}\gamma_k < \delta(t_k)/2\pi .$$

[1] See p. 63 for the definition of the function $f(t,\mu)$.

Set $\Psi = \Psi_{t_1,\gamma_1}\dots\Psi_{t_n,\gamma_n}$, where $\Psi_{t\gamma}$ is defined by (2.2). Since $c(t_k)/c(t_k+0) = \Psi(t_k)/\Psi(t_k+0)$ $(k = 1,\dots,n)$, the function $d := c/\Psi$ is continuous on Γ . Furthermore, the equality $c = d\Psi$ implies that the function Ψ is $\{p,\rho\}$-non-singular and that $\operatorname{ind} c^{p,\rho} = \operatorname{ind} d + \operatorname{ind} \Psi^{p,\rho}$. From Theorem 3.4 of Chapter 8 and Theorem 2.4 one concludes now that the function c is factorizable in $L_p(\Gamma,\rho)$ and that $\operatorname{ind} c|L_p(\Gamma,\rho) = \operatorname{ind} d + \operatorname{ind} \Psi^{p,\rho}$. By Theorem 3.1 of Chapter 8, the sufficiency of condition (3.1) follows. The necessity will be shown once the following proposition has been proved.

Proposition 3.1. *Let $a, b \in PC(\Gamma)$ with Γ being a closed curve. If the operator $A = aP_\Gamma + bQ_\Gamma$ is a Φ- or $\Phi_\pm$-operator then the functions a and b are subject to condition (3.1).*

Proof. Suppose A is either a Φ- or $\Phi_\pm$-operator. Then, by Theorem 4.1 of Chapter 7, $a(t \pm 0) \neq 0$ and $b(t \pm 0) \neq 0$ $(t \in \Gamma)$. The function $c = a/b$ can be represented in the form $c = \Psi d$ with the function d being continuous on Γ and $\Psi = \Psi_{t_1,\gamma_1}\dots\Psi_{t_n,\gamma_n}$. With this representation we may write the operator A as

$$A = b(\Psi P_\Gamma + Q_\Gamma)(dP_\Gamma + Q_\Gamma) + T \tag{3.2}$$

with a certain compact operator T . Since the function d is continuous and does not vanish on Γ , the operator $dP_\Gamma + Q_\Gamma$ is a Φ-operator. This implies that the operator $\Psi P_\Gamma + Q_\Gamma$ is a Φ- or $\Phi_\pm$-operator (in accordance with A). But then, by Theorem 2.5., the function Ψ is $\{p,\rho\}$-non-singular, and this leads to the $\{p,\rho\}$-non-singularity of the function c . Thus, condition (3.1) is satisfied. ■

Recall Theorem 6.1 of Chapter 7 to finish the proof of Theorem 3.1. ■

From Theorem 3.1 and Theorem 3.1 in Chapter 8 one derives the following

Corollary 3.1. *Let $a \in PC(\Gamma)$. For the factorizability in $L_p(\Gamma,\rho)$ of the function a it is necessary and sufficient that a is $\{p,\rho\}$-non-singular. If the function a is $\{p,\rho\}$-non-singular then* $\operatorname{ind} a|L_p(\Gamma,\rho) = \operatorname{ind} a^{p,\rho}$.

In Section 8.4 we obtained the following result. If the function a/b is factorizable in $L_p(\Gamma,\rho)$ then one can get an explicit expression for the inverse (or the one-sided inverse) operator of $A = aP_\Gamma + bQ_\Gamma$ in terms of this factorization (likewise for the operator $P_\Gamma aI + Q_\Gamma bI)$. Furthermore, the factorization enables us to describe a basis of the kernel and the cokernel for the operator A and to derive conditions for the solvability of the equality $A\varphi = \psi$. By what we have shown above this can be done if the function $c = a/b$ is $\{p,\rho\}$-non-singular along a closed curve. For composed (not nessarily closed) curves we shall establish analogous results in the next section.

Before doing so let us consider an example. The problem is to find solvability conditions and the explicit solution of the equation $(aP_{\mathbf{T}} + Q_{\mathbf{T}})\varphi = \psi$ in the space $L_p(\mathbf{T}, |t-1|^\alpha |t+1|^\beta)$ $(1 < p < \infty,\ -1 < \alpha < p-1,\ -1 < \beta < p-1)$, where

$$a(t) = \begin{cases} 1 & \text{if} \quad t \in \mathbf{T}\,, \operatorname{Im} t > 0 \\ -1 & \text{if} \quad t \in \mathbf{T}\,, \operatorname{Im} t < 0\,. \end{cases}$$

One easily checks that the function a is $\{p,\rho\}$–non–singular if and only if

$$\alpha \neq p/2 - 1\,, \quad \beta \neq p/2 - 1\,.$$

Now define functions b and c on $\mathbf{T}$ by

$$b(e^{i\theta}) = e^{-i\theta/2} \quad (0 \le \theta < 2\pi), \quad c(e^{i\theta}) = e^{-i\theta/2}\ (-\pi \le \theta < \pi)\,.$$

A straightforward computation yields

$$b(t)c(t) = a(t)t^{-1},\ b^{-1}(t)c^{-1}(t) = a(t)t,\ b(t)c^{-1}(t) = b^{-1}(t)c(t) = a(t)\,.$$

By Theorem 2.3, the functions c, b, c^{-1}, b^{-1} admit factorizations in $L_p(\mathbf{T}, |t-1|^\alpha |t+1|^\beta)$ of the following form:

$$\begin{aligned} b(t) &= (t-1)^{-1/2}\,(1-\frac{1}{t})^{1/2} \quad (-1 < \alpha < \frac{p}{2} - 1, -1 < \beta < p-1)\,, \\ b^{-1}(t) &= (t-1)^{1/2}\,(1-\frac{1}{t})^{-1/2} \quad (\frac{p}{2} - 1 < \alpha < p-1, -1 < \beta < p-1)\,, \\ c(t) &= (t+1)^{-1/2}\,(1+\frac{1}{t})^{1/2} \quad (-1 < \alpha < p-1, -1 < \beta < \frac{p}{2} - 1)\,, \\ c^{-1}(t) &= (t+1)^{1/2}\,(1+\frac{1}{t})^{-1/2} \quad (-1 < \alpha < p-1,\ p/2 - 1 < \beta < p-1)\,. \end{aligned}$$

Since the functions b and c do not possess common discontinuities, the function a permits the following factorizations depending on the position of the numbers α and β:

$$\begin{aligned} a(t) &= (t-1)^{-1/2}(t+1)^{1/2}(1-\tfrac{1}{t})^{1/2}(1+\tfrac{1}{t})^{-1/2} && \text{if} \quad -1 < \alpha < p/2 - 1 < \beta < p-1\,, \\ a(t) &= (t-1)^{1/2}(t+1)^{-1/2}(1-\tfrac{1}{t})^{-1/2}(1+\tfrac{1}{t})^{1/2} && \text{if} \quad -1 < \beta < p/2 - 1 < \alpha < p-1\,, \\ a(t) &= (t^2-1)^{-1/2}t(1-\tfrac{1}{t^2})^{1/2} && \text{if} \quad -1 < \alpha, \beta < p/2 - 1\,, \\ a(t) &= (t^2-1)^{1/2}t^{-1}(1-\tfrac{1}{t^2})^{-1/2} && \text{if} \quad p/2 - 1 < \alpha, \beta < p-1\,. \end{aligned} \tag{3.3}$$

Here the function $(z^2-1)^{1/2}$ is considered in the complex plane with cut along $(-\infty,-1] \cup [1,+\infty)$, and the function $(1-1/z^2)^{1/2}$ in the plane with cut along the interval $[-1,1]$.

From Theorem 4.2 we deduce:

If $-1 < \alpha < p/2 - 1 < \beta < p - 1$, then the operator $A = aP_{\mathbb{T}} + Q_{\mathbb{T}}$ is invertible with

$$A^{-1} = \sqrt{\frac{t-1}{t+1}}\, A \sqrt{\frac{t+1}{t-1}}\,. \tag{3.4}$$

If $-1 < \beta < p/2 - 1 < \alpha < p - 1$ then the operator A is also invertible, and

$$A^{-1} = \sqrt{\frac{t+1}{t-1}}\, A \,\sqrt{\frac{t-1}{t+1}}\,. \tag{3.5}$$

In case $p/2 - 1 < \alpha < p - 1,\ p/2 - 1 < \beta < p - 1$ the operator A is invertible from the right and the equality

$$R_r = (1 - \frac{1}{t^2})^{-1/2}\, A\, (1 - \frac{1}{t^2})^{1/2} \tag{3.6}$$

provides us with one of its right inverses. Moreover

$$\ker A = \operatorname{span}\{(t^2 - 1)^{1/2}\, \chi(t)\}\,,$$

where χ stands for the characteristic function of the lower semi-circle.

Finally, if $-1 < \alpha < p/2 - 1, \quad -1 < \beta < p/2 - 1$ then the operator A is invertible from the left. One of its left inverses is given by

$$R_l = (1 - \frac{1}{t^2})\, A\, (1 - \frac{1}{t^2})^{-1/2}\,, \tag{3.7}$$

and the equation $A\varphi = \psi$ is solvable if and only if

$$\int_{\Gamma_0} \psi(t)(1 - \frac{1}{t^2})^{1/2}\, \frac{dt}{t} = 0\,.$$

9.4 Composed curves

The results of the preceding section carry over to the case of an arbitrary composed curve Γ. In this case we complete the curve Γ to a closed curve $\widetilde{\Gamma}$, and we extend the coefficients a and b $(\in PC(\Gamma))$ of the singular integral operator $A = aP_\Gamma + bQ_\Gamma$ to functions $\widetilde{a}$ and $\widetilde{b}$ defined on $\widetilde{\Gamma}$ by

$$\widetilde{a}(t) := \begin{cases} a(t) & \text{if } \ t \in \Gamma \\ 1 & \text{if } \ t \in \widetilde{\Gamma}\backslash\Gamma \end{cases}, \quad \widetilde{b}(t) := \begin{cases} b(t) & \text{if } \ t \in \Gamma \\ 1 & \text{if } \ t \in \widetilde{\Gamma}\backslash\Gamma\,. \end{cases} \tag{4.1}$$

Applying the results of the preceding section to the operator $\widetilde{A} = \widetilde{a}P_{\widetilde{\Gamma}} + \widetilde{b}Q_{\widetilde{\Gamma}}$ with coefficients $\widetilde{a}$ and $\widetilde{b}$ in $PC(\widetilde{\Gamma})$ we obtain via Theorem 1.1 of Chapter 7 the corresponding results for the operator A acting on the space $L_p(\Gamma, \rho)$. To do this we assume that $0 \notin \Gamma$

and that the curve $\widetilde{\Gamma}$ is chosen in such a way that the point $z = 0$ belongs to the region $F^+_{\widetilde{\Gamma}}$ which is bounded by $\widetilde{\Gamma}$.

From here on we assume that the functions in $PC(\Gamma)$ (Γ being a composed curve) are continuous at the end points of the non–closed arcs of the curve Γ. Further we denote by $\tau_1, \ldots, \tau_m$ and $\tau_{m+1}, \ldots, \tau_{2m}$ the starting and the end points of all non–closed arcs of the curve Γ, respectively. To each function $a \in PC(\Gamma)$, number p, and weight ρ we associate the function $a^{p,\rho} = a^{p,\rho}(t, \mu)$ defined by the following equalities:

If the point t is different from each of the end points $\tau_1, \ldots, \tau_{2m}$ of Γ then

$$a^{p,\rho}(t,\mu) := a(t+0)f(t,\mu) + a(t)(1 - f(t,\mu)) \quad (0 \le \mu \le 1),$$

where $f(t,\mu) := f_{\delta(t)}(\mu)$ and $f_{\delta(t)}$ stands for the function defined in (1.2) and (1.3); otherwise, if t coincides with one of the end points then we set

$$a^{p,\rho}(\tau_k,\mu) = a(\tau_k)f(\tau_k,\mu) + 1 - f(\tau_k,\mu) \quad (k = 1, \ldots, m)$$

and

$$a^{p,\rho}(\tau_k,\mu) = a(\tau_k)(1 - f(\tau_k,\mu)) + f(\tau_k,\mu) \quad (k = m+1, \ldots, 2m).$$

The range of the function $a^{p,\rho}$ is the union of the range of the function a with the finitely many circular arcs (or straight lines if $\delta(t_k) = \pi)l(a(t_k), a(t_k+0); \delta(t_k))$ joining the limits $a(t_k)$ and $a(t_k+0)$ of a at its points of discontinuity t_k, along with the $2m$ circular arcs (or straight lines if $\delta(t_k) = \pi$) $l(1, a(\tau_k); \delta(\tau_k))$ $(k = 1, \ldots, m)$ and $l(a(\tau_k), 1; \delta(\tau_k))$ $(k = m+1, \ldots, 2m)$ joining the points $a(\tau_k)$ to the point $z = 1$.

Clearly, the range of the function $a^{p,\rho}$ $(t \in \Gamma,\ 0 \le \mu \le 1)$ coincides with the range of the function $\widetilde{a}^{p,\rho}$ $(t \in \widetilde{\Gamma},\ 0 \le \mu \le 1)$, with the function $\widetilde{a}$ being defined on the closed curve $\widetilde{\Gamma}$ via equality (4.1).

In the case of a composed curve Γ we call a function a in $PC(\Gamma)$ $\{p,\rho\}$–non–singular if $a^{p,\rho}(t,\mu) \ne 0$ $(t \in \Gamma, 0 \le \mu \le 1)$. If a is a $\{p,\rho\}$–non–singular function then the number

$$\operatorname{ind} a^{p,\rho} := \operatorname{ind} \widetilde{a}^{p,\rho}$$

is referred to as its $\{p,\rho\}$–index. [1]

Theorem 4.1. *Let Γ be a composed curve and $a, b \in PC(\Gamma)$. Then the operator $A = aP_\Gamma + bQ_\Gamma$ $(P_\Gamma aI + Q_\Gamma bI)$ is at least one-sided invertible on $L_p(\Gamma,\rho)$ if and only if the following conditions are satisfied:*

1. $a(t+0)b(t)f(t,\mu) + a(t)b(t+0)(1 - f(t,\mu)) \ne 0 \qquad (t \in \Gamma \backslash \{\tau_1, \ldots, \tau_{2m}\})$,
2. $a(\tau_k)f(\tau_k,\mu) + b(\tau_k)(1 - f(\tau_k,\mu)) \ne 0 \qquad (k = 1, \ldots, m)$,
3. $a(\tau_k)(1 - f(\tau_k,\mu)) + b(\tau_k)f(\tau_k,\mu) \ne 0 \qquad (k = m+1, \ldots, 2m)$.

[1] For closed curves, the notion of the $\{p,\rho\}$–index was introduced in Section 9.1.

If these conditions are fulfilled and if $c := a/b$ *then the operator* A ***is invertible, invertible*** *only from the left or invertible only from the right depending on* ***whether the number*** $\kappa = \text{ind } c^{p,\rho}$ *is equal to zero, positive or negative. If* $\kappa > 0$ *, then* $\dim \text{coker } A = \kappa$, ***whereas*** $\kappa < 0$ *implies* $\dim \ker A = |\kappa|$.

This theorem is an immediate consequence of Theorem 1.1 of Chapter 7 and of Theorem 3.1. One only has to take into account that the conditions 1-3 coincide with condition (3.1), where the functions a and b in (3.1) are replaced by $\widetilde{a}$ and $\widetilde{b}$ defined on the closed curve $\widetilde{\Gamma}$ by equality (4.1).

Let $a \in L_\infty(\Gamma)$, $b \in GL_\infty(\Gamma)$, and suppose $c := a/b$ is a $\{p,\rho\}$-non-singular function in $PC(\Gamma)$. Then the function $\widetilde{c}$ is $\{p,\rho\}$-non-singular on the closed curve $\widetilde{\Gamma}$ and, hence, it admits a factorization $\widetilde{c} = \widetilde{c}_- t^\kappa \widetilde{c}_+$ in $L_p(\widetilde{\Gamma},\rho)$, where $\kappa = \text{ind } c^{p,\rho}$. The operators $\widetilde{A} = \widetilde{a}P_{\widetilde{\Gamma}} + \widetilde{b}Q_{\widetilde{\Gamma}}$ and $A = aP_\Gamma + bQ_\Gamma$ are at least one-sided invertible, and Theorem 1.1 of Chapter 7 implies that the inverse A^{-1} of the operator A is nothing other than the restriction of the operator $\chi\widetilde{A}^{-1}$, $\chi(t)(t \in \widetilde{\Gamma})$ standing for the characteristic function of the curve Γ , to the subspace $L_p(\Gamma,\rho)(\subset L_p(\widetilde{\Gamma},\rho))$. Furthermore we obtain from Theorem 4.1 of Chapter 8:

Theorem 4.2. *Let* Γ *be a composed curve,* $a \in L_\infty(\Gamma)$, $b \in GL_\infty(\Gamma)$ *, and suppose the function* $c := a/b$ *to be* $\{p,\rho\}$*-non-singular with* $\kappa := \text{ind } c^{p,\rho}$ *. Then the at least one-sided inverses* A^{-1} *and* B^{-1} *of the operators* $A = aP_\Gamma + bQ_\Gamma$ *and* $B = P_\Gamma aI + Q_\Gamma bI$ *are given by*

$$A^{-1} = c_- a^{-1}(bP_\Gamma + aQ_\Gamma)b^{-1}c_-^{-1}I\,, \quad B^{-1} = b^{-1}c_+^{-1}(P_\Gamma b + Q_\Gamma a)^{-1}c_+ I\,, \tag{4.2}$$

where $c_\pm$ *denote the restrictions onto* Γ *of the terms* $\widetilde{c}_\pm$ *in the factorization* $\widetilde{c} = \widetilde{c}_- t^\kappa \widetilde{c}_+$ *of the function* $\widetilde{c}$ *in the space* $L_p(\widetilde{\Gamma},\rho)$ *.*

If $\kappa < 0$ *then*

$$\ker A = \text{span }\{g, gt, \ldots, gt^{-\kappa-1}\} \quad \textit{and} \quad \ker B = \text{span }\{h, ht, \ldots, ht^{-\kappa-1}\}$$

with $g := c_+^{-1} - c_- t^\kappa$ *and* $h := c_+^{-1}b^{-1}$ *.*

In case that $\kappa > 0$ *one has*

$$\begin{aligned} \text{coker } A &= \text{span }\{bc_-, bc_-t, \ldots, bc_-t^{\kappa-1}\}\,, \\ \text{coker } B &= \text{span }\{d^{-1}, d^{-1}t, \ldots, d^{-1}t^{\kappa-1}\}\,, \end{aligned}$$

where $d = c_-^{-1} - c_+ t^\kappa$ *.*

The equation $A\varphi = f$ *is solvable if and only if the conditions*

$$\int_\Gamma f(t)b^{-1}(t)c_-^{-1}(t)t^{-j}\,dt = 0 \quad (j = 1, \ldots, \kappa)$$

are satisfied; whereas for the solvability of $B\varphi = f$ *it is necessary and sufficient that*

$$\int_\Gamma f(t)d(t)t^{-j}\,dt = 0 \quad (j = 1, \ldots, \kappa)\,.$$

Now we examine a special case when the factorization of the function c in $L_p(\Gamma, \rho)$ can be effectively done.

Let c denote a $\{p, \rho\}$–non–singular function which is piecewise Hölder continuous on Γ. The latter means that the curve Γ can be divided into finitely many curves on each of which the function c satisfies a Hölder condition. In this case, the function $\widetilde{c}$ proves to be $\{p, \rho\}$–non–singular and piecewise Hölder continuous on the closed curve $\widetilde{\Gamma}$.

Let $\tau_1, \ldots, \tau_n$ denote all points of discontinuity of the function $\widetilde{c}$. By Theorem 1.2 we can write the quotients $\widetilde{c}(\tau_k)/\widetilde{c}(\tau_k + 0)$ in the form $\exp{(2\pi i \gamma_k)}$ with $\delta(\tau_k) - 2\pi < 2\pi\text{Re}\,\gamma_k < \delta(\tau_k)$. Put $\widetilde{\Psi} = \Psi_{\tau_1,\gamma_1} \ldots \Psi_{\tau_n,\gamma_n}$, with the functions $\Psi_{\tau,\gamma}$ defined on $\widetilde{\Gamma}$ by the equalities (2.2). Due to Theorem 2.4, the function $\widetilde{\Psi}$ admits the factorization $\widetilde{\Psi} = \widetilde{\Psi}_-\widetilde{\Psi}_+$ in $L_p(\widetilde{\Gamma}, \rho)$, the factors $\widetilde{\Psi}_\pm$ being the functions defined by (2.5) with $f_k^\pm = 1$.

Set $\widetilde{d} := \widetilde{c}/\widetilde{\Psi}$. Since, clearly,

$$\widetilde{\Psi}(\tau_k)/\widetilde{\Psi}(\tau_k + 0) = \exp{(2\pi i \gamma_k)} = \widetilde{c}(\tau_k)/\widetilde{c}(\tau_k + 0)\,,$$

the function $\widetilde{d}$ satisfies a Hölder condition along the whole curve $\widetilde{\Gamma}$. But then, by Theorem 3.1 of Chapter 3, the function $\widetilde{d}$ permits a factorization $\widetilde{d} = \widetilde{d}_- t^\kappa \widetilde{d}_+$ where $\kappa = \text{ind}\,\widetilde{d}$, and with factors $\widetilde{d}_\pm$ defined by equalities (3.3) of Chapter 3. The functions c_+ and c_- are just the restrictions of the functions $\widetilde{\Psi}_+\widetilde{d}_+$ and $\widetilde{\Psi}_-\widetilde{d}_-$ onto the curve Γ, respectively.

Let us consider an example: Let $\mathbf{T} := \{t : |t| = 1\}$ and $\mathbf{T}_1 := \{t : |t| = 4\}$ be two circles oriented counterclockwise, and denote by $a(t)$ a branch of the function $(t-3)^{1/2}$ which is defined in the complex plane $\mathbf{C}$ except for the cut along the ray $\text{Im}\; z = 0,\ \text{Re}\; z \geq 3$ and which satisfies the condition $a(4 + i0) = 1$. The function a is continuous at every point of the curve $\Gamma = \mathbf{T} \cup \mathbf{T}_1$ except for the point $t_0 = 4$. Furthermore, define the curve $\widetilde{\Gamma}$ as $\Gamma \cup \mathbf{T}_2$ with $\mathbf{T}_2$ an abbreviation for the circle $\mathbf{T}_2 = \{t : |t| = 2\}$ provided with a clockwise orientation, and set

$$\widetilde{\Psi}(t) := \begin{cases} (t-3)^{1/2} & \text{if} \quad t \in \mathbf{T}_1 \\ 1 & \text{if} \quad t \in \mathbf{T} \cup \mathbf{T}_2\,, \end{cases} \qquad \widetilde{a}(t) = \begin{cases} (t-3)^{1/2} & \text{if} \quad t \in \Gamma \\ 1 & \text{if} \quad t \in \mathbf{T}_2 \end{cases}$$

and $\widetilde{d} := \widetilde{a}/\widetilde{\Psi}$. The function d is continuous on $\widetilde{\Gamma}$, and $\text{ind}\,\widetilde{d} = 0$.

Let $\rho(t) := |t-4|^\beta$. By Theorem 2.1, the function $\widetilde{\Psi}$ is $\{p, \rho\}$–non–singular if and only if the number $1/2 - (1+\beta)/p$ is not an integer. Thus, because of $|1/2 - (1+\beta)/p| < 1/2$, the function $\widetilde{\Psi}$ is $\{p, \rho\}$–non–singular if and only if $p \neq 2(1+\beta)$. Invoking Theorem

2.1 once more we conclude that $\operatorname{ind}\widetilde{\Psi}^{p,\rho}=1$ if $p>2(1+\beta)$ and $\operatorname{ind}\Psi^{p,\rho}=0$ if $p<2(1+\beta)$.

In the case that $p<2(1+\beta)$ the function $\widetilde{\Psi}$ admits a generalized factorization $\widetilde{\Psi}=\widetilde{\Psi}_-\widetilde{\Psi}_+$ in $L_p(\widetilde{\Gamma},\rho)$ with factors given by (see (2.5))

$$\widetilde{\Psi}_-(t):=\begin{cases}\left(\frac{t-3}{t-4}\right)^{1/2} & \text{if } t\in\mathbf{T}_1\\ (t-4)^{-1/2} & \text{if } t\in\mathbf{T}\cup\mathbf{T}_2\end{cases},\ \widetilde{\Psi}_+(t):=(t-4)^{1/2}.$$

Hence the function $\widetilde{a}$ admits the factorization $\widetilde{a}=\widetilde{a}_-\widetilde{a}_+$ in the space $L_p(\widetilde{\Gamma},\rho)$, the factors of which are

$$\widetilde{a}_+(t):=\begin{cases}(t-4)^{1/2}(t-3)^{1/2} & \text{if } t\in\mathbf{T}\\ (t-4)^{1/2} & \text{if } t\in\mathbf{T}_1\cup\mathbf{T}_2,\end{cases}$$

$$\widetilde{a}_-(t):=\begin{cases}\left(\frac{t-3}{t-4}\right)^{1/2} & \text{if } t\in\mathbf{T}_1\\ (t-4)^{-1/2} & \text{if } t\in\mathbf{T}\cup\mathbf{T}_2.\end{cases}$$

The operator $A=aP_\Gamma+Q_\Gamma$ is invertible in $L_p(\Gamma,\rho)$, and for its inverse we find $A^{-1}=a_+^{-1}(P_\Gamma+aQ_\Gamma)a_-^{-1}I$ with

$$a_+(t):=\begin{cases}(t-4)^{1/2}(t-3)^{1/2} & \text{if } t\in\mathbf{T}\\ (t-4)^{1/2} & \text{if } t\in\mathbf{T}_1,\end{cases}$$

$$a_-(t):=\begin{cases}\left(\frac{t-3}{t-4}\right)^{1/2} & \text{if } t\in\mathbf{T}_1\\ (t-4)^{-1/2} & \text{if } t\in\mathbf{T}.\end{cases}$$

In case $p>2(1+\beta)$ one can easily represent the function $\widetilde{a}$ as the product $\widetilde{a}=\widetilde{\Psi}_1\widetilde{d}_1$ with

$$\widetilde{\Psi}_1(t):=\begin{cases}(t-3)^{-1/2} & \text{if } t\in\mathbf{T}_1\\ 1 & \text{if } t\in\mathbf{T}\cup\mathbf{T}_2\end{cases},\ \widetilde{d}_1(t):=\begin{cases}(t-3)^{1/2} & \text{if } t\in\mathbf{T}\\ 1 & \text{if } t\in\mathbf{T}_2\\ t-3 & \text{if } t\in\mathbf{T}_1\end{cases}.$$

The function $\widetilde{\Psi}_1$ is $\{p,\rho\}$-non-singular, $\operatorname{ind}\widetilde{\Psi}_1^{p,\rho}=0$, and $\widetilde{\Psi}_1$ admits the factorization $\widetilde{\Psi}_1=\widetilde{\Psi}_-\widetilde{\Psi}_+$ with (see (2.5))

$$\widetilde{\Psi}_-(t):=\begin{cases}\left(\frac{t-3}{t-4}\right)^{-1/2} & \text{if } t\in\mathbf{T}_1\\ (t-4)^{1/2} & \text{if } t\in\mathbf{T}\cup\mathbf{T}_2\end{cases},\quad \widetilde{\Psi}_+(t):=(t-4)^{-1/2}.$$

The function $\widetilde{d}_1$ is continuous on $\widetilde{\Gamma}$, $\operatorname{ind}\widetilde{d}_1=1$, and it admits the factorization $\widetilde{d}_1=\widetilde{d}_-t\widetilde{d}_+$ with

$$\widetilde{d}_-(t):=\begin{cases}\frac{t-3}{t} & \text{if } t\in\mathbf{T}_1\\ t^{-1} & \text{if } t\in\mathbf{T}\cup\mathbf{T}_2\end{cases},\quad \widetilde{d}_+(t):=\begin{cases}(t-3)^{1/2} & \text{if } t\in\mathbf{T}\\ 1 & \text{if } t\in\mathbf{T}_1\cup\mathbf{T}_2\end{cases}.$$

This implies that a factorization $\widetilde{a} = \widetilde{a}_- t \widetilde{a}_+$ in $L_p(\widetilde{\Gamma}, \rho)$ of the function $\widetilde{a}$ can be obtained by setting

$$\widetilde{a}_-(t) := \begin{cases} \left(\frac{t-3}{t-4}\right)^{-1/2} \cdot \frac{t-3}{t} & \text{if } t \in \mathbf{T}_1 \\ \frac{(t-4)^{1/2}}{t} & \text{if } t \in \mathbf{T} \cup \mathbf{T}_2 \end{cases},$$

$$\widetilde{a}_+(t) := \begin{cases} (t-3)^{1/2}(t-4)^{-1/2} & \text{if } t \in \mathbf{T} \\ (t-4)^{-1/2} & \text{if } t \in \mathbf{T}_1 \cup \mathbf{T}_2 \end{cases}.$$

The operator $A = aP_\Gamma + Q_\Gamma$ is invertible only from the left in $L_p(\Gamma, \rho)$. One of its left inverses is given by $A^{-1} = a_-(a^{-1}P_\Gamma + Q_\Gamma)a_-^{-1}I$ with

$$a_-(t) := \begin{cases} (t-4)^{1/2}t^{-1} & \text{if } t \in \mathbf{T} \\ \left(\frac{t-3}{t-4}\right)^{-1/2} \frac{t-3}{t} & \text{if } t \in \mathbf{T}_1 \end{cases},$$

and $\operatorname{coker}(aP_\Gamma + Q_\Gamma) = \operatorname{span}\{a_-\}$. The equation $(aP_\Gamma + Q_\Gamma)\varphi = f$ is solvable in $L_p(\Gamma, \rho)$ if and only if

$$\int_\Gamma f(t)a_-(t)t^{-1}\,dt = 0 .$$

We conclude this section by the following statements and by a few illustrative examples.

Theorem 4.3. *Let $a, b \in PC(\Gamma)$. Then the operator $A = aP_\Gamma + bQ_\Gamma$ $(A = P_\Gamma aI + Q_\Gamma bI)$ is a $\Phi-$ or $\Phi_\pm$-operator on the space $L_p(\Gamma, \rho)$ if and only if the conditions 1-3 of Theorem 4.1 are satisfied. If these conditions are satisfied then A is in fact a Φ-operator and* $\operatorname{Ind} A = -\operatorname{ind} c^{p,\rho}$ *with $c = a/b$.*

Proof. The sufficiency of the conditions of the theorem follows from Theorem 4.1, and their necessity is a consequence of Proposition 3.1 in the present chapter, as well as of Theorem 1.1 in Chapter 7. ■

Corollary 4.1. *Let $a, b \in PC(\Gamma)$ and $a(t \pm 0)b(t \pm 0) \neq 0$ $(t \in \Gamma)$. Then the operator $A = aP_\Gamma + bQ_\Gamma$ $(A = P_\Gamma aI + Q_\Gamma bI)$ is characterized by the following properties which are valid in each of the spaces $L_p(\Gamma, \rho)$:*

$$\dim \ker A < \infty \quad \textit{and} \quad \dim \ker A^* < \infty .$$

Proof. Because of $a(t \pm 0) \neq 0$ and $b(t \pm 0) \neq 0$ on Γ one can choose two numbers, p_1 and p_2, such that

1. $L_{p_1}(\Gamma, \rho) \subset L_p(\Gamma, \rho) \subset L_{p_2}(\Gamma, \rho)$,
2. the function $c := a/b$ is $\{p_1, \rho\}$- and $\{p_2, \rho\}$-non-singular.

Now Theorem 4.1 involves that A is a Φ-operator on each of the spaces $L_{p_1}(\Gamma,\rho)$ and $L_{p_2}(\Gamma,\rho)$. Due to the fact that $L_p(\Gamma,\rho) \subset L_{p_2}(\Gamma,\rho)$ we have $\dim\ker A < \infty$, whereas the inclusion $L_{p_1}(\Gamma,\rho) \subset L_p(\Gamma,\rho)$ implies that $\dim\ker A^* < \infty$. ■

Corollary 4.2. *Let Γ be a simple closed curve and $a \in PC(\Gamma)$. For the normal solvability of the operator $A = aP_\Gamma + Q_\Gamma$ $(A = P_\Gamma aI + Q_\Gamma I)$ on $L_p(\Gamma,\rho)$ it is necessary and sufficient that the function a is $\{p,\rho\}$-non-singular.*

Proof. The sufficiency of this condition is a consequence of Theorem 4.1. We explain its necessity. Let the operator A be normally solvable. Then, by Theorem 5.1 in Chapter 7, at least one of the numbers $\dim\ker A$ and $\dim\ker A^*$ is equal to zero. Thus, A is a Φ- or $\Phi_\pm$-operator, and now Theorem 4.3 states that the function a is $\{p,\rho\}$-non-singular. ■

Corollary 4.3. *Let Γ be a composed curve which is constituted by non-closed arcs only, and let $a \in PC(\Gamma)$. Then the operator $A = aP_\Gamma + Q_\Gamma$ $(A = P_\Gamma aI + Q_\Gamma)$ is normally solvable if and only if the function a is $\{p,\rho\}$-non-singular.*

Proof. Complete the curve Γ to a simple closed curve $\widetilde{\Gamma}$ and define $\widetilde{A} := \widetilde{a}P_{\widetilde{\Gamma}} + Q_{\widetilde{\Gamma}}$ $(\widetilde{A} = P_{\widetilde{\Gamma}}\widetilde{a}I + Q_{\widetilde{\Gamma}})$, where $\widetilde{a}$ is the continuation of the function a to $\widetilde{\Gamma}$ which is equal to 1 on $\widetilde{\Gamma}\backslash\Gamma$. By Theorem 1.1 of Chapter 7, the operator A is normally solvable if and only if the operator $\widetilde{A}$ is. Now it only remains to apply Corollary 4.2. ■

As an example we are going to investigate the equation

$$\frac{1}{\pi i}\int_\Gamma \frac{\varphi(\tau)\,d\tau}{t-\tau} = \psi(t) \qquad (-S_\Gamma\varphi = \psi) \tag{4.3}$$

in the space $L_p(\Gamma)$, where Γ refers to the semi-circle $\Gamma = \{z \in \mathbf{C} : |z| = 1,\ \operatorname{Im} z < 0\}$.

In accordance with the general scheme we complete the curve Γ to the circle $\mathbf{T}$, and on $L_p(\mathbf{T})$ we consider the operator $A = \widetilde{a}P_{\mathbf{T}} + Q_{\mathbf{T}}$ with

$$\widetilde{a}(t) = \begin{cases} 1 & \text{if } \operatorname{Im} t > 0 \\ -1 & \text{if } \operatorname{Im} t < 0\,. \end{cases}$$

We have already studied this operator in Section 9.3. Applying the results obtained there as well as Theorem 4.2 one concludes that

1. In case $p = 2$ the equation (4.3) is not normally solvable.
2. If $p > 2$ then the equation (4.3) is solvable if and only if

$$\int_\Gamma \psi(t)(t^2-1)^{1/2}t^{-2}\,dt = 0\,. \tag{4.4}$$

If condition (4.4) is fulfilled then the equation (4.3) possesses the unique solution

$$\varphi(t) = \frac{\sqrt{1-t^{-2}}}{\pi i} \int_\Gamma \frac{\psi(\tau)\, d\tau}{\sqrt{1-\tau^{-2}}\,(t-\tau)} . \tag{4.5}$$

3. If $1 < p < 2$, then the equation (4.3) is solvable for each right hand side, and its general solution is given by

$$\varphi(t) = \frac{1}{\pi i \sqrt{1-t^{-2}}} \int_\Gamma \frac{\sqrt{1-\tau^{-2}}\psi(\tau)\, d\tau}{t-\tau} + \alpha(t^2-1)^{-1/2}$$

where α is an arbitrary constant.

Let us look at one more example. Consider the equation

$$\lambda\varphi(t) - \frac{1}{\pi i} \int_\Gamma \frac{\varphi(\tau)\, d\tau}{\tau - t} = \psi(t) \quad (\lambda^2 \neq 1) \tag{4.6}$$

in $L_p(\Gamma)$ where Γ stands for the lower semi–circle, $\Gamma = \{z \in \mathbf{C} : |z| = 1,\ \operatorname{Im} z \leq 0\}$. Rewrite the equation (4.6) in the form

$$(\lambda+1)\left(\frac{\lambda-1}{\lambda+1} P_\Gamma + Q_\Gamma\right)\varphi = \psi.$$

Following the general scheme, what we need is the factorization in $L_p(\mathbf{T})$ of the function

$$\widetilde{a}(t) := \begin{cases} 1 & \text{if } \operatorname{Im} t > 0 \\ \frac{\lambda-1}{\lambda+1} & \text{if } \operatorname{Im} t < 0 . \end{cases}$$

Given an arbitrary complex number γ we consider the functions b and c on $\mathbf{T}$ defined by

$$b(e^{i\theta}) = e^{i\gamma\theta} \quad (0 \leq \theta < 2\pi), \quad c(e^{i\theta}) = e^{-i\gamma\theta} \quad (-\pi \leq \theta < \pi) .$$

One immediately checks that

$$b(t)c(t) = \begin{cases} 1 & \text{if } \operatorname{Im} t > 0 \\ e^{-2\pi i\gamma} & \text{if } \operatorname{Im} t < 0 . \end{cases}$$

Thus, choosing the number γ in such a way that

$$e^{-2\pi i\gamma} = \frac{\lambda-1}{\lambda+1} ,$$

we obtain

$$\widetilde{a}(t) = b(t)c(t) .$$

Since the functions b and c do not possess common discontinuities, the function $\widetilde{a}$ is factorizable in $L_p(\mathbf{T})$ if and only if both functions b and c are, that is, if and only if $\pm\mathrm{Re}\,\gamma - 1/p$ is not an integer (cf. Theorem 2.2). Let

$$\pm\,\mathrm{Re}\,\gamma - \frac{1}{p} \notin \mathbf{Z}\,. \tag{4.7}$$

Denote by κ_1 and κ_2 the integers determined uniquely by

$$-1 < \kappa_1 - \mathrm{Re}\,\gamma - \frac{1}{p} < 0\,, \quad -1 < \kappa_2 + \mathrm{Re}\,\gamma - \frac{1}{p} < 0\,,$$

and set $\alpha = \kappa_1 - \gamma$ and $\beta = \kappa_2 + \gamma$. Then the functions b and c are factorizable:

$$b(t) = (t-1)^{-\alpha} t^{\kappa_1} \left(\frac{t-1}{t}\right)^{\alpha}\,, \quad c(t) = (t+1)^{-\beta} t^{\kappa_2} \left(\frac{t+1}{t}\right)^{\beta}$$

(see Theorem 2.3), and consequently, the equality

$$\widetilde{a}(t) = \left(\frac{t-1}{t}\right)^{\alpha} \left(\frac{t+1}{t}\right)^{\beta} t^{\kappa} (t-1)^{-\alpha} (t+1)^{-\beta}\,, \tag{4.8}$$

holds, where $\kappa = \kappa_1 + \kappa_2$, providing us with a factorization of the function $\widetilde{a}$ in the space $L_p(\mathbf{T})$.

By Theorem 4.2, if $\kappa = 0$ $(\kappa > 0, \kappa < 0)$ then the operator $A = \lambda I - S_\Gamma$ is invertible (invertible from the left, invertible from the right) on $L_p(\Gamma)$. One of its (one-sided) inverses is given by

$$(Rf)(t) = \frac{1}{\lambda^2 - 1} \left(\frac{t-1}{t}\right)^{\alpha} \left(\frac{t+1}{t}\right)^{\beta} (\lambda I + S_\Gamma) \left(\frac{t-1}{t}\right)^{-\alpha} \left(\frac{t+1}{t}\right)^{-\beta} I\,. \tag{4.9}$$

In case $\kappa < 0$,

$$\ker A = \mathrm{span}\,\{g, gt, \ldots, gt^{-\kappa-1}\} \tag{4.10}$$

with $g(t) = (t-1)^{\alpha}(t+1)^{\beta}$.

If $\kappa > 0$ then the equation $A\varphi = \psi$ is solvable if and only if

$$\int_\Gamma \psi(t)(t-1)^{\alpha}(t+1)^{\beta} t^{l}\, dt = 0 \quad (l = 0, 1, \ldots, \kappa - 1)\,. \tag{4.11}$$

9.5 Singular integral operators with continuous coefficients on a composed curve

The results of the preceding section can be further improved in case the coefficients of the singular integral operator are continuous functions.

Let Γ again stand for a composed curve, τ_k $(k = 1, \ldots, m)$ for all starting points and τ_k $(k = m+1, \ldots, 2m)$ all end points of its non-closed arcs. In the formulation of the main results of this section we need a spatially closed curve $(\subset \mathbf{R}^3)$ denoted by $\Lambda = \Lambda(\Gamma)$. This curve consists of all points (x, y, z) being subject to the conditions

$$x + iy \in \Gamma, \; -1 \leq z \leq 1, \quad \text{and} \quad (1 - z^2) \prod_{k=1}^{2m} (x + iy - \tau_k) = 0 \, .$$

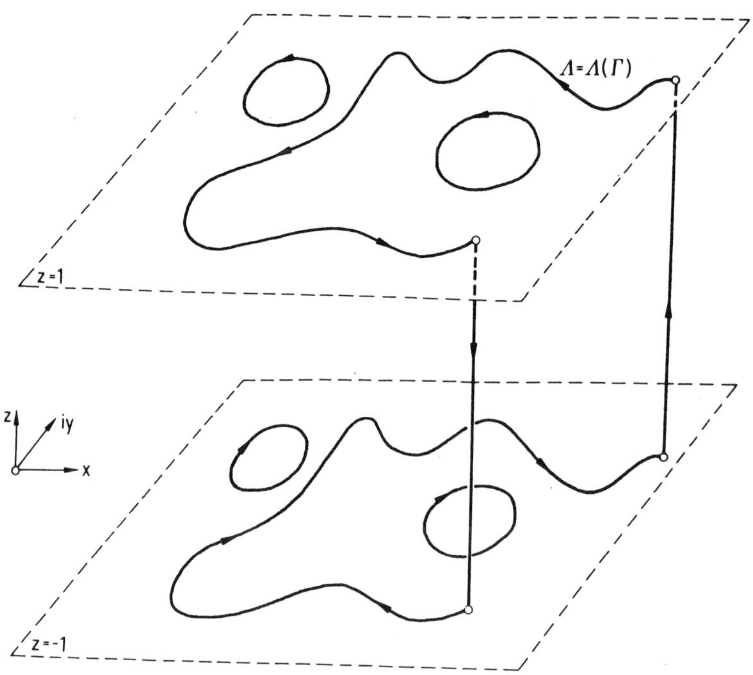

Figure 9.2

In other words: the curve $\Lambda(\Gamma)$ consists of two images of the curve Γ located in the planes $z = 1$ and $z = -1$ and of all straight lines which run parallel to the z–axis and which connect the starting points or the end points of the curve Γ . If, in particular, $\Gamma = [a, b]$ then $\Lambda(\Gamma)$ can be identified with the boundary of the rectangle $[a, b] \times [-1, 1]$.

We orient the curve $\Lambda(\Gamma)$ in such a way that its orientation coincides with the orientation of Γ in the plane $z = 1$ and that has the reverse orientation of Γ if $z = -1$.

By $\Omega_{p,\rho}$ we abbreviate the function which is defined on the curve $\Lambda(\Gamma)$ by

$$\Omega_{p,\rho}(t,z) := \begin{cases} \frac{z(1+a_k^2)-i(1-z^2)a_k}{1+z^2a_k^2} & \text{if} \quad t=\tau_k \text{ for some } k \le m \\ z & \text{if} \quad t \in \Gamma\backslash\{\tau_1,\ldots,\tau_{2m}\} \\ \frac{z(1+a_k^2)+i(1-z^2)a_k}{1+z^2a_k^2} & \text{if} \quad t=\tau_k \text{ for some } k \ge m+1\,, \end{cases}$$

where $a_k := \cot(\delta(\tau_k)/2)$, and the function $\delta(t)$ is defined by equality (1.3).

Now let c and d be continuous functions on Γ and $A := cI + dS_\Gamma$. The function $\mathcal{A}_{p,\rho}(t,z)$ defined on the curve $\Lambda(\Gamma)$ by

$$\mathcal{A}_{p,\rho}(t,z) := c(t) + d(t)\Omega_{p,\rho}(t,z)$$

is called the *symbol of the operator* A *acting on* $L_p(\Gamma,\rho)$. In other words: in the planes $z=1$ and $z=-1$, the function $\mathcal{A}_{p,\rho(t,z)}$ is defined by

$$\mathcal{A}_{p,\rho}(t,1) = c(t) + d(t), \quad \mathcal{A}_{p,\rho}(t,-1) = c(t) - d(t)\,,$$

and on each of the straight lines $t=\tau_k$ $(k=1,\ldots,2m)$ the range of the function $\mathcal{A}_{p,\rho}(\tau_k,z)$ is a certain curve s_k joining the points $\zeta_k = c(\tau_k)+d(\tau_k)$ and $w_k = c(\tau_k)-d(\tau_k)$ in the complex plane. We are going to show that $s_k = l(w_k,\zeta_k;\delta(\tau_k))$ if $k=1,2,\ldots,m$ and $s_k = l(\zeta_k,w_k;\delta(\tau_k))$ if $k=m+1,\ldots,2m$.

In Section 9.1 we derived the parameter representation

$$w = \zeta_k f(\tau_k,\mu) + w_k(1-f(\tau_k,\mu)) \quad (0 \le \mu \le 1) \tag{5.1}$$

of the circular arcs (resp. straight lines) $l(w_k,\zeta_k;\delta(\tau_k))$. The equality (5.1) can be rewritten as $z = c(\tau_k) + d(\tau_k)\,(2f(\tau,\mu)-1)$. The following identities can be easily checked:
If $\delta(\tau_k) \ne \pi$ then

$$\begin{aligned} 2f(\tau_k,\mu)-1 &= \frac{2\sin(\theta\mu)\exp(i\theta(\mu-1))}{\sin\theta} - 1 \\ &= \frac{\tan\frac{\theta(2\mu-1)}{2}\left(1+\tan^2\frac{\theta}{2}\right)}{\left(1+\tan^2\frac{\theta(2\mu-1)}{2}\right)\tan\frac{\theta}{2}} - i\frac{\tan^2\frac{\theta}{2}-\tan^2\frac{\theta(2\mu-1)}{2}}{\left(1+\tan^2\frac{\theta(2\mu-1)}{2}\right)\tan\frac{\theta}{2}} \end{aligned} \tag{5.2}$$

with $\theta = \pi - \delta(\tau_k)$. Put $z := (1/a_k)\tan(\theta(2\mu-1)/2)$ with $a_k := \cot(\delta(\tau_k)/2)$. If μ runs through the interval $[0,1]$ then z runs through $[-1,1]$. Now substitute $\tan\frac{\theta(2\mu-1)}{2} = a_k z$ and $\tan(\theta/2) = a_k$ in (5.2) to obtain

$$2f(\tau_k,\mu)-1 = \frac{z(1+a_k^2)-ia_k(1-z^2)}{1+a_k^2z^2} = \Omega_{p,\rho}(\tau_k,z) \quad (k=1,\ldots,m)\,.$$

In case $\delta(\tau_k)=\pi$, we set $z = 2\mu-1$ and find

$$2f(\tau_k,\mu)-1 = 2\mu-1 = z = \Omega_{p,\rho}(\tau_k,z)\,.$$

In the same way, in case $k \leq m$ we can rewrite equation (5.1) in the form

$$w = c(\tau_k) + d(\tau_k)\Omega_{p.\rho}(\tau_k, z) \quad (= \mathcal{A}_{p,\rho}(\tau_k, z)) \quad (-1 \leq z < 1)\,.$$

Hence, $s_k = l(w_k, \zeta_k; \delta(\tau_k))$.

Setting

$$z := \begin{cases} a_k^{-1}\tan\left((1-2\mu)\frac{\theta}{2}\right) & \text{if} \quad \delta(\tau_k) \neq \pi \\ 1 - 2\mu & \text{if} \quad \delta(\tau_k) = \pi \end{cases}$$

one can analogously show that $s_k = l(\zeta_k, w_k; \delta(\tau_k))$ if $k = m+1, \ldots, 2m$. Denote by ind $\mathcal{A}_{p,\rho}$ the increase of the argument of the function $\mathcal{A}_{p,\rho}(t, z)$ if the point (t, z) runs along the curve $\Lambda(\Gamma)$ modulo 2π .

Theorem 5.1. *Let* Γ *be a composed curve and assume the functions* c *and* d *to be continuous on* Γ *. Then the operator* $A = cI + dS_\Gamma$ $(A = cI + S_\Gamma dI)$ [1] *is at least one–sided invertible on* $L_p(\Gamma, \rho)$ *if and only if*

$$\mathcal{A}_{p,\rho}(t, \mu) \neq 0 \quad ((t, \mu) \in \Lambda(\Gamma))\,. \tag{5.3}$$

If the condition (5.3) is satisfied and $\kappa = \operatorname{ind} \mathcal{A}_{p,\rho}$ *then the operator* A *is invertible, invertible only from the right or invertible only from the left in dependence on whether the number* κ *is equal to zero, negative or positive. If* $\kappa > 0$ *then* $\dim \operatorname{coker} A = \kappa$ *, and if* $\kappa < 0$ *then* $\dim \ker A = |\kappa|$.

Proof. We shall establish the equivalence of condition (5.3) with the conditions 1-3 in Theorem 4.1. First let us formulate the conditions 1-3 in Theorem 4.1 for the operator $A = (c+d)P_\Gamma + (c-d)Q_\Gamma$. Due to the continuity of the functions c and d on Γ , the condition 1. means that $c(t)+d(t) \neq 0$ and $c(t)-d(t) \neq 0$, which in turn is equivalent to $\mathcal{A}_{p,\rho}(t, 1) \neq 0$ and $\mathcal{A}_{p,\rho}(t, -1) \neq 0$. The condition 2 can be rewritten as $c(\tau_k) + d(\tau_k)(2f(\tau, \mu) - 1) \neq 0$ which is tantamount to $\mathcal{A}_{p,\rho}(\tau_k, z) \neq 0$ $(k = 1, \ldots, m)$, as we have seen above. Finally, the condition 3 is equivalent to $\mathcal{A}_{p,\rho}(\tau_k, z) \neq 0$ $(k = m+1, \ldots, 2m)$. This proves the equivalence of the conditions (5.3) and 1-3. In order to finish the proof of the theorem it remains to verify that $\operatorname{ind} \mathcal{A}_{p,\rho} = \operatorname{ind} g^{p,\rho}$ with $g = (c+d)/(c-d)$.

Let $\Gamma_1 (\subset \Gamma)$ be a simple arc or a simple closed curve, and let Γ' and Γ'' denote the two images of the curve Γ_1 located in the planes $z = 1$ and $z = -1$, respectively. The orientation on Γ' and Γ'' might coincide with the orientation on the curve $\Lambda(\Gamma)$. Then

$$\begin{aligned} [\arg g(t)]_{t\in\Gamma_1} &= \left[\arg \frac{c(t) + d(t)}{c(t) - d(t)}\right]_{t\in\Gamma_1} \\ &= [\arg(c(t) + d(t))]_{\Gamma_1} - [\arg(c(t) - d(t))]_{\Gamma_1} \\ &= [\arg \mathcal{A}_{p,\rho}(t, 1)]_{\Gamma'} - [\arg \mathcal{A}_{p,\rho}(t, -1)]_{\Gamma''}\,. \end{aligned}$$

[1] In this section it is more convenient to represent singular integral operators in this form.

Let $\gamma_k(\subset \Lambda(\Gamma))$ stand for the straight line joining the point $(\tau_k, -1)$ to $(\tau_k, 1)$. Then one has for $k = 1, \ldots, m$:

$$\begin{aligned}[\arg \mathcal{A}_{p,\rho}(\tau_k, z)]_{\gamma_k} &= [\arg\,(c(\tau_k) + d(\tau_k)\Omega_{p,\rho}(\tau_k, z))]_{z=-1}^{1} \\ &= [\arg\,(c(\tau_k) + d(\tau_k)(2f(\tau_k, \mu) - 1))]_{\mu=0}^{1} \\ &= [\arg(g(\tau_k)f(\tau_k, \mu) + 1 - f(\tau_k, \mu))]_{\mu=0}^{1} \\ &= [\arg\, g^{p,\rho}(\tau_k, \mu)]_{\mu=0}^{1} .\end{aligned}$$

Analogously one checks the validity of the equality $[\arg \mathcal{A}_{p,\rho}(\tau_k, z)]_{\gamma_k} = [\arg\, g^{p,\rho}(\tau_k, \mu)]_{\mu=0}^{1}$ in case $k = m+1, \ldots, 2m$. Summing up all these increments one obtains the equality $\operatorname{ind} \mathcal{A}_{p,\rho} = \operatorname{ind} g^{p,\rho}$. ∎

Corollary 5.1. *For the operator A to be a Φ- or $\Phi_\pm$-operator on $L_p(\Gamma, \rho)$ it is necessary and sufficient that its symbol $\mathcal{A}_{p,\rho}$ does not vanish on $\Lambda(\Gamma)$. If this condition is satisfied then A is a Φ-operator, and* $\operatorname{Ind} A = -\operatorname{ind} \mathcal{A}_{p,\rho}$.

Indeed, it follows without much ado from Corollary 4.1 that if the operator A is a Φ- or $\Phi_\pm$-operator, then it is at least one-sided invertible. In this case, Theorem 5.1 implies that $A_{p,\rho}(t, z) \neq 0$ on $\Lambda(\Gamma)$ and, furthermore, that A is a Φ-operator with index $\operatorname{Ind} A = -\operatorname{ind} \mathcal{A}_{p,\rho}$.

For illustration, let us consider a simple example. Let $\Gamma := [-1, 1]$, $A := itI + S_\Gamma$, and consider the operator A on the space $L_p(\Gamma, \rho)$ with $\rho(t) := (t+1)^{\beta_1}(1-t)^{\beta_2}$. Then the curve $\Lambda(\Gamma)$ equals the boundary of the square $[-1, 1] \times [-1, 1]$, and the range of the symbol $\mathcal{A}_{p,\rho}(t, z)$ consists of the two straight lines $l_1 = \{it - 1\}$ and $l_2 = \{it + 1\}$ $(-1 \leq t \leq 1)$ and of the two arcs (resp. straight lines) $l_3 = l(-1-i, 1-i; 2\pi(1+\beta_1)/p)$ and $l_4 = l(1+i, -1+i; 2\pi(1+\beta_2)/p)$.

In no case do the straight lines l_1 and l_2 meet the point $\lambda = 0$, whereas the arcs $l_k (k = 3, 4)$ run through this point $\lambda = 0$ if and only if $2\pi(1+\beta_{k-2})/p = 2\pi - (\pi/2)$, that is, if $4(1+\beta_{k-2}) = 3p$. Thus, the operator A is at least one-sided invertible on $L_p(\Gamma, \rho)$ if and only if the two conditions $4(1+\beta_1) \neq 3p$ and $4(1+\beta_2) \neq 3p$ are satisfied. Further it is not hard to see that:

1) If $4(1+\beta_1) < 3p$ and $4(1+\beta_2) < 3p$ then $\operatorname{ind} \mathcal{A}_{p,\rho} = 1$. Hence, the operator A is invertible from the left, and $\dim \operatorname{coker} A = 1$.

2) If the numbers $4(1+\beta_1) - 3p$ and $4(1+\beta_2) - 3p$ have different signs then $\operatorname{ind} \mathcal{A}_{p,\rho} = 0$, and so the operator A is invertible.

3) If $4(1+\beta_1) > 3p$ and $4(1+\beta_2) > 3p$ then $\operatorname{ind} \mathcal{A}_{p,\rho} = -1$, the operator A is invertible from the right, and $\dim \ker A = 1$.

If A is considered as an operator acting on $L_p(-1, 1)$ (without weight) then there is no p for which A is invertible. Indeed, if $1 < p < 4/3$ then A is invertible only from the right, if $4/3 < p < \infty$ is invertible from the left, and in case $p = 4/3$ it is neither right nor left invertible.

Sometimes the following criterion for the one-sided invertibility proves to be useful.

Theorem 5.2. *Let $\tau_1, \ldots, \tau_m$ refer to the starting and $\tau_{m+1}, \ldots, \tau_{2m}$ end points of all non-closed arcs of the composed curve Γ ; furthermore let $c, d \in C(\Gamma), c^2(t) - d^2(t) \neq 0$ on Γ; $g := (c+d)/(c-d)$; $\rho(t) := \prod_{k=1}^{2m} |t-\tau_k|^{\beta_k}$; $\gamma_k = -(1/2\pi i)\ln g(\tau_k)$ $(k = 1, \ldots, m)$ and $\gamma_k = (1/2\pi i)\ln g(\tau_k)$ $(k = m+1, \ldots, 2m)$. Then the operator $A = cI + dS_\Gamma$ $(A = cI + S_\Gamma dI)$ is at least one-sided invertible on $L_p(\Gamma, \rho)$ if and only if none of the numbers*

$$\operatorname{Re} \gamma_k - (1 + \beta_k)/p$$

is an integer.

Proof. The operator A can be written as $A = (c+d)P_\Gamma + (c-d)Q_\Gamma = (c-d)(gP_\Gamma + Q_\Gamma)$. Let $\widetilde{\Gamma}(\supset \Gamma)$ denote a closed composed curve and $\widetilde{g}$ the continuation of g onto $\widetilde{\Gamma}$ satisfying $\widetilde{g}|(\widetilde{\Gamma}\backslash\Gamma) = 1$ By Theorem 4.1, the operator A is at least one-sided invertible if and only if the function g is $\{p, \rho\}$-non-singular, and the latter is equivalent to the $\{p, \rho\}$-non-singularity of the function $\widetilde{g}$.

Because

$$\begin{aligned} \widetilde{g}(\tau_k - 0)/\widetilde{g}(\tau_k + 0) &= \frac{1}{g(\tau_k)} \quad (k = 1, \ldots, m), \\ \widetilde{g}(\tau_k - 0)/\widetilde{g}(\tau_k + 0) &= g(\tau_k) \quad (k = m+1, \ldots, 2m), \end{aligned}$$

Theorem 1.2 yields that the function $\widetilde{g}$ is $\{p, \rho\}$-non-singular if and only if the numbers $\operatorname{Re} \gamma_k - (1 + \beta_k)/p$ are not integers. ■

Consider, for example, the operator $A = S_\Gamma$. Then $g = -1$ and $\gamma_k = 1/2 + n$ for a certain integer n, and Theorem 5.1 states that the operator S_Γ is at least one-sided invertible if and only if

$$2(1 + \beta_k) \neq p \quad (k = 1, \ldots, 2m).$$

The following theorem completes this result.

Theorem 5.3. *For the one-sided invertibility of the operator S_Γ it is necessary and sufficient that $2(1 + \beta_k) \neq p$ for all $k = 1, \ldots, 2m$. If $2(1 + \beta_k) \neq p$ and if n denotes the number of those points τ_k for which $2(1 + \beta_k) < p$ then :*

1. *in case $m > n$ the operator S_Γ is right invertible and* $\dim \ker S_\Gamma = m - n$;
2. *in case $n > m$ the operator S_Γ is left invertible and* $\dim \operatorname{coker} S_\Gamma = n - m$;
3. *in case $n = m$ the operator S_Γ is invertible.*

Proof. It follows immediately from the definition of the symbol $\mathcal{A}_{p,\rho}(t, z)$ of the operator A that its range consists of $2m$ circular arcs or straight lines joining the points

-1 and 1. If $2(1+\beta_k) \neq p$ $(k = 1, \ldots, 2m)$ then no straight lines appear in the range of the symbol, and thus $\mathcal{A}_{p,\rho}(t,z) \neq 0$ on $\Lambda(\Gamma)$. In this case,

$$\text{ind } \mathcal{A}_{p,\rho} = \frac{1}{2\pi} \sum_{k=1}^{m} [\arg \Omega_{p,\rho}(\tau_k, z)]_{z=-1}^{1} + \frac{1}{2\pi} \sum_{k=m+1}^{2m} [\arg \Omega_{p,\rho}(\tau_k, z)]_{z=1}^{-1} .$$

It is not hard to see that if $2(1+\beta_k) < p$ then $[\arg \Omega_{p,\rho}(\tau_k, z)]_{-1}^{1} = \pi$ in case $k \leq m$, and $[\arg \Omega_{p,\rho}(\tau_k, z)]_{1}^{-1} = \pi$ in case $k > m$. If $2(1+\beta_k) > p$ then $[\arg \Omega_{p,\rho}(\tau_k, z)]_{-1}^{1} = -\pi$ in case $k \leq m$, and $[\arg \Omega_{p,\rho}(\tau_k, z)]_{1}^{-1} = -\pi$ if $k > m$. This shows that $\text{ind } \mathcal{A}_{p,\rho} = n - m$ with n denoting the number of those points τ_k $(k = 1, \ldots, 2m)$ at which $2(1+\beta_k) < p$. Now it remains to apply Theorem 5.1. ■

Next we state another formula for the inversion of singular operators. Let $c, d \in PC(\Gamma)$ with $c^2 - d^2 \neq 0$, suppose the function $g := (c+d)(c-d)^{-1}$ to be $\{p, \rho\}$-non-singular, and let $\widetilde{g}_- t^\kappa \widetilde{g}_+$ stand for the generalized factorization of the function $\widetilde{g}$ with respect to the curve $\widetilde{\Gamma}$ in $L_p(\widetilde{\Gamma}, \rho)$. In this situation, the operators $A := cI + dS_\Gamma$ and $B := cI + S_\Gamma dI$ are at least one-sided invertible, and their corresponding inverses (see (4.2)) can be written in the following form:

$$\begin{aligned} A^{-1} &= ((c+d)P_\Gamma + (c-d)Q_\Gamma)^{-1} \\ &= g_-(c+d)^{-1}((c-d)P_\Gamma + (c+d)Q_\Gamma)(c-d)^{-1} g_-^{-1} I \\ &= g_-(c+d)^{-1}(cI - dS_\Gamma)(c-d)^{-1} g_-^{-1} I . \end{aligned}$$

Thus,

$$A^{-1} = c^* I - d^* z S_\Gamma z^{-1} I \tag{5.4}$$

with $c^* := c(c^2 - d^2)^{-1}$, $d^* := d(c^2 - d^2)^{-1}$, and $z := g_-(c-d)$. Analogously

$$B^{-1} = c^* I - z_1^{-1} S_\Gamma d^* z_1 I \tag{5.5}$$

with $z_1 := g_+(c-d)$.

Now we are going to describe the structure of the factors z and z_1 in the special case when the functions c and d satisfy a Hölder condition on Γ . Let $\tau_1, \ldots, \tau_m$ and $\tau_{m+1}, \ldots, \tau_{2m}$ denote the starting and end points of all non-closed subarcs of the curve Γ , suppose the functions c and d to satisfy a Hölder condition on Γ , and define

$$\begin{aligned} g &:= (c+d)(c-d)^{-1}; \\ \gamma_k &:= \frac{-1}{2\pi i} \ln g(\tau_k) \quad (k = 1, \ldots, m) , \\ \gamma_k &:= \frac{1}{2\pi i} \ln g(\tau_k) \quad (k = m+1, \ldots, 2m) , \end{aligned} \tag{5.6}$$

and

$$\rho(t) = \prod_{k=1}^{2m} |t - \tau_k|^{\beta_k} \quad \text{with} \quad \frac{1+\beta_k}{p} - 1 < \operatorname{Re} \gamma_k < \frac{1+\beta_k}{p} .$$

Under these restricitions we have

Theorem 5.4. *The operators $A = cI + dS_\Gamma$ and $B = cI + S_\Gamma dI$ are at least one-sided invertible on $L_p(\Gamma, \rho)$. Their inverses are given by (5.4) and (5.5) with the factors*

$$z(t) := \prod_{k=1}^{2m} (t-\tau_k)^{-\gamma_k} \cdot w(t) , \quad z_1(t) := \prod_{k=1}^{2m} (t-\tau_k)^{\gamma_k} \cdot w_1(t), \tag{5.7}$$

where the functions $w(t)$ and $w_1(t)$ together with their inverses satisfy a Hölder condition on the curve Γ.

Proof. The invertibility of the operators A and B is a consequence of Theorem 5.2. Let $\widetilde{\Gamma}(\supset \Gamma)$ be a closed curve, and $\widetilde{\Psi} := \Psi_{\tau_1,\gamma_1} \cdots \Psi_{\tau_{2m},\gamma_{2m}}$.[1] As we have seen in the preceding section, the function $\widetilde{g}$ admits the factorization $\widetilde{g} = \widetilde{g}_- t^\kappa g_+$ in $L_p(\widetilde{\Gamma}, \rho)$ with the factors $\widetilde{g}_+ = \widetilde{\Psi}_+ \widetilde{d}_+$ and $\widetilde{g}_- = \widetilde{\Psi}_- \widetilde{d}_-$ where the functions $d_\pm$ together with their inverses satisfy a Hölder condition on Γ. The functions $g_\pm$ are just the restrictions onto Γ of the functions $\widetilde{g}_\pm$ in the factorization of the function g. Now set $\Psi_\pm := \widetilde{\Psi}_\pm | \Gamma$. Then formula (2.5) yields

$$\Psi_+(t) = h_1(t) \prod_{k=1}^{2m} (t-\tau_k)^{\gamma_k}, \quad \Psi_-(t) = h_2(t) \prod_{k=1}^{2m} (t-\tau_k)^{-\gamma_k},$$

with the functions h_1 and h_2 together with their inverses satisfying a Hölder condition on Γ. This shows that the functions z and z_1 can be represented in the form (5.7). ∎

If the coefficients of the singular integral are constant then the function $z(t)$ can be explicitly determined and the inverse operator can be effectively computed. To illustrate this fact we consider the singular integral operator with constant coefficients on the space $L_p[\alpha, \beta]$.

Theorem 5.5. *Let $c, d \in \mathbf{C}$, $\Gamma = [\alpha, \beta]$ and $A = cI + dS_\Gamma$. For the one-sided invertibility of the operator A on $L_p[\alpha, \beta]$ it is necessary and sufficient that the inequalities*

$$c^2 - d^2 \neq 0 \tag{5.8}$$

and

$$\operatorname{Re} \gamma \neq 1/p , \quad \operatorname{Re} \gamma \neq 1/q \tag{5.9}$$

[1] The functions $\Psi_{\tau\gamma}$ are defined by (2.2).

are satisfied, where

$$\gamma = \frac{1}{2\pi i} \ln \frac{c+d}{c-d}$$

with

$$\max(1/p, 1/q) - 1 < \operatorname{Re}\gamma \leq \max(1/p, 1/q)$$

and $p^{-1} + q^{-1} = 1$.

Let the conditions (5.8) and (5.9) be satisfied. If

$$1/q < \operatorname{Re}\gamma < 1/p$$

then the operator A *is invertible from the right-hand side,*

$$\ker A = \operatorname{span}\{(t-\alpha)^{-1+\gamma}(\beta - t)^{\gamma}\}, \tag{5.10}$$

and the operator

$$(Rf)(t) = c^* f(t) - \frac{d^*(t-\alpha)^{\gamma-1}}{(\beta - t)^{\gamma}\pi i} \int_{\alpha}^{\beta} \frac{(\beta-\tau)^{\gamma} f(\tau}{(\tau-\alpha)^{\gamma-1}(\tau - t)}\, d\tau , \tag{5.11}$$

with $c^* = c(c^2 - d^2)^{-1}$ *and* $d^* = d(c^2 - d^2)^{-1}$, *is one of its right inverses. If*

$$1/p < \operatorname{Re}\gamma < 1/q$$

then the operator A *is invertible from the left, the operator*

$$(Rf)(t) = c^* f(t) - \frac{d^*(t-\alpha)^{\gamma}}{(\beta - t)^{\gamma-1}\,\pi i} \int_{\alpha}^{\beta} \frac{(\beta-\tau)^{\gamma-1} f(\tau)\, d\tau}{(\tau-\alpha)^{\gamma}(\tau - t)} \tag{5.12}$$

is one of its left inverses A , *and the equation* $A\varphi = f$ *is solvable if and only if*

$$\int_{\alpha}^{\beta} \frac{f(t)(\beta - t)^{\gamma-1}}{(t-\alpha)^{\gamma}}\, dt = 0 . \tag{5.13}$$

In the remaining cases, i.e. if

$$\max(1/p, 1/q) - 1 < \operatorname{Re}\gamma < \min(1/p, 1/q) ,$$

the operator A *is invertible and*

$$(A^{-1}f)(t) = c^* f(t) - \frac{d^*}{\pi i} \int_{\alpha}^{\beta} \left(\frac{t-\alpha}{\beta - t}\right)^{\gamma} \left(\frac{\beta-\tau}{\tau-\alpha}\right)^{\gamma} \frac{f(\tau)\, d\tau}{\tau - t} . \tag{5.14}$$

Proof. The necessity of the conditions (5.8) and (5.9) follows from Theorem 4.1. Let $\widetilde{\Gamma}$ ($[\alpha,\beta] \subset \widetilde{\Gamma}$) be a simple closed curve surrounding the point $z = i$, and write Γ_α and Γ_β for the rays starting at the point i and running through the points α and β, respectively.

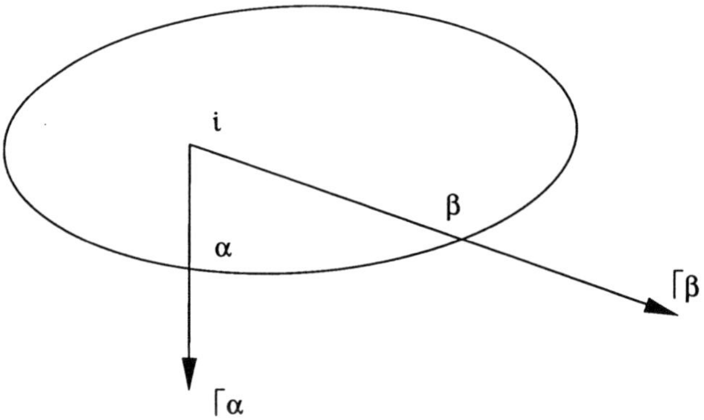

Figure 9.3

Our goal is the explicit factorization of the function

$$g(t) = \begin{cases} \frac{c+d}{c-d} & \text{if} \quad t \in [\alpha,\beta] \\ 1 & \text{if} \quad t \in \widetilde{\Gamma}\backslash[\alpha,\beta] \,. \end{cases}$$

Once this factorization has been found then all assertions of the theorem are consequences of Theorem 4.1 and of equality (5.4).

We consider the two functions $a(z) = (z-i)^{-\gamma}$ and $b(z) = (z-i)^{\gamma}$ which are analytic in the complex plane except for the cut along the rays Γ_α and Γ_β, respectively. Their branches are selected in such a way that the equality $a(t)b(t) \equiv 1$ holds along $\widetilde{\Gamma}\backslash[\alpha,\beta]$. In this situation one has $a(t)b(t) = \exp(2\pi i\gamma) = (c+d)/(c-d)$ along the interval $[\alpha,\beta]$. Thus, $g(t) = a(t)b(t)$ ($t \in \widetilde{\Gamma}$).

If $\max(1/p, 1/q) - 1 < \operatorname{Re}\gamma < \min(1/p, 1/q)$ then

$$\frac{1}{p} - 1 < \operatorname{Re}\gamma < \frac{1}{p} \quad \text{and} \quad \frac{1}{p} - 1 < -\operatorname{Re}\gamma < \frac{1}{p} \,.$$

Under these restrictions, both functions a and b admit a factorization in $L_p(\widetilde{\Gamma})$ with vanishing indices. Since these functions do not possess common discontinuities, the function g also admits a factorization with vanishing index:

$$g(t) = g_-(t)g_+(t)$$

with

$$g_+(z) = \left(\frac{\beta - z}{z - \alpha}\right)^{\gamma}, \quad g_-(z) = \left(\frac{z-i}{\beta - z}\right)^{\gamma} \left(\frac{z-\alpha}{z-i}\right)^{\gamma}.$$

For the points $t, \tau \in [\alpha, \beta]$ we find in particular

$$z(t)z^{-1}(\tau) = \frac{(t-\alpha)^\gamma(\beta-\tau)^\gamma}{(\beta-t)^\gamma(\tau-\alpha)^\gamma},$$

from which (5.14) follows.

If $1/p < \operatorname{Re}\gamma < 1/q$ then

$$\frac{1}{p} - 1 < \operatorname{Re}\gamma - 1 < \frac{1}{p} \quad \text{and} \quad \frac{1}{p} - 1 < -\operatorname{Re}\gamma < \frac{1}{p}$$

and

$$g(t) = g_-(t)(t-i)g_+(t)$$

with

$$g_-(z) = \left(\frac{z-i}{\beta-z}\right)^{\gamma-1}\left(\frac{z-\alpha}{z-i}\right)^\gamma, \quad g_+(z) = \frac{(\beta-z)^{\gamma-1}}{(z-\alpha)^\gamma}.$$

Taking into account Theorem 4.2 this yields condition (5.13), and the equality (5.4) shows that the operator

$$(R_1 f)(t) = c^* f(t) - \frac{d^*(t-\alpha)^\gamma}{\pi i(\beta-t)^{\gamma-1}(t-i)}\int_\alpha^\beta \frac{(\beta-\tau)^{\gamma-1}(\tau-i)f(\tau)\,d\tau}{(\tau-\alpha)^\gamma(\tau-t)}$$

is one of the left inverses of the operator A. The operator R_1 can be rewritten as

$$(R_1 f)(t) = (Rf)(t) - \frac{d^*(t-\alpha)^\gamma}{\pi i(\beta-t)^{\gamma-1}(t-i)}\int_\alpha^\beta \frac{(\beta-\tau)^{\gamma-1}}{(\tau-\alpha)^\gamma} f(\tau)\,d\tau,$$

where R refers to the operator defined by (5.12). Now we conclude from (5.13) that the operators R_1 and R coincide on the image of the operator A; hence R is a left inverse for A, too.

Finally, if $1/q < \operatorname{Re}\gamma < 1/p$ then

$$g(t) = g_-(t)(t-i)^{-1}\, g_+(t)$$

with

$$g_-(z) = \left(\frac{z-\alpha}{z-i}\right)^{\gamma-1}\left(\frac{z-i}{\beta-z}\right)^\gamma, \quad g_+(z) = \frac{(\beta-z)^\gamma}{(z-\alpha)^{\gamma-1}}.$$

Using this factorization, (5.10) follows from Theorem 4.2, and from (5.4) we conclude that the operator

$$(R_2 f)(t) = c^* f(t) - \frac{d^*(t-\alpha)^{\gamma-1}(t-i)}{\pi i(\beta-t)^\gamma}\int_\alpha^\beta \frac{(\beta-\tau)^\gamma f(\tau)\,d\tau}{(\tau-\alpha)^{\gamma-1}(\tau-i)(\tau-t)}$$

is one of the right inverses of the operator A . Comparing the operator R_2 with the operator R defined by (5.11) we see that

$$(Rf)(t) - (R_2 f)(t) = \frac{d^*(t-\alpha)^{\gamma-1}}{\pi i(\beta - t)^{\gamma}} \int_{\alpha}^{\beta} \frac{(\beta-\tau)^{\gamma} f(\tau)\, d\tau}{(\tau-\alpha)^{\gamma-1}(\tau - i)} \in \ker A$$

for all $f \in L_p(\Gamma)$. This shows that the operator R is a right inverse of A , too. ∎

As an example we consider the equation

$$\varphi(t) + \frac{1}{\pi}\int_0^1 \frac{\varphi(s)\, ds}{s-t} = f(t) \tag{5.15}$$

in the space $L_p(0,1)$. Here, $(c+d)(c-d)^{-1} = i = \exp(\pi i/2)$ and $\gamma = 1/4$. If $4/3 < p < 4$ then

$$\gamma < \min(1/p, 1/q) .$$

For these values of p , the equation (5.15) is uniquely solvable for an arbitrary right–hand side, and its solution is of the form

$$\varphi(t) = \frac{1}{2} f(t) - \frac{1}{2\pi}\sqrt[4]{\frac{t}{1-t}} \int_0^1 \sqrt[4]{\frac{1-\tau}{\tau}} \frac{f(\tau)}{\tau - t}\, d\tau .$$

If $1 < p < 4/3$ then the equation (5.15) is solvable in $L_p(0,1)$ for any right–hand side, and its general solution is

$$\varphi(t) = \frac{1}{2} f(t) - \frac{1}{2\pi \sqrt[4]{t^3(1-t)}} \int_0^1 \frac{\sqrt[4]{\tau^3(1-\tau)}}{(\tau - t)} f(\tau)\, d\tau + \lambda \sqrt[4]{t^3(1-t)} ,$$

with λ standing for an arbitrary constant.

If $p > 4$ then the equation (5.15) is solvable if and only if

$$\int_0^1 \frac{f(t)\, dt}{\sqrt[4]{t^3(1-t)}} = 0 .$$

If this condition is satisfied then the solution is unique and given by

$$\varphi(t) = \frac{1}{2} f(t) - \frac{\sqrt[4]{t(1-t)^3}}{2\pi} \int_0^1 \frac{f(\tau)\, d\tau}{\sqrt[4]{\tau(1-\tau)^3}(\tau - t)} .$$

9.6 The case of the real axis

The results obtained in the preceding sections can be generalized to unbounded curves in a natural manner. For the sake of simplicity we restrict ourselves to the case when the curve under consideration is the real axis.

Denote by $PC(\overline{\mathbf{R}})$ the collection of all functions a which are left-sided continuous on the real axis $\mathbf{R}$, possess finitely many discontinuities of the first kind and for which the limits $a(+\infty)$ and $a(-\infty)$ exist. Here $\overline{\mathbf{R}}$ stands for the extended (by ∞) real axis. Further we let $L_p(\mathbf{R}, \rho_0)$ refer to the space L_p over the axis $\mathbf{R}$ provided with the weight

$$\rho_0(t) := (1+t^2)^{\beta/2} \prod_{k=1}^{N} |t-t_k|^{\beta_k}$$

with $t_1, \dots, t_N \in \mathbf{R}$, $-1 < \beta_k < p-1$, $1 < p < \infty$ and

$$-1 < \sum_{k=1}^{N} \beta_k + \beta < p-1 .$$

Given $a \in PC(\overline{\mathbf{R}})$ we define the function $a_{p,\rho_0}(t,\mu)$ on the set $\overline{\mathbf{R}} \times [0,1]$ by

$$a_{p,\rho_0}(t,\mu) := a(t+0)f(t,\mu) + a(t)(1-f(t,\mu)) \quad \text{if} \quad t \in \mathbf{R},\ 0 \le \mu \le 1 ,$$

and

$$a_{p,\rho_0}(\infty,\mu) := a(-\infty)f_\delta(\mu) + a(+\infty)(1-f_\delta(\mu)) \quad (0 \le \mu \le 1) .$$

Herein, $\delta := 2\pi(p-1-\beta-\sum_{k=1}^{N}\beta_k)/p$, and the functions $f_\delta(\mu)$ and $f(t,\mu)$ are defined in Section 9.1.

The range of the function a_{p,ρ_0} is a closed curve which results from adding to the range of the function a the $(N+1)$ circular arcs (resp. straight lines) $(a(t_k), a(t_k+0);\ \delta(t_k))$ $(k = 1, \dots, N)$ and $l(a(+\infty), a(-\infty);\ \delta)$. If $a_{p,\rho_0}(t,\mu) \neq 0$ $(t \in \overline{\mathbf{R}},\ 0 \le \mu \le 1)$ then the function a is said to be $\{p, \rho_0\}$–non–singular, and the winding number of the curve a_{p,ρ_0} with respect to the point $\lambda = 0$ will be designated by $\operatorname{ind} a_{p,\rho_0}$.

Theorem 6.1. *The operator* $A = aP_{\mathbf{R}} + bQ_{\mathbf{R}}$ $(a, b \in PC(\overline{\mathbf{R}}))$ *is at least one-sided invertible on the space* $L_p(\mathbf{R}, \rho_0)$ *if and only if the functions* a *and* b *satisfy the conditions*

$$\begin{aligned} a(t+0)b(t)f(t,\mu) + a(t)b(t+0)(1-f(t,\mu)) &\neq 0 \quad (t \in \mathbf{R}, 0 \le \mu \le 1) , \\ a(-\infty)b(+\infty)f_\delta(\mu) + a(+\infty)b(-\infty)(1-f_\delta(\mu)) &\neq 0 \quad (0 \le \mu \le 1) . \end{aligned} \tag{6.1}$$

If the conditions (6.1) are fulfilled and $\kappa := \operatorname{ind} a_{p,\rho_0}$ *then the operator* A *is invertible, left invertible or right invertible depending on whether the number* κ *is equal to zero, positive or negative. In case* $\kappa < 0$ *one has* $\dim\ker A = |\kappa|$ *, and if* $\kappa > 0$ *then* $\dim\operatorname{coker} A = \kappa$.

Proof. Let $\mathbf{T}$ stand for the unit circle, and define $\rho(\zeta) := \prod_{k=0}^{N} |\zeta - \zeta_k|^{\beta_k}$ with $\zeta_k := (t_k + i)(t_k - i)^{-1}$ $(k = 1, \ldots, N)$, $\zeta_0 := 1$ and $\beta_0 := p - 2 - \beta - \sum_{k=1}^{N} \beta_k$. Furthermore, we denote by B the operator mapping the space $L_p(\mathbf{R}, \rho_0)$ onto $L_p(\mathbf{T}, \rho)$ via

$$(B\varphi)(\zeta) := \frac{1}{\zeta - 1} \varphi\left(i\frac{\zeta + 1}{\zeta - 1}\right) .$$

As we have seen in Section 6.3 (see also Section 1.5), the linear operator B is bounded, and the equality $BS_{\mathbf{R}}B^{-1} = -S_{\mathbf{T}}$ holds with

$$(S_{\mathbf{R}}\varphi)(t) := \frac{1}{\pi i} \int_{-\infty}^{\infty} \frac{\varphi(s)}{s - t}\, ds\,, \quad (S_{\mathbf{T}}f)(\zeta) := \frac{1}{\pi i} \int_{\mathbf{T}} \frac{f(\tau)}{\tau - \zeta}\, d\tau\,.$$

It is easy to see that $BaB^{-1} = a(i(\zeta + 1)/(\zeta - 1))I$ for each function $a \in L_\infty(\overline{\mathbf{R}})$. Let us denote the function $c(i(\zeta + 1)/(\zeta - 1))$ by $c_0(\zeta)$. Then $BAB^{-1} = a_0 P_{\mathbf{T}} + b_0 Q_{\mathbf{T}}$. Let $c \in PC(\overline{\mathbf{R}})$ and $\rho(\zeta) = \prod_{k=0}^{N} |\zeta - \zeta_k|^{\beta_k}$, with $\zeta_0 = 1$, $\zeta_k = (t_k + i)(t_k - i)^{-1}$ $(k = 1, \ldots, N)$ and $\beta_0 = \beta + \sum_{k=1}^{N} \beta_k$. Because of $c(t_k \pm 0) = c_0(\zeta_k \pm 0)$ and $c(\mp\infty) = c_0(1 \pm 0)$, the ranges of the functions c_{p,ρ_0} and c_0^{p,ρ_0} coincide.

So it only remains to apply Theorem 3.1 to the operator BAB^{-1}. ∎

Now suppose the conditions (6.1) for the operator A are satisfied. Then the equality $aP_{\mathbf{R}} + bQ_{\mathbf{R}} = B^{-1}(a_0 P_{\mathbf{T}} + b_0 Q_{\mathbf{T}})B$ in combination with Theorem 4.2 yield a formula for the inverse or one–sided inverse operator of A without difficulty. Analogously one can desribe the bases of the kernel and cokernel of A as well as the condition for the solvability of the singular integral equation. The same holds true for the operator $P_{\mathbf{R}} aI + Q_{\mathbf{R}} bI$.

Notice further that the Theorem 1.1 in Chapter 7 remains true in the case of an unbounded curve Γ. As one consequence of this, the problem of inversion of a singular integral operator on $L_p(\mathbf{R}_+, \rho_0)$ with $\mathbf{R}_+$ referring to the positive half–axis can be reduced to the corresponding problem on the space $L_p(\mathbf{R}, \rho_0)$.

Let us consider an example. Our goal is a condition for the one–sided invertibility of the operator

$$(S_{\mathbf{R}_+}\varphi)(t) = \frac{1}{\pi i} \int_0^\infty \frac{\varphi(\tau)}{\tau - t}\, d\tau \quad (0 \le t \le \infty)$$

thought of as acting $L_p(0, +\infty)$. This problem is equivalent to the problem of one–sided invertibility of the operator $A = cP_{\mathbf{R}} + Q_{\mathbf{R}}$ acting on $L_p(-\infty, \infty)$ where $c(t) = -1$ if $t > 0$, and $c(t) = 1$ if $t < 0$.

A criterion for the one–sided invertibility of the operator A was established in Theorem 6.1. The first one of the conditions (6.1) states that $2f(0, \mu) - 1 \ne 0$. Because of $f(0, \mu) = f_{2\pi/p}(\mu)$, the range of the function $2f(0, \mu) - 1$ is a circular arc (if $p \ne 2$) or a straight line (if $p = 2$) between -1 and 1. This shows that $2f(0, \mu) - 1$ can be zero only if $p = 2$, and so the first one of the conditions (6.1) is satisfied if and only if $p \neq 2$. The same result is implied by the second condition in (6.1). Thus, the operator $S_{\mathbf{R}_+}$ is one–sided invertible if and only if $p \neq 2$.

In the second example we seek the spectrum of the operator $S_{\mathbf{R}_+}$ on the space $L_p(\mathbf{R}_+, t^\beta)$ with $-1 < \beta < p-1$. The invertibility of the operator $S_{\mathbf{R}_+} - \lambda I$ on $L_p(\mathbf{R}_+, t^\beta)$ is equivalent to the invertibility of the operator $A := aP_{\mathbf{R}} + bQ_{\mathbf{R}}$ on $L_p(\mathbf{R}, |t|^\beta)$ where

$$a(t) := \begin{cases} 1-\lambda & \text{if } t > 0 \\ 1 & \text{if } t < 0\,, \end{cases} \qquad b(t) := \begin{cases} -1-\lambda & \text{if } t > 0 \\ 1 & \text{if } t < 0\,. \end{cases}$$

Now we look for all of those points at which the conditions (6.1) are violated. For $0 < t < \infty$ this means that $(1-\lambda)(1+\lambda) = 0$; in case $t = 0$ we obtain the curve $\lambda = 2f(0,\mu) - 1$, and if $t = \infty$ the curve $\lambda = 1 - 2f_\delta(\mu)$. Since $\delta = 2\pi - 2\pi(1+\beta)/p = 2\pi - \delta(0)$, these two curves coincide. Thus, condition (6.1) is violated only if $\lambda \in l(-1,1;2\pi(1+\beta)/p)$. Due to the connectedness of the complement G of this curve and to the one–sided invertibility of the operator A for each $\lambda \in G$ (by Corollary 5.3 in Chapter 2, [GK 1]), the spectrum of $S_{\mathbf{R}_+}$ on $L_p(\mathbf{R}_+, t^\beta)$ is nothing other than the arc $l(-1,1;2\pi(1+\beta)/p)$.

Let us finally remark that the results of this section carry over to the case of arbitrary admissible (and possibly unbounded) curves Γ (see Section 1.5).

9.7 Another method of inversion

We consider the singular integral operator $A = cI + dS_{\mathbf{R}}$ acting on the space $L_p(\mathbf{R}, |t|^\beta)$ with the coefficients

$$c(t) := \begin{cases} c_1 & \text{if } -\infty < t < 0 \\ c_2 & \text{if } 0 < t < \infty\,, \end{cases} \qquad d(t) := \begin{cases} d_1 & \text{if } -\infty < t < 0 \\ d_2 & \text{if } 0 < t < \infty \end{cases}$$

$(c_1, c_2, d_1, d_2 \in \mathbf{C})$.

Recalling formula (6.1) it is easy to see that the operator A is invertible if and only if the arc (resp. the straight line) $l((c_2+d_2)(c_1-d_1),\ (c_1+d_1)(c_2-d_2);\ 2\pi(1+\beta)/p)$ does not contain the point $z = 0$. If, on the other hand, this arc runs through the point $z = 0$ that one sees from Corollary 4.3 the operator A is not even normally solvable. In case the operator A is invertible a formula for its inverse A^{-1} was established in Section 9.4.

In this section we present another method to invert the operator A.

For let $\nu : L_p(\mathbf{R}, |t|^\beta) \to L_p(\mathbf{R}_+, t^\beta) \dot{+} L_p(\mathbf{R}_+, t^\beta)$ be the mapping defined by $(\nu\varphi)(t) := (\varphi(t),\ \varphi(-t))(t \geq 0)$. The operator $\nu A \nu^{-1}$ acting on the space $L_p^2(\mathbf{R}_+, t^\beta)$ $(= L_p(\mathbf{R}_+, t^\beta) \times$

$L_p(\mathbf{R}_+, t^\beta))$ can be rewritten as an matrix

$$\nu A \nu^{-1} = \begin{pmatrix} A_{11} & A_{12} \\ A_{21} & A_{22} \end{pmatrix}$$

with entries in $L(L_p(\mathbf{R}_+, t^\beta))$. We are going to desribe the operators A_{jk} now. Let $\varphi \in L_p(\mathbf{R}, |t|^\beta)$ and $\psi := A\varphi$. The latter equality is equivalent to

$$\begin{cases} c_2\varphi(t) & +\frac{d_2}{\pi i}\int\limits_{-\infty}^{0} \frac{\varphi(\tau)}{\tau - t}\, d\tau & +\frac{d_2}{\pi i}\int\limits_{0}^{\infty} \frac{\varphi(\tau)}{\tau - t}\, d\tau & = \psi(t) & (t > 0)\,, \\ c_1\varphi(-t) & +\frac{d_1}{\pi i}\int\limits_{-\infty}^{0} \frac{\varphi(\tau)}{\tau + t}\, d\tau & +\frac{d_1}{\pi i}\int\limits_{0}^{\infty} \frac{\varphi(\tau)}{\tau + t}\, d\tau & = \psi(-t) & (t < 0)\,, \end{cases}$$

or, equivalently,

$$\begin{cases} c_2\varphi_1 + d_2 S\varphi_1 - d_2 N\varphi_2 & = \psi_1\,, \\ d_1 N\varphi_1 + c_1\varphi_2 - d_1 S\varphi_2 & = \psi_2 \end{cases}$$

with $\varphi_1(t) = \varphi(t),\ \varphi_2(t) = \varphi(-t)\ (t > 0)$, and

$$(Sf)(t) := \frac{1}{\pi i}\int\limits_0^\infty \frac{f(\tau)}{\tau - t}\, d\tau\,, \quad (Nf)(t) := \frac{1}{\pi i}\int\limits_0^\infty \frac{f(\tau)}{\tau + t}\, d\tau\,. \quad (0 < t < \infty)\,.$$

Thus we obtain

$$\nu A \nu^{-1} = \begin{pmatrix} c_2 I + d_2 S & -d_2 N \\ d_1 N & c_1 I - d_1 S \end{pmatrix}. \tag{7.1}$$

Letting $W_\gamma : L_p(\mathbf{R}_+, t^\beta) \to L_p(\mathbf{R}_+, t^{-1})$ stand for the isometry

$$(W_\gamma u)(t) = t^\gamma u(t) \quad (\gamma = (1+\beta)/p),$$

we find

$$(W_\gamma S W_\gamma^{-1}\varphi)(t) = \frac{1}{\pi i}\int\limits_0^\infty \frac{\varphi(\tau)t^\gamma}{(\tau - t)\tau^\gamma}\, d\tau = \frac{1}{\pi i}\int\limits_0^\infty \varphi(\tau)\frac{(t\tau^{-1})^\gamma}{1 - t\tau^{-1}} \cdot \frac{d\tau}{\tau} \tag{7.2}$$

and

$$(W_\gamma N W_\gamma^{-1}\varphi)(t) = \frac{1}{\pi i}\int\limits_0^\infty \varphi(\tau)\frac{(t\tau^{-1})^\gamma}{1 + t\tau^{-1}} \cdot \frac{d\tau}{\tau}\,. \tag{7.3}$$

Now we interprete the positive half axis $\mathbf{R}_+$ as an abelian group (with respect to the multiplication) with the Haar measure $d\tau/\tau$ (cf.GELFAND, RAIKOW, SHILOW [1], p.156). Then the integrals on the right hand sides of (7.2) and (7.3) can be identified with the convolutions

$$(W_\gamma S W_\gamma^{-1}\varphi)(t) = \frac{1}{\pi i}\frac{t^\gamma}{1-t} * \varphi(t), \ (W_\gamma N W_\gamma^{-1}\varphi)(t) = \frac{1}{\pi i}\frac{t^\gamma}{1+t} * \varphi(t)\,.$$

Denote by M the Fourier transform on the group $\mathbf{R}_+$ (see GELFAND, RAIKOW, SHILOW [1], Chapter IV) and by $s(\xi)$ and $n(\xi)$ the images of the functions $(1/\pi i)\, t^\gamma/(1-t)$ and $(1/\pi i)t^\gamma/(1+t)$ under the Fourier transform, and define $R := MW_\gamma\nu$. Then the operator RAR^{-1} proves to be the operator of multiplication by the function matrix

$$\begin{pmatrix} c_2 + d_2 s(\xi) & -d_2 n(\xi) \\ d_1 n(\xi) & c_1 - d_1 s(\xi) \end{pmatrix}.$$

Looking for the concrete form of the functions $n(\xi)$ and $s(\xi)$, one finds

$$n(\xi) = \frac{1}{\pi i}\int_0^\infty \frac{t^{\gamma-i\xi-1}}{1+t}\,dt = \frac{2e^{\pi(\xi+i\gamma)}}{e^{2\pi(\xi+i\gamma)}-1},$$

(see, e.g., RYSHIK,GRADSTEIN[1], p. 307) and, by a straightforward calculation,

$$s(\xi) = \frac{e^{2\pi(\xi+i\gamma)}+1}{e^{2\pi(\xi+i\gamma)}-1}\,. \tag{7.4}$$

The latter result could be also obtained in the following manner. Because of $S_{\mathbf{R}}^2 = I$ and

$$RS_{\mathbf{R}}R^{-1} = \begin{pmatrix} s(\xi) & -n(\xi) \\ n(\xi) & -s(\xi) \end{pmatrix}$$

one has

$$\begin{pmatrix} s^2 - n^2 & 0 \\ 0 & s^2 - n^2 \end{pmatrix} = \begin{pmatrix} 1 & 0 \\ 0 & 1 \end{pmatrix}$$

and, thus,

$$s(\xi)^2 = 1 + n(\xi)^2\,. \tag{7.5}$$

Moreover, the spectrum of the operator S on $L_p(\mathbf{R}_+, t^\beta)$ coincides with the arc $l(-1,1;2\pi\gamma)$ (see the concluding example in the previous section). Since the range of the function $s(\xi)$ is contained in the spectrum of the operator S , it is

$$\operatorname{Im} s(\xi) > 0 \quad \text{if} \quad \gamma > 1/2\,, \quad \operatorname{Im} s(\xi) < 0 \quad \text{if} \quad \gamma < 1/2\,. \tag{7.6}$$

The function $s(\xi)$ is uniquely determined by the conditions (7.5) and (7.6). Abbreviating $s(\xi)$ to ζ then $n(\xi) = \sqrt{\zeta^2 - 1}$ where $\sqrt{\zeta^2-1}$ refers to the unique continuous branch taking at the point $\zeta = -i\cot\,(\pi\gamma)$ the value $1/i\,\sin\pi\gamma$. Thus the operator RAR^{-1} is just the operator of multiplication by the function matrix

$$\mathcal{A}(\zeta) = \begin{pmatrix} c_2 + d_2\zeta & -d_2\sqrt{\zeta^2-1} \\ d_1\sqrt{\zeta^2-1} & c_1 - d_1\zeta \end{pmatrix}$$

with the variable $\zeta = s(\xi)$ running over the arc $l(-1,1;2\pi\gamma)$.

Let us consider a simple example. In the preceding section we have seen that the operator S is invertible in $L_p(\mathbf{R}_+)$ $(1 < p < \infty)$ if and only if $p \neq 2$. Now we look for its inverse. Let $R_0 := MW_{1/p}$ with the operators M and $W_{1/p}$ being defined as above; and given $\varphi \in L_p(\mathbf{R}_+)$, set $\psi = S\varphi$ and $\widehat{\varphi} = R_0\varphi$. Then, $R_0SR_0^{-1}\widehat{\varphi} = \widehat{\psi}$ and, as we have just seen, $s(\xi)\widehat{\varphi}(\xi) = \widehat{\psi}(\xi)$.

Hence,

$$\widehat{\psi}(\xi) = \frac{e^{2\pi(\xi+i\gamma)}+1}{e^{2\pi(\xi+i\gamma)}-1}\widehat{\varphi}(\xi)$$

with $\gamma = 1/p$. Taking into account that equality (7.4) holds for all $0 < \lambda < 1$ we can rewrite the function $\widehat{\psi}(\xi)$ in the following form:

$$\widehat{\psi}(\xi) = \frac{e^{2\pi(\xi+i\gamma+i/2)}-1}{e^{2\pi(\xi+i\gamma+i/2)}+1}\,\widehat{\varphi}(\xi) \quad \text{if} \quad p > 2$$

and

$$\widehat{\psi}(\xi) = \frac{e^{2\pi(\xi+i\gamma-i/2)}-1}{e^{2\pi(\xi+i\gamma-i/2)}+1}\,\widehat{\varphi}(\xi) \quad \text{if} \quad p < 2\,.$$

Now take the inverse Fourier transform and remember that $(W_{1/p}f)(t) = f(t)t^{1/p}$ to obtain for $1 < p < 2$,

$$t^{1/p}\psi(t) = \frac{1}{\pi i}\int_0^\infty \varphi(\tau)\tau^{1/p}\frac{(t\tau-1)^{1/p-1/2}}{1-t\tau^{-1}}\frac{d\tau}{\tau}\,,$$

and thus,

$$\psi(t) = \frac{1}{\pi i}\int_0^\infty \sqrt{\frac{\tau}{t}}\,\frac{\varphi(\tau)}{\tau - t}\,d\tau\,.$$

This shows that if $1 < p < 2$, then

$$(S^{-1}\varphi)(t) = \frac{1}{\pi i}\int_0^\infty \sqrt{\frac{\tau}{t}}\,\frac{\varphi(\tau)}{\tau - t}\,d\tau\,. \tag{7.7}$$

Analogously, if $2 < p < \infty$,

$$(S^{-1}\varphi)(t) = \frac{1}{\pi i}\int_0^\infty \sqrt{\frac{t}{\tau}}\,\frac{\varphi(\tau)}{\tau - t}\,d\tau\,. \tag{7.8}$$

We remark that the operator standing on the right hand side of (7.7) also proves to be the inverse of the operator S acting on $L_p(\mathbf{R}_+, t^\beta)$ if only $2(1+\beta) > p$, and the operator on the right hand side of (7.8) is the inverse of S whenever $2(1+\beta) < p$. In case $2(1+\beta) = p$, the operator S is not one–sided invertible (and not even normally solvable).

9.8 Singular integral operators with regel functions coefficients

In the present section we are going to generalize the results of the previous sections by applying the local principle explained in Section 6.8 (see also [GK 1, 5.1]) to the case where the coefficients of the singular integral operator belong to the closure of $PC(\Gamma)$.

For this we need the following lemma.

Proposition 8.1. *Let* $A_\tau = a_\tau P_\Gamma + b_\tau Q_\Gamma$ $(\tau \in \Gamma, a_\tau \in L_\infty(\Gamma),\ b_\tau \in L_\infty(\Gamma))$ *be a family of* Φ*-operators on* $L_p(\Gamma,\rho)$ *and* $a, b \in L_\infty(\Gamma)$. *If for each point* $\tau \in \Gamma$ *and for each* $\varepsilon > 0$ *there is a neighbourhood* $U(\tau)$ *so that*

$$\operatorname*{ess\ sup}_{t\in U(\tau)} |a(t) - a_\tau(t)| < \varepsilon, \quad \operatorname*{ess\ sup}_{t\in U(\tau)} |b(t) - b_\tau(t)| < \varepsilon,$$

then the operator $A = aP_\Gamma + bQ_\Gamma$ *is also a* Φ*-operator in* $L_p(\Gamma,\rho)$.

Proof. Since the operators A_τ are Φ–operators, the cosets $\widehat{A}_\tau$ must be invertible in the quotient algebra $\widehat{L} = L(\mathcal{B})/\mathcal{T}(\mathcal{B})$, where $\mathcal{B} = L_p(\Gamma,\rho)$. Now introduce a system of localizing classes $\{M_\tau\}_{\tau\in\Gamma}$ in the algebra $\widehat{L}$ as follows: Let Z_τ denote the class of all continuous non–negative functions on Γ which are equal to one in a certain (depending on the function) neighbourhood of the point τ . Furthermore we let $\check{M}_\tau$ refer to the class of all operators of multiplication by a function in Z_τ . Finally we write M_τ for the collection of all cosets $\widehat{R} \in \widehat{L}$ generated by the operators $R \in \check{M}_\tau$.

One easily checks that

1. the system $\{M_\tau\}(\tau \in \Gamma)$ is covering,
2. the cosets $\widehat{A}$ and $\widehat{A}_\tau$ commute with all cosets in M_τ , and
3. $\widehat{A} \overset{M_\tau}{\sim} \widehat{A}_\tau$ at each point $\tau \in \Gamma$.

By Theorem 1.1 of [GK 1, 5.1], the coset $\widehat{A}$ is invertible in the algebra $\widehat{L}$ and, consequently, $A \in \Phi(\mathcal{B})$. ■

Let Γ be a closed curve without intersections. By $\overline{PC(\Gamma)}$ we denote the closure in the $L_\infty(\Gamma)$–norm of the set $PC(\Gamma)$. The functions belonging to $\overline{PC(\Gamma)}$ will be called *regel functions.* It is well–known [see, e.g., BOURBAKI [1], p. 75] that the class $\overline{PC(\Gamma)}$ consists of all functions which possess finite left–sided and right–sided limits at each point $\tau \in \Gamma$, and such that the set of all points of discontinuity of a function a in $\overline{PC(\Gamma)}$ is at most countable.

Let $\rho(t) = \prod_{k=1}^n |t - t_k|^{\beta_k}$ $(t_k \in \Gamma,\ -1 < \beta_k < p - 1,\ 1 < p < \infty)$ and $a \in \overline{PC(\Gamma)}$. The function a is said to be $\{p,\rho\}$–non–singular if

$$a(t+0)f(t,\mu) + a(t-0)(1 - f(t,\mu)) \neq 0$$

for all $t \in \Gamma$ and $\mu \in [0,1]$. Herein, the function $f(t,\mu)$, which depends on p and ρ, is defined as in (1.3).

A little thought shows that if a ($\in \overline{PC(\Gamma)}$) is a $\{p,\rho\}$–non–singular function and

$$\|a - a_n\|_{L_\infty(\Gamma)} \to 0$$

for certain functions $a_n \in PC(\Gamma)$ then, beginning at an n, the functions a_n are also $\{p,\rho\}$–non–singular, and their $\{p,\rho\}$–indices $\operatorname{ind} a_n^{p,\rho}$ stabilize. Define

$$\operatorname{ind} a^{p,\rho} := \lim_{n\to\infty} \operatorname{ind} a_n^{p,\rho}.$$

Theorem 8.1. *Let $a, b \in \overline{PC(\Gamma)}$. The operator $A = aP_\Gamma + bQ_\Gamma$ is a Φ–operator in the space $L_p(\Gamma,\rho)$ if and only if*

$$a(t \pm 0) \neq 0 \quad (t \in \Gamma),\ b(t \pm 0) \neq 0 \quad (t \in \Gamma) \tag{8.1}$$

and the function $c(t) := a(t)/b(t)$ is $\{p,\rho\}$–non–singular. If these conditions are satisfied and $\kappa := \operatorname{ind} c^{p,\rho}$, then in case $\kappa \geq 0$, the operator A is left invertible and $\dim \operatorname{coker} A = \kappa$*, and if $\kappa \leq 0$, A is right invertible and* $\dim \ker A = -\kappa$.

Proof. Given $\tau \in \Gamma$ we choose functions a_τ and b_τ which are continuous at each point $t \neq \tau$ and do not vanish there, and which, at the point τ, are subject to the conditions

$$\begin{aligned} a_\tau(\tau+0) &= a(\tau+0), \quad a_\tau(\tau-0) = a(\tau-0), \\ b_\tau(\tau+0) &= b(\tau+0), \quad b_\tau(\tau-0) = b(\tau-0). \end{aligned} \tag{8.2}$$

If the conditions (8.1) are fulfilled and if c is a $\{p,\rho\}$–non–singular function then each of the operators $A_\tau := a_\tau P_\tau + b_\tau Q_\tau$ is a Φ–operator (by Theorem 4.1), and Proposition 8.1 states that A is a Φ–operator, too.

Conversely, assume the operator A to be a Φ–operator. If the point τ is fixed then, for each point $\tilde{t} \neq \tau$ and for each $\varepsilon > 0$, there exists a neighbourhood $U(\tilde{t})$ such that

$$|a_\tau(t) - a_\tau(\tilde{t})| < \varepsilon, \quad |b_\tau(t) - b_\tau(\tilde{t})| < \varepsilon \quad (t \in U(\tilde{t})),$$

and the operators $B_{\tilde{t},\tau} := a_\tau(\tilde{t})P_\Gamma + b_\tau(\tilde{t})Q_\Gamma$ with constant coefficients are invertible on $L_p(\Gamma,\rho)$. Furthermore, for each point τ there is a neighbourhood $U(\tau)$ such that

$$\operatorname*{ess\,sup}_{t\in U(\tau)} |a_\tau(t) - a(t)| < \varepsilon, \quad \operatorname*{ess\,sup}_{t\in U(\tau)} |b_\tau(t) - b(t)| < \varepsilon,$$

and the operator $aP_\Gamma + bQ_\Gamma$ is a Φ–operator. From Proposition 8.1 we conclude that the operator A_τ is a Φ–operator, and this fact implies via Theorem 3.1 and equality (8.2) that

$$a(\tau+0)b(\tau-0)f(\tau,\mu) + a(\tau-0)b(\tau+0)(1 - f(\tau,\mu)) \neq 0$$

$(0 \le \mu \le 1)$. Since the choice of the point τ was arbitrary this proves the necessity of the conditions in the theorem.

Finally, let $(A_n)_1^\infty$ be a sequence of Φ-operators of the form $a_n P_\Gamma + b_n Q_\Gamma$ which converges uniformly to A , and set $c_n = a_n/b_n$. Since $\operatorname{Ind} A = \lim_{n\to\infty} \operatorname{Ind} A_n$ and $\lim_{n\to\infty} \operatorname{ind} c_n^{p,\rho} = \operatorname{ind} c^{p,\rho}$, one has $\operatorname{ind} A = -\operatorname{ind} c^{p,\rho}$. The remaining assertions of the theorem follow from the general Theorem 4.1. ∎

9.9 Estimates for the norms of the operators P_Γ, Q_Γ and S_Γ

Thought of as an application of the results of the present chapter, we are now going to derive lower estimates for the quotient norms, and thus for the norms, of the operators P_Γ, Q_Γ and S_Γ acting on the space $L_p(\Gamma, \rho)$.

Let Γ be a composed curve, $\rho(t) := \prod_{k=1}^N |t - t_k|^{\beta_k}$ $(-1 < \beta_k < p-1,\ 1 < p < \infty)$, $r_k := p(1+\beta_k)^{-1}$ $(k = 1, \dots, N)$, $r_0 := p$, $\overline{r_k} := \max\left(r_k, r_k(r_k - 1)^{-1}\right)$, and $r := \max\left(\overline{r}_0, \overline{r}_1, \dots, \overline{r}_N\right)$.

Theorem 9.1. *If Γ is a closed composed curve then the following estimates hold*

$$\inf_{T \in \mathcal{T}(L_p(\Gamma,\rho))} \|P_\Gamma + T\| \ge \frac{1}{\sin\frac{\pi}{r}}, \tag{9.1}$$

$$\inf_{T \in \mathcal{T}(L_p(\Gamma,\rho))} \|Q_\Gamma + T\| \ge \frac{1}{\sin\frac{\pi}{r}}, \tag{9.2}$$

$$\inf_{T \in \mathcal{T}(L_p(\Gamma,\rho))} \|S_\Gamma + T\| \ge \cot\frac{\pi}{2r}. \tag{9.3}$$

Proof. Assume that $\inf \|P_\Gamma + T\| < 1/\sin(\pi/r)$. Then we choose a function a in $PC(\Gamma)$ which takes the values $\cos(\pi/r)\exp(i\pi/r)$ and $\cos(\pi/r)\exp(-i\pi/r)$ on Γ and for which

$$a(t_m + 0) = \begin{cases} \cos\frac{\pi}{r}\exp(i\pi/r) & \text{if } \ \overline{r}_m = r_m(r_m-1)^{-1} \\ \cos\frac{\pi}{r}\exp(-i\pi/r) & \text{if } \ \overline{r}_m = r_m \end{cases}$$

and

$$a(t_m - 0) = \begin{cases} \cos\frac{\pi}{r}\exp(-i\pi/r) & \text{if } \ \overline{r}_m = r_m(r_m-1)^{-1} \\ \cos\frac{\pi}{r}\exp(i\pi/r) & \text{if } \ \overline{r}_m = r_m \end{cases},$$

where t_m is a point at which $r = \overline{r}_m$. Because of $a(t_m - 0)/a(t_m + 0) = \exp(2\pi i/r_m)$, the function a is $\{p,\rho\}$-singular (i.e. is not $\{p,\rho\}$-non-singular) by Theorem 1.2, and thus, by Theorem 4.3, the operator $A = aP_\Gamma + Q_\Gamma$ is not a Φ-operator in $L_p(\Gamma, \rho)$. On the other hand, $|a(t)-1| = |\cos^2(\pi/r) \pm i\sin(\pi/r)\cos(\pi/r) - 1| = \sin\pi/r$ and, by our assumption,

$$\inf_{T\in\mathcal{T}(L_p(\Gamma,\rho))} \|B+T\| < 1$$

with $B := (a-1)P_\Gamma$. Since $A = aP_\Gamma + Q_\Gamma = I + B$, Theorem 7.2 of Chapter 4, [GK1] implies that A is a Φ–operator. This contradiction proves the estimate (9.1).

Analogously one verifies (9.2): Pick a function a taking the two values $[\cos(\pi/r)]^{-1}\exp(i\pi/r)$ and $[\cos(\pi/r)]^{-1}\exp(-i\pi/r)$ only. This function is also $\{p,\rho\}$–singular, and one furthermore has $|(1-a)/a| = \sin(\pi/r)$ and $A = a(I + (1-a)a^{-1}Q_\Gamma)$, which easily gives (9.2). Finally, to prove (9.3), consider a function a taking the two values $\exp(i\pi/r)$ and $\exp(-i\pi/r)$ and then write

$$aP_\Gamma + Q_\Gamma = \frac{a+1}{2}(I + \frac{a-1}{a+1}S_\Gamma)$$

with $|(a-1)(a+1)^{-1}| = \tan(\pi/2r)$. ■

In case the composed curve Γ also contains non–closed arcs the above proof does not work since the point t_m at which $r = \overline{r}_m$ is possibly an end point of a certain non–closed arc. But if t_m is an inner point of the composed curve Γ then Theorem 9.1 remains true. In particular, for the space $L_p(\Gamma)$ (without weight) one has $r = \max(p,q)$ $(p^{-1}+q^{-1}=1)$, and the point t_m can be arbitrarily selected from the curve Γ . Thus we have

Theorem 9.2. *Let Γ be a composed curve and $1 < p < \infty$. Then*

$$\inf_{T\in\mathcal{T}(L_p(\Gamma))} \|P_\Gamma + T\|_p \geq \frac{1}{\sin\pi/p}\,, \inf_{T\in\mathcal{T}(L_p(\Gamma))} \|Q_\Gamma + T\|_p \geq \frac{1}{\sin\pi/p}\,, \tag{9.4}$$

$$\inf_{T\in\mathcal{T}(L_p(\Gamma))} \|S_\Gamma + T\|_p \geq \cot\frac{\pi}{2p} \quad \text{if} \quad 2 \leq p < \infty\,, \tag{9.5}$$

and

$$\inf_{T\in\mathcal{T}(L_p(\Gamma))} \|S_\Gamma + T\|_p \geq \tan\frac{\pi}{2p} \quad \text{if} \quad 1 < p \leq 2\,. \tag{9.6}$$

The estimates (9.5), (9.6) and (2.1) of Chapter 1, [GK1] imply the following:

Theorem 9.3. *Let $\mathbf{T}$ denote the unit circle. Then the equalities*

$$\|S\|_{L_p(\mathbf{T})} = \cot\frac{\pi}{2p} \quad (p = 2^n, n = 1,2,\ldots)$$

and

$$\|S\|_{L_p(\mathbf{T})} = \tan\frac{\pi}{2p} \quad (p = (2^n-1)^{-1},\ n = 1,2,\ldots)$$

hold for the operator S acting on $L_p(\mathbf{T})$.

Let us finally mention that the Theorems 9.2 and 9.3 carry over without changes to the case when Γ is the real axis or a part of it. Further, Theorem 9.1 possesses an analogue for the space $L_p(\mathbf{R},\rho_0)$, where

$$\rho_0(t) := (1+t^2)^{\beta/2}\prod_{k=1}^N |t-t_k|^{\beta_k} \quad (-1<\beta_k<p-1,\ 1<p<\infty, -1<\sum_{k=1}^N \beta_k+\beta<p-1)\,.$$

For the formulation of the corresponding theorem one has only to replace the r in Theorem 9.1 by $r := \max(\overline{r}_0,\overline{r}_1,\ldots,\overline{r}_N,\overline{s})$ with

$$r_0=p,\quad r_k=p(1+\beta_k)^{-1}\quad (k=1,\ldots,N)$$

and

$$s=p(p-1-\beta-\sum_{k=1}^N\beta_k)^{-1}\,.$$

9.10 Singular operators on spaces $H^\circ_\mu(\Gamma,\rho)$

Let Γ be a composed curve. By $H_\mu(\Gamma;t_1,\ldots,t_n)$ $(0<\mu<1)$ we mean the class of all functions on Γ fulfilling a Hölder condition with exponent μ on Γ , with the possible exception of the points $t_1,\ldots,t_n$ at which the functions are continuous from the left and have a discontinuity of the first kind. Remember that (see the definition in Section 1.6) $H^\circ_\mu(\Gamma,\rho)$ with $\rho(t) := \prod_{k=1}^n|t-t_k|^{\alpha_k}$ stands for the Banach space of all functions ψ satisfying $\rho\psi\in H^\circ_\mu(\Gamma;t_1,\ldots,t_n)$, where $H^\circ_\mu(\Gamma;t_1,\ldots,t_n)$ denotes the subspace of $H_\mu(\Gamma)$ consisting of all functions which vanish at the points $t_1,\ldots,t_n$. We further suppose that the set $\{t_1,\ldots,t_n\}$ contains all starting and end points of the non–closed subarcs of the curve Γ .

Now let $\mu,a_1,\ldots,\alpha_n$ be real numbers such that

$$0<\mu<1\,,\quad \mu<\alpha_k<\mu+1\quad (k=1,\ldots,n)\,, \tag{10.1}$$

and put $\rho(t) := \prod_{k=1}^n|t-t_k|^{\alpha_k}$. To each function $a\in H_\mu(\Gamma;t_1,\ldots,t_n)$ we associate the function $a^{\mu,\rho} : \Gamma\times[0,1]\to \mathbf{C}$ which is defined as follows: If the point $t\in\Gamma$ is neither a starting nor an end point of a non–closed arc, then we set $a^{\mu,\rho}(t,x) := a(t)$ in case $t\notin\{t_1,\ldots,t_n\}$, and $a^{\mu,\rho}(t_k,x) := a(t_k+0)f_{\lambda_k}(x)+a(t_k)(1-f_{\lambda_k}(x))$ $(0\le x\le 1)$, where $\lambda_k := 2\pi(\alpha_k-\mu)$ and $f_{\lambda_k}(x)$ is the function introduced in Section 9.1. At the starting and end points t_k of the non-closed subarcs of the curve Γ we set

$$a^{\mu,\rho}(t_k,x) := f_{\lambda_k}(x)a(t_k)+1-f_{\lambda_k}(x)$$

and

$$a^{\mu,\rho}(t_k,x) := f_{\lambda_k}(x)+a(t_k)(1-f_{\lambda_k}(x))\,,$$

respectively.

The function $a(\in H_\mu(\Gamma; t_1, \dots, t_n))$ is called $H_{\mu,\rho}$–non–singular if $a^{\mu,\rho}(t,x) \neq 0$ $(t \in \Gamma, 0 \leq x \leq 1)$. If the function a is $H_{\mu,\rho}$–non–singular then the winding number of the range curve of the function $a^{\mu,\rho}$ around the point $z = 0$ is referred to as the $H_{\mu,\rho}$–index of a .

Theorem 10.1 *Let* $a, b \in H_\mu(\Gamma; t_1, \dots, t_n)$. *Then the operator* $A = aP_\Gamma + bQ_\Gamma$ $(A = P_\Gamma aI + Q_\Gamma bI)$ *is Fredholm on* $H_\mu^\circ(\Gamma, \rho)$ *if and only if*

1. $b(t \pm 0) \neq 0$ $(t \in \Gamma)$, *and*

2. the function a/b *is* $H_{\mu,\rho}$*–non–singular.*

If these conditions are satisfied and κ *is the* $H_{\mu,\rho}$*–index of the function* a/b *then the operator* A *is invertible, left invertible or right invertible depending on whether the number* κ *is equal to zero, positive or negative. If* $\kappa < 0$ *then* $\dim \ker A = |\kappa|$, *and in case* $\kappa > 0$ *one has* $\dim \operatorname{coker} A = \kappa$.

Proof. To start with we consider the case when Γ is a closed curve. Suppose the conditions 1 and 2 are satisfied, and define $c := a/b$. Further we set $p := 2, \beta_k := 2(\alpha_k - \mu) - 1$ and $\rho_1(t) := \prod_{k=1}^n |t - t_k|^{\beta_k}$. It is easy to see that the ranges of the functions c^{2,ρ_1} (defined in Section 9.1) and $c^{\mu,\rho}$ coincide. Hence, the function c is $\{2, \rho_1\}$–non–singular. Furthermore, due to the piecewise Hölder continuity of c , one can represent this function in the form $c = \psi d$ with $d \in H_\mu(\Gamma)$, $\psi = \psi_{t_1,\gamma_1} \cdot \ldots \cdot \psi_{t_n,\gamma_n}$, $\alpha_k - \mu - 1 < \operatorname{Re}\gamma_k < \alpha_k - \mu$; where the function ψ_{t_k,γ_k} is defined by (2.2). The function ψ admits a factorization (in the space $L_2(\Gamma, \rho_1)$) with factors of the form

$$\psi_+(t) := g_1(t) \prod_{k=1}^n (t - t_k)^{\gamma_k} , \quad \psi_-(t) := g_2(t) \prod_{k=1}^n (t - t_k)^{-\gamma_k}$$

where $g_1^\pm \in H_\mu(\Gamma), g_2^\pm \in H_\mu(\Gamma)$. Moreover, the function $d(\in H_\mu(\Gamma))$ is factorizable: $d = d_+ t^\kappa d_-$ with $(d_\pm)^{\pm 1} \in H_\mu(\Gamma)$ (see Section 3.3). Hence, the function c permits a factorization $c = c_- t^\kappa c_+$ in $L_2(\Gamma, \rho_1)$ whose factors are subject to the conditions

$$c_+(t) = g_3(t) \prod_{k=1}^n (t - t_k)^{\gamma_k} , \quad c_-(t) = g_4(t) \prod_{k=1}^n (t - t_k)^{-\gamma_k} \tag{10.2}$$

for certain functions $g_3^{\pm 1} \in H_\mu(\Gamma)$ and $g_4^{\pm 1} \in H_\mu(\Gamma)$. By Theorem 3.1 (see also Theorem 4.1 in Chapter 8), the operator A is at least one–sided invertible on $L_2(\Gamma, \rho_1)$, and

$$\begin{aligned} (aP_\Gamma + bQ_\Gamma)^{-1} &= c_- a^{-1}(bP_\Gamma + aQ_\Gamma) b^{-1} c_-^{-1} I , \\ (P_\Gamma aI + Q_\Gamma bI)^{-1} &= b^{-1} c_+^{-1} (P_\Gamma bI + Q_\Gamma aI) a^{-1} c_+ I . \end{aligned}$$

With regard to (10.2) the operator A^{-1} can be rewritten as

$$A^{-1} = f_1 I + f_2 \prod_{k=1}^n (t - t_k)^{-\gamma_k} S_\Gamma f_3 \prod_{k=1}^n (t - t_k)^{\gamma_k} I$$

for certain functions f_1, f_2 $f_3 \in H_\mu(\Gamma)$.

We are going to show that the operator A^{-1} is bounded on the space $H_\mu^\circ(\Gamma, \rho)$. Obviously, the boundedness of A^{-1} is equivalent to be boundedness of the operator $\rho_0 A^{-1} \rho_0^{-1} I$ acting on the space $H_\mu^\circ(\Gamma, \rho\rho_0^{-1})$ where $\rho_0(t) := \prod_{k=1}^n (t - t_k)^{\gamma_k}$. But since

$$\rho_0 A^{-1} \rho_0^{-1} I = f_1 I + f_2 S_\Gamma f_3 I \, , \quad |\rho(t)\rho_0^{-1}(t)| = \prod_{k=1}^n |t - t_k|^{\alpha_k - \gamma_k}$$

and $\mu < \alpha_k - \mathrm{Re}\gamma_k < \mu + 1$ the boundedness of the operator $\rho_0 A^{-1} \rho_0^{-1} I$ on $H_\mu^\circ(\Gamma, \rho\rho_0^{-1})$ follows easily by means of Theorem 6.2 of Chapter 1, [GK1]. Thus, the operator A^{-1} is indeed at least a one-sided inverse to the operator A . Now Theorem 4.2 immediately gives

$$\ker A|L_2(\Gamma, \rho_1) = \ker A|H_\mu^\circ(\Gamma, \rho)$$

and

$$\dim \operatorname{coker} A|L_2(\Gamma, \rho_1) = \dim \operatorname{coker} A|H_\mu^\circ(\Gamma, \rho) \, .$$

This proves the sufficiency of the conditions of the theorem in case the curve Γ is closed.

The general case can be traced back to the situation just considered by means of Theorem 1.1 in Chapter 7, which obviously holds true for the spaces $H_\mu^\circ(\Gamma, \rho)$ and $H_\mu^\circ(\widetilde{\Gamma}, \rho)$.

For the proof of the necessity of the conditions of the theorem see DUDUCHAVA's paper [4]. ■

9.11 Singular operators on symmetric spaces

Let Γ again be a composed curve. A Banach space $E(\Gamma)$ of complex valued measurable functions on Γ is called a *symmetric space* if it is subject to the following two conditions:

1. if $x \in E(\Gamma)$ and y is a measurable function such that $|y(t)| \le |x(t)|$ for all $t \in \Gamma$ then $y \in E(\Gamma)$ and $\|y\| \le \|x\|$,

2. if $x \in E(\Gamma), y$ is measurable and $\mathrm{mes}\{t \in \Gamma : |x(t)| > \sigma\} = \mathrm{mes}\{t \in \Gamma : |y(t)| > \sigma\}$ for each $\sigma > 0$ then $y \in E(\Gamma)$ and $\|x\| = \|y\|$.

For the symmetric space $E = E(\Gamma)$ we denote by $\varphi_E(t)$ its *fundamental function,* that is, the function $\varphi_E(t) = \|\chi_U\|$ where χ_U is the characteristic function of an arbitrary arc $U \subseteq \Gamma$ of length t . Furthermore we put

$$m(E) := \underline{\lim}_{t \to 0} \frac{\varphi_E(2t)}{\varphi_E(t)} \, , \quad M(E) := \overline{\lim}_{t \to 0} \frac{\varphi_E(2t)}{\varphi_E(t)} \, .$$

Let $x \in E(\Gamma)$. A function $f : [0, \operatorname{mes}\Gamma] \to \mathbf{R}$ which is measurable on the interval $[0, \operatorname{mes}\Gamma]$ is said to be *equi-measurable with the function* x if

$$\operatorname{mes}\{t \in \Gamma : |x(t)| > \sigma\} = \operatorname{mes}\{s \in [0, \operatorname{mes}\Gamma] : f(s) > \sigma\}$$

for each $\sigma > 0$. We let E_0 stand for the class of all functions $f(s)$ $(0 \le s \le \operatorname{mes}\Gamma)$ which are equi-measurable with a certain function x_f in $E(\Gamma)$. The set E_0 is a symmetric Banach space under the norm $\|f\|_{E_0} = \|x_f\|_{E(\Gamma)}$. On E_0 we define the linear operator W_μ $(0 < \mu < 1)$ by

$$(W_\mu f)(s) := \begin{cases} f(\frac{s}{\mu}) & \text{if} \quad 0 \le s \le \mu \operatorname{mes}\Gamma \\ 0 & \text{if} \quad s > \mu \operatorname{mes}\Gamma . \end{cases}$$

Finally, we suppose that the condition

$$\|W_\mu\|_{E_0} \le C \sup_s \frac{\varphi_E(\mu s)}{\varphi_E(s)} \quad (0 < \mu < 1) \tag{11.1}$$

is satisfied. Then the interpolation theorem of SEMENOV [2] holds:

Theorem 11.1. *Let* $0 \le \alpha_j \le 1$ $(j = 1, 2)$ *and suppose that*

$$2^{\alpha_1} < m(E), \quad M(E) < 2^{\alpha_2} .$$

If the linear operator A *is bounded on* $L_p(\Gamma)$ *for every* $p \in (\alpha_2^{-1}, \alpha_1^{-1})$ *then the operator* A *is also bounded on* $E(\Gamma)$.

If the symmetric space $E(\Gamma)$ is separable then $C(\Gamma)$ is dense in $E(\Gamma)$ (see SEMENOV [2], Theorem 1.4), and thus the linear manifold $R(\Gamma)$[1] is also dense in $E(\Gamma)$. If we further suppose the curve Γ is closed and composed then the operator S_Γ maps the linear manifold $R(\Gamma)$ into itself (see Section 1.1). Thus, provided that the operator S_Γ is bounded on $R(\Gamma)$ in the norm of $E(\Gamma)$, it can be extended continuously onto all of $E(\Gamma)$.

In the sequel we confine ourselves to the consideration of symmetric spaces on which the operator S_Γ is bounded. The boundedness of the operator S_Γ on all spaces $L_p(\Gamma)$ $(1 < p < \infty)$ implies via Theorem 11.1 its boundedness on all symmetric spaces $E(\Gamma)$ satisfying

$$1 < m(E), \; M(E) < 2 .$$

In particular, the operator S_Γ is bounded on every reflexive Orlicz space as well as on every uniformly convex Lorentz space Λ (see BOYD [1]). All the results obtained in Chapters 3 and 4 for singular integral operators with continuous coefficients on closed curves remain valid on these spaces, and the results of the present chapter permit the following generalization:

[1] $R(\Gamma)$ denotes the class of all rational functions without poles on Γ.

Theorem 11.2. *Let* Γ *be a closed composed curve,* $a, b \in PC(\Gamma)$ *, and let* $E(\Gamma)$ *denote a separable symmetric space satisfying*

$$1 < m(E) = M(E) < 2 . \tag{11.2}$$

Then the operator $aP_\Gamma + bQ_\Gamma$ $(A = P_\Gamma aI + Q_\Gamma bI)$ *is at least one sided invertible on* $E(\Gamma)$ *if and only if*

$$a(t+0)b(t)f_\delta(\mu) + a(t)b(t+0)(1 - f_\delta(\mu)) \neq 0 \quad (t \in \Gamma, 0 \leq \mu \leq 1) \tag{11.3}$$

with $\delta := 2\pi \log_2 M(E)$ *.*

If the condition (11.3) is satisfied and $c := a/b$ *then the operator* A *is invertible, invertible only from the left or invertible only from the right on* $E(\Gamma)$ *depending on whether the number* $\kappa = \operatorname{ind} c^{p_0,1}$ $(p_0 := 1/\log_2 M(E))$ *is equal to zero, positive or negative, respectively. If* $\kappa > 0$ *then* $\dim \operatorname{coker} A = \kappa$ *, and in case* $\kappa < 0$ *one has* $\dim \ker A = |\kappa|$ *.*

The proof of this theorem proceeds analogously as that one of Theorem 3.1. One only has to explain the boundedness on $E(\Gamma)$ of the operator $\Psi_+^{-1} P_\Gamma \Psi_-^{-1}$ which appears in the definition of the factorization of the function $\Psi = \Psi_{t_1,\gamma_1} \cdot \ldots \cdot \Psi_{t_n,\gamma_n}$ in the space $L_{p_0}(\Gamma)$ (with $p_0 = 1/\log_2 M(E)$ in the present situation).

From condition (11.2) one concludes that $b(t \pm 0) \neq 0$ and that the function $c = a/b$ is $\{p_0, 1\}$–non–singular $(p_0 = 1/\log_2 M(E))$. Further it is easy to see that each $\{p_0, 1\}$–non–singular function is also $\{p, 1\}$–non–singular for all p belonging to the interval $(p_0 - \varepsilon, p_0 + \varepsilon)$ with a certain positive number ε . Now, by Theorem 2.4, the function Ψ admits a factorization in each space $L_p(\Gamma)$ $(p \in (p_0 - \varepsilon, p_0 + \varepsilon))$ and, consequently, the operator $\Psi_+^{-1} P_\Gamma \Psi_-^{-1}$ is bounded on each of these spaces. So Theorem 11.1 implies the boundedness of the operator $\Psi_+^{-1} P_\Gamma \Psi_-^{-1}$ on $E(\Gamma)$. ■

In case $m(E) \neq M(E)$ we can only derive sufficient conditions for the one–sided invertibility of the operator A . For it is sufficient that condition (11.3) is satisfied for all δ belonging to the interval

$$2\pi \log_2 m(E) \leq \delta \leq 2\pi \log_2 M(E) . \tag{11.4}$$

The latter means that $b(t \pm 0) \neq 0$ and that the function $c = a/b$ is p–non–singular for all values p belonging to the interval $[\log_2^{-1} M(E), \log_2^{-1} m(E)]$.

Let us finally remark that, by means of the factors $c_\pm$ appearing in the factorization of the p–non–singular function c in $L_p(\Gamma)$, one can get explicit representations of the kernel and the cokernel of the operator A on $E(\Gamma)$, a solvability condition for the equation $A\varphi = \psi$, and even a formula for the at least one–sided inverse operator of A .

9.12 Fredholm conditions in the case of arbitrary weights

Let Γ be a simple closed curve, $0 \in \Gamma_\Gamma^+$, and $\rho \in W_p(\Gamma)$. The latter is tantamount to the boundedness of the operator S_Γ on the space $L_p(\Gamma, \rho)$.

In the present section we are going to establish criteria for the Fredholmness of singular integral operators with piecewise continuous coefficients on spaces $L_p(\Gamma, \rho)$ with arbitrary weight $\rho \in W_p(\Gamma)$. For this we need the following two properties of the class $W_p(\Gamma)$ $(1 < p < \infty)$ (see HUNT, MUCKENHOUPT, WHEEDEN [1]):

12.1°. *If $\rho \in W_p(\Gamma)$ then there exists a neighbourhood $U(1)$ of the number 1 such that $\rho^x \in W_p(\Gamma)$ for all $x \in U(1)$.*

12.2°. *If $\rho_1, \rho_2 \in W_p(\Gamma)$ then $\rho := \rho_1^y \rho_2^{1-y} \in W_p(\Gamma)$ for all $y \in (0,1)$.*

Define

$$I_\tau = \{\alpha \in \mathbf{R} : |\tau - t|^{\alpha p} \rho(t) \in W_p(\Gamma)\}. \tag{12.1}$$

From properties 12.1° and 12.2° we conclude that I_τ is an open connected subset of the real axis $\mathbf{R}$, that is, an interval, and this interval cannot be empty since $0 \in I_\tau$. Denote the left and right end of this interval by $-\nu_-(\tau)$ and $1 - \nu_+(\tau)$, respectively. Then, clearly, $\nu_-(\tau) > 0$ and $\nu_+(\tau) < 1$. We shall show that $\nu_-(\tau) \le \nu_+(\tau)$, or in other words, that the length of the interval I_τ is not more than 1.

Assume for contrary that there is a number $\alpha_0 \in \mathbf{R}$ such that $\alpha_0 \in I_\tau$ and $\alpha_0 + 1 \in I_\tau$. Then $|\tau - t|^{\alpha_0} \rho^{1/p} \in L_p(\Gamma)$ and $|\tau - t|^{-1-\alpha_0} \rho^{-1/p} \in L_q(\Gamma)$ (cf. Theorem 4.5 in Chapter 1, [GK1]), from which it follows that $|\tau - t|^{-1} \in L_1(\Gamma)$, which is evidently false.

The basic result of this section is

Theorem 12.1. *Let $a \in PC(\Gamma)$. Then the operator $A = aP_\Gamma + Q_\Gamma$ is a Φ-operator on the space $L_p(\Gamma, \rho)$ if and only if*

$$a(t \pm 0) \neq 0 \tag{12.2}$$

at all points $t \in \Gamma$ and if

$$\frac{1}{2\pi} \arg \frac{a(t-0)}{a(t+0)} \notin [\nu_-(t), \nu_+(t)], \tag{12.3}$$

with the argument being chosen in the interval $[0, 2\pi)$.

If, in particular, $\rho(t) = |t - t_0|^\beta$ $(-1 < \beta < p - 1)$ then

$$I_{t_0} = \{\alpha \in \mathbf{R} : |t - t_0|^{\beta + \alpha p} \in W_p(\Gamma)\} = \left(-\frac{1+\beta}{p}, 1 - \frac{1+\beta}{p}\right). \tag{12.4}$$

In this case,

$$\nu_-(t_0) = \nu_+(t_0) = \frac{1+\beta}{p}\,, \tag{12.5}$$

and condition (12.3) means that

$$\arg\frac{a(t_0-0)}{a(t_0+0)} \neq \frac{2\pi(1+\beta)}{p} \tag{12.6}$$

and

$$\arg\frac{a(t-0)}{a(t+0)} \neq \frac{2\pi}{p} \quad (t \neq t_0)\,. \tag{12.7}$$

The condition (12.3) can be geometrically reinterpreted as follows: The point $\lambda = 0$ does not belong to the subset of the complex plane which consists of the range of the function a and certain planar sets bounded by two circular arcs (or by one arc and one straight line).

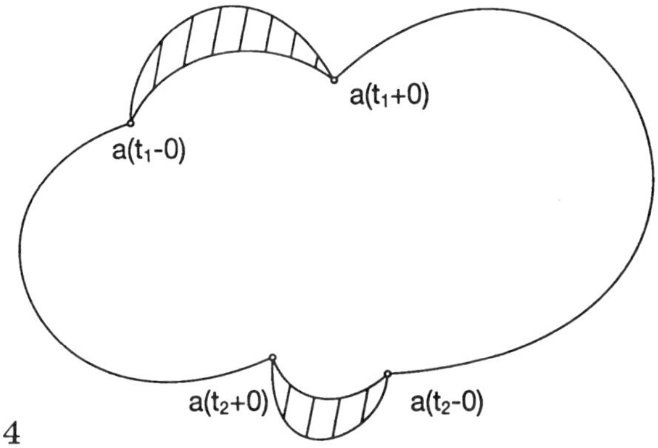

Figure 9.4

Proof of Theorem 12.1. First let $a(t) = t^\gamma$ $(0 \leq \mathrm{Re}\gamma < 1)$ be the function which is continuous at each point $t \neq t_0$ $(t_0 \in \Gamma)$. Factoring the function a, one immediately gets that

$$\dim\ker(aP_\Gamma + Q_\Gamma) = 0$$

and

$$\dim\ker(aP_\Gamma + Q_\Gamma)^* = \kappa\,,$$

where $\kappa = 0$ or $\kappa = 1$ in dependence on γ.

Let the operator $A = aP + Q$ be a Φ-operator. Then $a(t) = a_-(t)t^\kappa a_+(t)$ and $\rho|a_+|^{-p} \in W_p(\Gamma)$ (see Theorem 3.1 in Chapter 8). In addition, the function a can be represented as

$$a(t) = (t-t_0)^\gamma(1-\frac{t_0}{t})^\gamma \tag{12.8}$$

as well as

$$a(t) = (t-t_0)^{\gamma-1}\,(1-\frac{t_0}{t})^{1-\gamma}t\,. \tag{12.9}$$

Hence, in case $\kappa = 0$ one can suppose that $a_+(t) = (t - t_0)^\gamma$, and if $\kappa = 1$ that $a_+(t) = (t - t_0)^{\gamma - 1}$. In the first case we obtain

$$|t - t_0|^{-p\mathrm{Re}\gamma} \rho \in W_p(\Gamma) \, ,$$

from which it follows that

$$\nu_+(t_0) - 1 < \mathrm{Re}\gamma < \nu_-(t_0) \, .$$

In the second case, the condition

$$|t - t_0|^{(1 - \mathrm{Re}\gamma)p} \rho \in W_p(\Gamma)$$

leads to

$$\mathrm{Re}\gamma > \nu_+(t_0) \, .$$

Combining these results we find that

$$\mathrm{Re}\gamma \notin [\nu_-(t_0), \ \nu_+(t_0)] \, ,$$

and, since $\arg(a(t_0 - 0)/a(t_0 + 0)) = 2\pi \mathrm{Re}\gamma$, that

$$\frac{1}{2\pi} \arg \frac{a(t_0 - 0)}{a(t_0 + 0)} \notin [\nu_-(t_0), \ \nu_+(t_0)] \, . \tag{12.10}$$

For the points $t \neq t_0$ the condition (12.3) follows from $\nu_-(t) > 0$.

Conversely, assume the function $a(t) = t^\gamma$ is subject to the condition (12.3). Then one of the inclusions

$$|t - t_0|^{-p\mathrm{Re}\gamma} \rho \in W_p(\Gamma), \ |t - t_0|^{p(1 - \mathrm{Re}\gamma)} \rho \in W_p(\Gamma)$$

holds, and consequently, one of the equalities (12.8) and (12.9) describes the factorization of the function a in the space $L_p(\Gamma, \rho)$. By Theorem 3.1 of Chapter 8,

$$t^\gamma P_\Gamma + Q_\Gamma \in \Phi(L_p(\Gamma, \rho)) \, .$$

This proves the theorem for the function $\psi(t) = t^\gamma$. The general case can be found from this in the manner outlined in Section 9.3. ■

Theorem 12.2. *Let a ($\in PC(\Gamma)$) be a function satisfying the conditions (12.2) and (12.3); furthermore let $t_1, \ldots, t_N$ denote all of its discontinuities , l the number of the points of discontinuity at which*

$$\arg \frac{a(t_m - 0)}{a(t_m + 0)} > 2\pi \nu_+(t_m) \, ,$$

and Γ_j *the arcs into which* Γ *is divided by the points* t_j *. Then one has for the operator* $aP_\Gamma + Q_\Gamma$ *acting on* $L_p(\Gamma, \rho)$

$$\text{Ind}\,(aP_\Gamma + Q_\Gamma) = -l + \frac{1}{2\pi}\left(-\sum_m [\arg a]_{\Gamma_m} + \sum_{m=1}^{N} \arg\frac{a(t_m - 0)}{a(t_m + 0)} \right) . \tag{12.11}$$

Proof. By means of the principle of separation of singularities (cf. Section 7.3) one can reduce the problem of computing the index of the operator $aP_\Gamma + Q_\Gamma$ $(a \in PC(\Gamma))$ to the corresponding problem for the operator $\psi_{\tau,\gamma}P_\Gamma + Q_\Gamma$. But if $0 \le \text{Re}\gamma < 1$ then the index of these simple operators is either equal to zero or to minus one, and this immediately yields the equality (12.11). ∎

As in Section 9.1 one can find a geometric interpretation of the $\{p, \rho\}$–index of a given function a satisfying the conditions (12.2) and (12.3). This is left to the reader.

9.13 Technical lemmas

In this section we give some results in connection with the boundedness on l_p of Toeplitz operators generated by piecewise continuous functions and of their inverses. These results will be needed in the forthcoming section which deals with criteria for the Fredholmness of such operators.

Proposition 13.1. *Let* β *and* p *be numbers satisfying*

$$-1/p < \beta < 1 - 1/p\,, \quad 1 \le p \le \infty\,,$$

and set

$$c_{nk} = \frac{n^\beta k^{-\beta}}{n+k}\,.$$

Then the operator C *defined by its matrix representation* $(c_{nk})_{n,k=1}^\infty$ *is bounded on* l_p *.*

Proof. Let $\xi = (\xi_k)_1^\infty \in l_p$ and $\eta = (\eta_k)_1^\infty \in l_q$ $(1/p + 1/q = 1)$. Then

$$\begin{aligned}
\left| \sum_{n=1}^\infty \overline{\eta}_n \sum_{k=1}^\infty c_{nk}\xi_k \right| &\le \sum_{n=1}^\infty\sum_{k=1}^\infty \frac{n^\beta k^{-\beta}}{n+k} |\xi_k||\eta_n| \\
&= \sum_{n=1}^\infty\sum_{k=1}^\infty \left(\frac{n^\beta k^{-\beta}}{n+k}\right)^{1/p} \left(\frac{k}{n}\right)^{1/pq} |\xi_k| \left(\frac{n^\beta k^{-\beta}}{n+k}\right)^{1/q} \left(\frac{n}{k}\right)^{1/pq} |\eta_n| \\
&\le \left(\sum_{k=1}^\infty |\xi_k|^p \sum_{n=1}^\infty \frac{n^\beta k^{-\beta}}{n+k}\left(\frac{k}{n}\right)^{1/q} \right)^{1/p} \left(\sum_{n=1}^\infty |\eta_n|^q \sum_{k=1}^\infty \frac{n^\beta k^{-\beta}}{n+k}\left(\frac{n}{k}\right)^{1/p} \right)^{1/q} .
\end{aligned}$$

Since $\beta - 1/q < 0$ the function $f(x) = x^{\beta-1/q}(1+x)^{-1}$ is increasing. Thus,

$$\sum_{n=1}^{\infty} \frac{n^\beta k^{-\beta}}{n+k}\left(\frac{k}{n}\right)^{1/q} = \sum_{n=1}^{\infty} f\left(\frac{n}{k}\right)\cdot\frac{1}{k} < \int_0^\infty \frac{x^{\beta-1/q}}{1+x}\,dx\,.$$

Analogously one shows that the series

$$\sum_{k=1}^{\infty} \frac{n^\beta k^{-\beta}}{n+k}\left(\frac{n}{k}\right)^{1/p}$$

is bounded by a number independent of n.

So we arrive at

$$\left|\sum_{n=1}^{\infty} \overline{\eta}_n \sum_{k=1}^{\infty} c_{nk}\xi_k\right| \le r\|\xi\|_{l_p}\|\eta\|_{l_q}$$

with a constant r, and this estimate implies the boundedness of the operator C on all spaces l_p $(1 < p < \infty)$. In the cases $p = 1$ and $p = \infty$ the assertion of the proposition is evident. ■

Remember (see [GK 1, 5.2]) that $\widetilde{l}_p$ stands for the Banach space of all sequences $x = (\xi_n)_{n=-\infty}^{\infty}$ provided with the norm

$$\|x\|_p = \left(\sum_{n=-\infty}^{\infty} |\xi_n|^p\right)^{1/p} \quad (1 \le p < \infty)\,,$$

$$\|x\|_\infty = \sup |\xi_n|$$

and that, given a function $a \in L_\infty(\mathbf{T})$, the notation R_a refers to the operator being defined on the collection $\mathcal{L}$ of all finite sequences by the infinite matrix

$$R_a = (a_{j-k})_{j,k=-\infty}^{\infty}$$

where the a_j are the Fourier coefficients of the function a. Further we denote by P and Q the projection operators

$$P(\xi_k)_{-\infty}^{\infty} = (\delta_k\xi_k)_{-\infty}^{\infty} \quad \text{and} \quad Q(\xi_k)_{-\infty}^{\infty} = ((1-\delta_k)\xi_k)_{-\infty}^{\infty}$$

with

$$\delta_k = \begin{cases} 1 & \text{if } k \ge 0 \\ 0 & \text{if } k < 0\,. \end{cases}$$

Finally, we define $S = P - Q$.

Theorem 13.1 *Let the complex number* α *satisfy the estimate* $-1/p < \operatorname{Re}\alpha < 1 - 1/p$ *and define the function* $a(t)$ *on* $\mathbf{T}$ *by*

$$a(t) = (1-t)^{-\alpha} = (1-\exp(i\theta))^{-\alpha} \quad (0 < \theta \le 2\pi)\,.$$

Then the operator

$$R = R_a S R_{a^{-1}}$$

is bounded on $\widetilde{l}_p$.

Proof. The functions $a(t) = (1-t)^{-\alpha}$ and $b(t) = (1-t)^{\alpha}$ belong to the space $L_1^+(\mathbf{T})$ and, as is well known, their Fourier coefficients are

$$a_n = \begin{cases} \binom{-\alpha}{n} & \text{if } n \geq 0 \\ 0 & \text{if } n < 0 \,, \end{cases} \qquad b_n = \begin{cases} \binom{\alpha}{n} & \text{if } n \geq 0 \\ 0 & \text{if } n < 0 \,, \end{cases}$$

where $\binom{\alpha}{n} = \alpha(\alpha-1)\ldots(\alpha-n+1)/n!$ if $n > 0$ and $\binom{\alpha}{0} = 1$. This shows that

$$R_a = \begin{pmatrix} A_1 & 0 \\ A_3 & A_2 \end{pmatrix}, \quad R_b = \begin{pmatrix} B_1 & 0 \\ B_3 & B_2 \end{pmatrix},$$

with

$$A_1 = \begin{pmatrix} \ldots & \ldots & \\ \ldots & \ldots & \\ \ldots & a_0 & 0 \\ \ldots & a_1 & a_0 \end{pmatrix}, \quad A_2 = \begin{pmatrix} a_0 & 0 & \ldots \\ a_1 & a_0 & \ldots \\ \ldots & & \ldots \\ \ldots & & \ldots \end{pmatrix} \quad A_3 = \begin{pmatrix} \ldots & a_2 & a_1 \\ \ldots & a_3 & a_2 \\ \ldots & \ldots & \\ \ldots & \ldots & \end{pmatrix}$$

and analogously,

$$B_1 = \begin{pmatrix} \ldots & \ldots & \\ \ldots & \ldots & \\ \ldots & b_0 & 0 \\ \ldots & b_1 & b_0 \end{pmatrix}, \quad B_2 = \begin{pmatrix} b_0 & 0 & \ldots \\ b_1 & b_0 & \ldots \\ \ldots & & \ldots \\ \ldots & & \ldots \end{pmatrix} \quad B_3 = \begin{pmatrix} \ldots & b_2 & b_1 \\ \ldots & b_3 & b_2 \\ \ldots & \ldots & \\ \ldots & \ldots & \end{pmatrix}.$$

Since $ab = 1$,

$$\begin{pmatrix} I & 0 \\ 0 & I \end{pmatrix} = \begin{pmatrix} A_1 & 0 \\ A_3 & A_2 \end{pmatrix} \begin{pmatrix} B_1 & 0 \\ B_3 & B_2 \end{pmatrix} = \begin{pmatrix} A_1 B_1 & 0 \\ A_3 B_1 + A_2 B_3 & A_2 B_2 \end{pmatrix},$$

and consequently,

$$R = \begin{pmatrix} A_1 & 0 \\ A_3 & A_2 \end{pmatrix} \begin{pmatrix} -I & 0 \\ 0 & I \end{pmatrix} \begin{pmatrix} B_1 & 0 \\ B_3 & B_2 \end{pmatrix}$$

$$= \begin{pmatrix} -A_1 B_1 & 0 \\ -A_3 B_1 + A_2 B_3 & A_2 B_2 \end{pmatrix} = \begin{pmatrix} -I & 0 \\ 2 A_2 B_3 & I \end{pmatrix}$$

For the operator $A_2 B_3$ one finds

$$A_2 B_3 = \begin{pmatrix} a_0 & 0 & 0 & \ldots \\ a_1 & a_0 & 0 & \ldots \\ a_2 & a_1 & a_0 & \ldots \\ . & . & . & \end{pmatrix} \begin{pmatrix} \ldots & b_3 & b_2 & b_1 \\ \ldots & b_4 & b_3 & b_2 \\ \ldots & b_5 & b_4 & b_3 \\ \ldots & . & . & . \end{pmatrix}$$

$$= \begin{pmatrix} \cdots & \gamma_{0,-3} & \gamma_{0,-2} & \gamma_{0,-1} \\ \cdots & \gamma_{1,-3} & \gamma_{1,-2} & \gamma_{1,-1} \\ & \vdots & \vdots & \vdots \end{pmatrix},$$

with

$$\gamma_{j,-k} = \sum_{m=0}^{j} a_{j-m} b_{k+m} = \sum_{m=0}^{j} \binom{-\alpha}{j-m} \binom{\alpha}{k+m} \quad (j \geq 0,\ k \geq 1)\,. \tag{13.1}$$

It is not hard to check (for instance via induction over j) the following identity

$$\sum_{m=0}^{j} \binom{-\alpha}{j-m} \binom{\alpha}{k+m} = \binom{\alpha-1}{k-1} \binom{-\alpha-1}{j} \frac{\alpha}{k+j}\,. \tag{13.2}$$

Write $\alpha = \beta + i\gamma$. Since $\ln(a+x) \leq \ln a + \frac{x}{a}$ whenever $a > 0$ and $x \geq 0$ one has

$$\ln\left[\left(1 - \frac{\beta}{m}\right)^2 + \frac{\gamma^2}{m^2}\right] \leq \ln\left(1 + \frac{\beta}{m}\right)^2 + \frac{\gamma^2}{(m-\beta)^2}\,,$$

and, consequently,

$$\begin{aligned} \left|\binom{\alpha-1}{k-1}\right| &= \prod_{m=1}^{k-1} \left|1 - \frac{\alpha}{m}\right| \leq c_1 \prod_{m=1}^{k-1} \left|1 - \frac{\beta}{m}\right| \\ &\leq c_1 \exp\left(-\beta \sum_{m=1}^{k-1} \frac{1}{m}\right) \\ &\leq c_1 \exp(-\beta \ln k + c_2) = c_3 k^{-\beta}\,, \end{aligned} \tag{13.3}$$

where $c_1 = \sum_{m=1}^{\infty} \frac{\gamma^2}{(m-\beta)^2}$.

From (13.1), (13.2) and (13.3) one concludes that

$$|\gamma_{j,-k}| \leq c\, \frac{k^{-\beta}(1+j)^{\beta}}{k+j}$$

with a constant c being independent of k and j . Now apply Proposition 13.1 to derive the boundedness of the operator $A_2 B_3$ on l_p and, thus, that of R on $\widetilde{l}_p$. ■

As an immediate consequence of this theorem we remark

Corollary 13.1. *Let t_0 be an arbitrary point on the unit circle and γ be a complex number satisfying*

$$-1/p < \operatorname{Re} \gamma < 1 - 1/p\,. \tag{13.4}$$

Then the operator

$$R_{(t-t_0)^{-\gamma}} S R_{(t-t_0)^{\gamma}}$$

is bounded on $\widetilde{l}_p$.

Further we shall need

Corollary 13.2. *Let the function* $\psi(t) = t^\gamma$ *be continuous at every point* $t \neq t_0$ *of the unit circle. If the condition (13.4) is fulfilled, then the operator* $A = R_\psi P + Q$ *is invertible on the space* $\widetilde{l}_p$ $(1 < p < \infty)$.

Indeed, define $u := \frac{\psi^{-1}+t}{2}$, $v := \frac{\psi^{-1}-1}{2}$ and

$$B := R_u + R_v R_{\psi^{-1}} S R_\psi .$$

The results in Section 9.4 entail the equality

$$AB\xi = BA\xi = \xi$$

for all finite sequences ξ . Moreover we know from Theorem 2.2 of Chapter 5, [GK1] that $\psi^\pm \in \mathcal{R}_p$ [1] for all $p \in (1,\infty)$, from which the boundedness of the operators A, R_u, R_v on the space $\widetilde{l}_p$ follows. Since the operator $R_{\psi^{-1}} S R_\psi$ is also bounded, as we have seen above, this yields the boundedness of the operators A and B on $\widetilde{l}_p$. ■

9.14 Toeplitz and paired operators with piecewise continuous coefficients on the spaces l_p and $\widetilde{l}_p$

Let a be a function in $L_\infty(\mathbf{T})$ and $(a_n)_{n=-\infty}^{\infty}$ its Fourier coefficient sequence. Remember (see [GK 1, 5.3]) that the operator defined on l^p by the Toeplitz matrix

$$T_a = (a_{j-k})_{j,k=0}^{\infty}$$

is called a *Toeplitz operator*, the operator $\Pi(a,b) := PR_a + QR_b$ $(a, b \in L_\infty(\mathbf{T}))$ acting on $\widetilde{l}_p$ is said to be a *paired operator*, and the operator $\prod'(a,b) = R_a P + R_b Q$ is referred to as the *transpose of a paired operator.*

By $PC_{<p>}$ we denote the class of all piecewise continuous functions a on $\mathbf{T}$ which belong to the set $\mathcal{R}_{<p>}$, that is, they generate bounded multipliers R_a on the spaces $\widetilde{l}_r$ for all r in a certain neighbourhood of the point p . In the present section we shall derive criteria for the Fredholmness and descriptions of the kernel and cokernel of the operators $T_a, \Pi(a,b)$ and $\Pi'(a,b)$ with functions a, b in $PC(\mathbf{T})$. In contrast to singular integral operators on $L_p(\Gamma)$ where the p-non-singularity of the function a/b played the decisive role, it will now turn out that the q-non-singularity of this function is responsible for the

[1] $\mathcal{R}_p$ denotes the class of all functions $a \in L_\infty(\mathbf{T})$ for which the operator R_a extends to a linear bounded operator in $\widetilde{l}_p$, i.e. all multipliers in $\widetilde{l}_p$.

Fredholmness of the operator T_a in l_p $(p^{-1}+q^{-1}=1)$. Notice that any q–non–singular function permits a q–factorization (cf. Corollary 3.1).

Theorem 14.1. *Let* $a \in PC_{<p>}(\mathbf{T})$ $(1 < p < \infty)$. *For the operator* T_a *to be a* $\Phi-$ *or* $\Phi_{\pm}$*-operator on* l_p *it is necessary and sufficient that the function* $a(t)$ *is* q*–non–singular* $(1/p + 1/q = 1)$.

Let a *be a* q*–non–singular function and* $a = a_- t^\kappa a_+$ *be its factorization in* $L_q(\mathbf{T})$.

1.° *If* $\kappa \le 0$ *then the operator* T_a *is right invertible, the operator*

$$T_{a_+^{-1}} T_{a_-^{-1}} V^{|\kappa|} ,$$

with $V = T_t$ *denoting the forward shift operator, is one of its right inverses of* T_a, *and*

$$\ker T_a = \operatorname{span}\{g, Vg, \ldots, V^{-\kappa+1} g\} , \tag{14.1}$$

where $g = (g_k)_0^\infty$ *stands for the Fourier coefficient sequence of the function* a_+^{-1}.

2.° *If* $\kappa > 0$ *then the operator* T_a *is invertible from the left and the operator*

$$W^\kappa T_{a_+^{-1}} T_{a_-^{-1}} ,$$

with $W = T_{t^{-1}}$ *referring to the backward shift operator, is one of its left inverses. The equation* $T_a \xi = \eta$ *is solvable if and only if the equalities*

$$\sum_{n=0}^{\infty} \eta_n f_{-m-n} = 0 \quad (m = 0, \ldots, \kappa - 1) , \tag{14.2}$$

with $\eta = (\eta_n)_0^\infty$ *and* f_n *abbreviating the* n*–th Fourier coefficient of the function,* a_-^{-1} *are fulfilled.*

In preparation for the proof of this theorem we formulate two lemmas.

Proposition 14.1. *Let* C *be a bounded operator on* $\widetilde{l}_p$ *for all* p *belonging to the interval* (p_1, p_2). *If this operator is compact on the space* $\widetilde{l}_{p_0}$ *with* $p_0 \in (p_1, p_2)$, *then* C *is compact on every space* $\widetilde{l}_p$ $(p \in (p_1, p_2))$.

Proof. Define

$$P_n(\xi)_{-\infty}^{\infty} = (\ldots, 0, \xi_{-n}, \ldots, \xi_n, 0, \ldots) .$$

This sequence $\{P_n\}$ of projections converges strongly to I, and thus, the sequence $\{P_n C\}$ converges uniformly to the operator C on $\widetilde{l}_{p_0}$. The interpolation theorem of M.Riesz applied to the triple (r, s, p_0) (with either $p_1 < s < r < p_0$ or $p_0 < r < s < p_2$) yields

$$\|P_n C - C\|_{\widetilde{l}_r} \le \|P_n C - C\|_{\widetilde{l}_s}^{\theta} \, \|P_n C - C\|_{\widetilde{l}_{p_0}}^{1-\theta}$$

This shows the uniform convergence of the finite rank operators P_nC to the operator C on $\tilde{l}_r$, hence, the operator C is compact on $\tilde{l}_r$. Since r is an arbitrary number in the interval (p_1, p_2), this proves the assertion. ■

Proposition 14.2. *Let $a, b \in PC_{<p>}(\mathbf{T})$. If the functions a and b do not possess common discontinuities then the operator $T = T_aT_b - T_{ab}$ is compact on l_p.*

Proof. The operator T is bounded on l_r for all r in a certain interval $(p_1 - \varepsilon, p_2 + \varepsilon)$ with $p_1 = \min(p, q)$ and $p_2 = \max(p, q)$ (see Section 2, Chapter 5, [GK1]). From Proposition 3.1 and the inclusion $\mathcal{R}_{<p>} \subset \mathcal{R}_{<2>}$ we conclude that T is compact on l_2, and the preceding proposition states its compactness on l_p. ■

Proof of Theorem 14.1. Let a be a q–non-singular function. Then as in Proposition 3.1, it can be represented in the form

$$a = \psi_1\psi_2 \cdot \ldots \cdot \psi_n a_0 , \tag{14.3}$$

where each of the functions ψ_m has only one discontinuity. These discontinuities are pairwise distinct, the functions ψ_m satisfy the conditions in Corollary 13.2, and the function a_0 is continuous on $\mathbf{T}$. By Theorem 2.2 Chapter 5, [GK1], $\psi_k \in \mathcal{R}_{(1,\infty)}$. [1]

Then by Proposition 14.2, the operator T_a can be rewritten as

$$T_a = T_{\psi_1}T_{\psi_2} \cdot \ldots \cdot T_{\psi_n}T_{a_0} + T \tag{14.4}$$

with a compact operator T. We know from Corollary 13.2 that each of the operators T_{ψ_m} is invertible. Furthermore, the function a_0 belongs to $C_{<p>}(\mathbf{T})$, and by Theorem 3.4 of Chapter 5, [GK1], the operator T_{a_0} is a Φ–operator with index $\operatorname{Ind} T_{a_0} = -\operatorname{ind} a_0$. Now remember that in Section 9.1, we showed that

$$\operatorname{ind} a^q = \sum_m \operatorname{ind} \psi_m^q + \operatorname{ind} a_0 = \operatorname{ind} a_0 .$$

Hence we conclude that T_a is a Φ–operator with index $\operatorname{Ind} T_a = -\operatorname{ind} a^q = -\kappa$.

The one–sided invertibility of the operator T_a is a consequence of Proposition 3.1 of Chapter 5, [GK1]. The equality (14.1) can be checked in a straightforward manner, having only to note that $\dim \ker T_a = -\kappa$ in case $\kappa \le 0$. Finally, the condition (14.2) follows by passing to the adjoints.

The proof of the necessity of the conditions of the theorem will be split into several steps.

1. At first we show that if $\psi(t) = t^\gamma$ $(-1/p < \operatorname{Re}\gamma \le 1 - 1/p)$ and if T_ψ is a Φ– (or $\Phi_\pm$–)operator then

$$\operatorname{Re} \gamma \ne 1 - 1/p , \tag{14.5}$$

[1] $\mathcal{R}_\Omega := \bigcap_{p \in \Omega} \mathcal{R}_p$.

that is, that the function ψ is q–non–singular.

Assume that $\mathrm{Re}\,\gamma = 1 - 1/p$. Then the arc l associated with $\{\psi(t)\}$ runs through the point $z = 0$. Choosing two complex numbers z_1 and z_2 with sufficiently small absolute values and which are located at different sides of the arc l , one can conclude that the functions $\psi - z_1$ and $\psi - z_2$ are q–non–singular, but that their q–indices differ from each other. By the above,

$$\mathrm{Ind}\, T_{\psi - z_1} \neq \mathrm{Ind}\, T_{\psi - z_2} \ .$$

On the other hand,

$$\mathrm{Ind}\, T_{\psi - z_1} = \mathrm{Ind}\, T_{\psi} = \mathrm{Ind}\, T_{\psi - z_2}$$

due to the invariance of the index under small perturbation. This contradiction proves (14.5).

2. Our next claim is that the one–sided invertibility of the operator T_a implies that

$$a(t \pm 0) \neq 0 \tag{14.6}$$

for all points $t \in \mathbf{T}$.

Assume (14.6) to be violated. If needed, we add to the operator T_a the scalar operator εI with a sufficiently small $|\varepsilon| (\neq 0)$ so that $a(t_0) = 0$ at a certain point t_0 .

We introduce a system of localizing classes (see [GK 1, 5.1]) in the algebra $\mathcal{A} = L(l_p)/\mathcal{T}(l_p)$: Let M_τ stand for the collection of all cosets $\widehat{T}_x$ with $x \in C_{<p>}$, $0 \leq x(t) \leq 1$ and $x(\tau) = 1$. It is immediate that $\{M_\tau\}$ $(\tau \in \mathbf{T})$ is actually a covering system of localizing classes, and that $\widehat{T}_a \widehat{T}_x = \widehat{T}_x \widehat{T}_a$ for all $a \in PC_{<p>}(\mathbf{T})$.

The one–sided invertibility of T_a involves the one–sided invertibility of $\widehat{T}_a$ in the algebra $\mathcal{A}$. Since $\widehat{T}_a \overset{M_{t_0}}{\sim} \widehat{0}$ this implies, via Theorem 1.1 in [GK 1, 5.1], that the zero element $\widehat{0}$ is one–sided invertible in $\mathcal{A}$ which clearly is impossible.

3. To finish the proof of the theorem notice first that Proposition 3.1 in [GK 1, 5.3] states that whenever the operator T_a is a Φ–(or $\Phi_\pm$–)operator then it is one–sided invertible and, as we have just seen, $a(t \pm 0) \neq 0$ for all $t \in \mathbf{T}$. Let $t_1, \ldots, t_n$ denote all points of discontinuity for the function a , and let $\psi_k(t) = t^{\gamma_k}$ be functions which are discontinuous only at $t = t_k$ and satisfying

$$\frac{\psi_k(t_k + 0)}{\psi_k(t_k - 0)} = \frac{a(t_k + 0)}{a(t_k - 0)} \quad (k = 1, \ldots, n)$$

and

$$-1/p < \mathrm{Re}\gamma_k < 1 - 1/p \ .$$

Since the operators $T_{\psi_1}, \ldots, T_{\psi_n}, T_{a_0}$ appearing in (14.4) commute modulo compact operators, they are all Φ– (or $\Phi_\pm$–)operators (in depending on T_a). We have already seen that then

$$\mathrm{Re}\ \gamma_k \neq 1 - 1/p$$

which yields the q-non-singularity of the function a. ■

Let us consider an example. The problem is to find out all $\alpha \in \mathbf{R}\backslash\mathbf{Z}$ for which the equation

$$\begin{pmatrix} \frac{1}{\alpha} & \frac{1}{\alpha+1} & \frac{1}{\alpha+2} & \cdot \\ \frac{1}{\alpha-1} & \frac{1}{\alpha} & \frac{1}{\alpha+2} & \cdot \\ \frac{1}{\alpha-2} & \frac{1}{\alpha-1} & \frac{1}{\alpha} & \cdot \\ \cdot & \cdot & \cdot & \cdot \end{pmatrix} \begin{pmatrix} \xi_0 \\ \xi_1 \\ \xi_2 \\ \vdots \end{pmatrix} = \begin{pmatrix} \eta_0 \\ \eta_1 \\ \eta_2 \\ \vdots \end{pmatrix} \tag{14.7}$$

is uniquely solvable in l_p $(1 < p < \infty)$ for arbitrary right side. Furthermore we look for the solution of this equation when $\eta = (\eta_n) = (1, 0, 0, \ldots)$.

Solution. Because

$$\int_{\mathbf{T}} t^\alpha t^{-n} \, |dt| = \frac{2 \sin \pi\alpha \exp(\pi i \alpha)}{\alpha - n} ,$$

the equation (14.7) can be rewritten in the form

$$\tfrac{\exp(-\pi i \alpha)}{2 \sin \pi\alpha} T_{t^\alpha} \xi = \eta \, . \tag{14.8}$$

By Theorem 14.1, the operator T_{t^α} is invertible if and only if $\alpha \in (-1/p, 1 - 1/p)$, and consequently, the operator on the left-hand side of equation (14.8) is invertible if and only if

$$\alpha \in (-1/p, 0) \cup (0, 1 - 1/p) \, .$$

Also, the identity

$$\exp(-\pi i \alpha) t^\alpha = (1 - t)^{-\alpha} (1 - \frac{1}{t})^{-\alpha} ,$$

being just the q-factorization of the function $t^\alpha \exp(-\pi i \alpha)$, implies that the vector

$$\xi = 2 \sin(\pi\alpha) \; T_{(1-t)^\alpha} \; T_{(1-\frac{1}{t})^\alpha} \eta \tag{14.9}$$

is the solution of the equation (14.8), i.e.

$$\begin{pmatrix} \xi_0 \\ \xi_1 \\ \xi_2 \\ \xi_3 \\ \cdot \end{pmatrix} = 2\sin(\pi\alpha) \begin{pmatrix} 1 & 0 & 0 & 0 & \cdot \\ \binom{\alpha}{1} & 1 & 0 & 0 & \cdot \\ \binom{\alpha}{2} & \binom{\alpha}{1} & 1 & 0 & \cdot \\ \binom{\alpha}{3} & \binom{\alpha}{2} & \binom{\alpha}{1} & 1 & \cdot \\ \cdot & \cdot & \cdot & \cdot & \cdot \end{pmatrix} \begin{pmatrix} 1 & \binom{\alpha}{1} & \binom{\alpha}{2} & \binom{\alpha}{3} & \cdot \\ 0 & 1 & \binom{\alpha}{1} & \binom{\alpha}{2} & \cdot \\ 0 & 0 & 1 & \binom{\alpha}{1} & \cdot \\ 0 & 0 & 0 & 1 & \cdot \\ \cdot & \cdot & \cdot & \cdot & \cdot \end{pmatrix} \begin{pmatrix} \eta_0 \\ \eta_1 \\ \eta_2 \\ \eta_3 \\ \cdot \end{pmatrix} . \tag{14.10}$$

In particular, for $\eta = (1, 0, \ldots)$ one obtains the solution

$$\xi = \left(1, \alpha, \frac{\alpha(\alpha-1)}{1 \cdot 2}, \frac{\alpha(\alpha-1)(\alpha-2)}{1 \cdot 2 \cdot 3}, \ldots\right) 2\sin(\pi\alpha) \, .$$

We examine one more example. Let $a(\in PC_{<p>}(\mathbf{T}))$ be a function for which the operator T_a is invertible on l_p, set $e_0 = (1, 0, \dots)$, and let the vector $x = (x_n)_0^\infty$ $(y = (y_n)_0^\infty)$ be the solution to the equation $T_a x = e_0$ $(T_{\overline{a}} y = e_0)$ on the space l_p (l_q). Then $x_0 \neq 0$ and, for each vector $\eta \in l_p$, the solution $\xi \in l_p$ of the equation $T_a \xi = \eta$ can be determined by $\xi = AB\eta$, with

$$A = \begin{pmatrix} x_0 & 0 & 0 & 0 & \dots \\ x_1 & x_0 & 0 & 0 & \dots \\ x_2 & x_1 & x_0 & 0 & \dots \\ . & . & . & . & \dots \end{pmatrix}, \quad B = \begin{pmatrix} 1 & b_1 & b_2 & \dots \\ 0 & 1 & b_1 & \dots \\ 0 & 0 & 1 & \dots \\ . & . & . & \dots \end{pmatrix}$$

and $b_k = \overline{y}_k / x_0$.

Indeed, the function a admits the q-factorization $a(t) = a_-(t)a_+(t)$. Without loss of generality we assume that $a_-(\infty) = 1$. The equality $T_{a_+^{-1}} T_{a_-^{-1}} e_0 = x$ implies that

$$a_+^{-1}(t) = \sum_{k=0}^{\infty} x_k t^k ,$$

meaning in particular, that $x_0 = a_+^{-1}(0) \neq 0$. Furthermore, the invertibility of the operator T_a on l_p causes the invertibility of the operator $T_{\overline{a}}$ on l_q, as well as the identity

$$T_{\overline{a}_-^{-1}} T_{\overline{a}_+^{-1}} e_0 = y .$$

This, in turn, implies that $T_{\overline{a}_-^{-1}}(\overline{x}_0 e_0) = y$, and hence,

$$a_-^{-1}(t) = 1 + \sum_{k=1}^{\infty} \frac{\overline{y}_k}{x_0} t^{-k} .$$

As a corollary one obtains $T_a^{-1} = AB$. ■

Concerning paired and transposes to paired operators, we prove now the following result.

Theorem 14.2. *Let $a, b \in PC_{<p>}(\mathbf{T})$. Then the operator $\Pi(a,b)$ $(\Pi'(a,b))$ is a Φ-operator on $\widetilde{l}_p$ $(1 < p < \infty)$ if and only if*

$$a(t \pm 0) \neq 0, \quad b(t \pm 0) \neq 0 \tag{14.11}$$

on $\mathbf{T}$ and the function $c(t) = a(t)/b(t)$ is q-non-singular $(1/p + 1/q = 1)$. If these conditions are satisfied and $c(t) = c_-(t)t^\kappa c_+(t)$ then, in case $\kappa \leq 0$, the operators $\Pi(a,b)$ and $\Pi'(a,b)$ are right invertible, the equalities

$$B_1 = R_g \Pi\left(\frac{1}{a}, \frac{1}{b}\right) R_{g^{-1}} \quad (g = c_- b) \tag{14.12}$$

and

$$B_2 = R_{f^{-1}}\Pi'(\frac{1}{a},\frac{1}{b})R_f \quad (f = c_+b) \tag{14.13}$$

provide us with right inverses of $\Pi(a,b)$ *and* $\Pi'(a,b)$, *respectively,*

$$\begin{aligned}\ker \Pi(a,b) &= \operatorname{span}\{R_he_0, R_he_1,\dots,R_he_{|\kappa|-1}\}\\ &= \operatorname{span}\{\xi, \widetilde{V}\xi,\dots,\widetilde{V}^{|\kappa|-1}\xi\}\,,\end{aligned} \tag{14.14}$$

with $h(t) = c_+^{-1}(t)(1-c(t)) = \sum_{-\infty}^{\infty} h_nt^n$ *and* $\xi = (h_n)_{-\infty}^{\infty}$, *and*

$$\begin{aligned}\ker \Pi'(a,b) &= \operatorname{span}\{R_ye_0, R_ye_1,\dots,R_ye_{|\kappa|-1}\}\\ &= \operatorname{span}\{\eta, \widetilde{V}_\eta,\dots,\widetilde{V}^{|\kappa|-1}\eta\}\,,\end{aligned} \tag{14.15}$$

with $y(t) = c_+^{-1}(t)b^{-1}(t) = \sum_{-\infty}^{\infty} y_kt^k$ *and* $\eta = (y_k)_{-\infty}^{\infty}$.

In case $\kappa \geq 0$ *the operators* $\Pi(a,b)$ *and* $\Pi'(a,b)$ *are left invertible on* $\widetilde{l}_p$ *and the operators* B_1 *and* B_2 *are left inverses of* $\Pi(a,b)$ *and* $\Pi'(a,b)$, *respectively. The equations*

$$\Pi(a,b)x = z \quad \text{and} \quad \Pi'(a,b)x = z$$

are solvable in $\widetilde{l}_p$ *if and only if their right-hand sides* $z = (z_n)_{-\infty}^{\infty}$ *are subject to the conditions*

$$\sum_{n=-\infty}^{\infty} z_nu_{k-n} = 0\,, \quad \sum_{n=-\infty}^{\infty} z_nv_{k-n} = 0 \quad (k = 0,1,\dots,\kappa-1) \tag{14.16}$$

where (u_n), (v_n) *refer to the Fourier coefficient sequences of the functions* $c_-^{-1}b$, $c_-^{-1}(1-c)$, *respectively.*

Proof. Let A stand for one of the operators $\Pi(a,b)$ or $\Pi'(a,b)$. Invoking the local principle one can show, as in Theorem 3.1 of [GK 1, 5.3], that whenever A is a Φ–operator and t is a continuity point of a and b , the functions a and b do not vanish at t . This implies without difficulties that the conditions (14.11) (see the proof of Theorem 14.1) are satisfied if A is a Φ–operator.

In addition, if the conditions (14.11) are fulfilled then the operator R_b is invertible on $\widetilde{l}_p$. To see this, represent b in the form $b(t) = \psi_1(t)\dots\psi_n(t)b_0(t)$ as in (14.3). The functions ψ_k^{-1} $(k = 1,\dots,n)$ belong to $\mathcal{R}_{<p>}$ since their total variations are bounded, and $b_0 \in \mathcal{R}_{<p>}$ due to Theorem 2.3 of [GK 1, Chapter 5]. Thus,

$$\Pi(a,b) = \Pi(c,1)R_b \quad \text{and} \quad \Pi'(a,b) = R_b\Pi'(c,1)\,,$$

with the operator R_b being invertible on $\widetilde{l}_p$.

Next we employ the standard identities

$$\begin{aligned}\Pi(c,1) &= (PR_cP+Q)(I+QR_cP)\,,\\ \Pi'(c,1) &= (I+PR_cQ)(PR_cP+Q)\,,\end{aligned}$$

in which the operators $I + QR_cP$ and $I + PR_cQ$ are invertible and the restriction of the operator $PR_cP + Q$ onto l_p coincides with T_c. Now Theorem 14.1 implies the necessity of the q-non-singularity of the function c appearing in Theorem 14.2. The other assertions of Theorem 14.2 can also be derived from Theorem 14.1. However once the factorization of the function c is available, one also may simply repeats the arguments from the proof of Theorem 3.3 in [GK 1, 5.3]. ■

9.15 Some applications

The present section is devoted to some applications of the theory of one-dimensional singular integral equations with piecewise continuous coefficients to boundary value problems, as well as to problems in mathematical physics.

1. Boundary value problem. As in [GK 1, 3.15], we consider the following problem, which also will be referred to as *problem (15.1)*.

Let Γ be a composed curve, $G \in L_\infty(\Gamma)$, and $g \in L_p(\Gamma)$. Then a function $\Phi(z)$ is to be found which admits in $\mathbf{C}\backslash\Gamma$ a representation by the Cauchy type integral

$$\Phi(z) = \frac{1}{2\pi i} \int_\Gamma \frac{\varphi(\tau)\, d\tau}{\tau - z}$$

with a density function $\varphi \in L_p(\Gamma)$, and which satisfies almost everywhere on Γ the boundary condition

$$\Phi_+(t) = G(t)\Phi_-(t) + g(t) . \tag{15.1}$$

Herein, t indicates an arbitrary point of Γ which is not an endpoint, and $\Phi_+(t)$ ($\Phi_-(t)$) stands for the limit of $\Phi(z)$ when z tends to t inside the region which is located at the left (right) of Γ.

Besides problem (15.1) we consider the singular integral equation

$$P_\Gamma\varphi + GQ_\Gamma\varphi = g . \tag{15.2}$$

As in [GK 1, 3.15], the problem (15.1) and the equation (15.2) are equivalent in the following sense: If the function $\Phi(z)$ is a solution to (15.1) then the function $\varphi(t) = \Phi_+(t) - \Phi_-(t)$ solves equation (15.2), and if φ is a solution of (15.2) in $L_p(\Gamma)$ then the function $\Phi(z)$ defined above gives a solution to (15.1).

In the remainder of this section we confine ourselves to piecewise continuous functions $G(t)$.

We know from Theorem 4.1 that the singular integral operator A standing on the left hand side of equation (15.2) is at least one-sided invertible on $L_p(\Gamma)$ if and only if the following conditions are satisfied:

1.

$$G(t-0)f(t,\mu)+G(t+0)(1-f(t,\mu))\neq 0 \tag{15.3}$$

for all $t\in\Gamma$ which are not end points, and for all $0\leq\mu\leq 1$.

2.

$$f(\tau_k,\mu)+G(\tau_k)(1-f(\tau_k,\mu))\neq 0 \tag{15.4}$$

for all starting points τ_k of non-closed subarcs of Γ, and

3.

$$1-f(\tau_k,\mu)+G(\tau_k)f(\tau_k,\mu)\neq 0 \tag{15.5}$$

for all end points τ_k of non-closed subarcs of Γ.

Clearly, these conditions provide that the function $G^{-1}(t)$ is p–non-singular. It is easy to see that the function $G^{-1}(t)$ is p–non-singular if and only if the function $G(t)$ is q–non-singular, and that in this case

$$\operatorname{ind}_p G^{-1}=\operatorname{ind}_q G\,.$$

Indeed, this can be immediately checked when G is a function of the form $\psi(t)=t^{\gamma}$, and in the general case one represents G as $G(t)=G_0(t)\psi_1(t)\ldots\psi_n(t)$ with a continuous function G_0 and the functions $\psi_k(t)=t^{\gamma_k}$ having exactly one point of discontinuity, with all points of discontinuity distinct. Hence, if the conditions (15.3)–(15.5) are fulfilled then

$$\operatorname{Ind} A=-\operatorname{ind}_p G^{-1}=\operatorname{ind}_q G \quad (p^{-1}+q^{-1}=1)\,.$$

Let Γ be a closed simple curve, $0\in F_\Gamma^+$, and let the equality

$$G(t)=G_-(t)t^{\kappa}G_+(t)$$

give the q–factorization of the function G. As in Section 3.15 we define

$$X(z)=\begin{cases} G_+(z) & \text{if } z\in F_\Gamma^+ \\ z^{-\kappa}G_-^{-1}(z) & \text{if } z\in F_\Gamma^-\,. \end{cases}$$

In this setting one has:

Theorem 15.1. *Let κ denote the q–index of the function G. Then*

1°. *If $\kappa=0$, problem (15.1) is uniquely solvable for each right–hand side, and its solution is given by*

$$\Phi(z)=\frac{X(z)}{2\pi i}\int_\Gamma \frac{g(\tau)\,d\tau}{G_+(\tau)(\tau-z)} \quad (z\in \mathbf{C}\backslash\mathbf{\Gamma})\,. \tag{15.6}$$

2°. *In case* $\kappa > 0$ *problem (15.1) possesses a solution for each right–hand side, and the general form of the solution is*

$$\Phi(z) = \frac{X(z)}{2\pi i} \int_\Gamma \frac{g(\tau)\, d\tau}{G_+(\tau)(\tau - z)} + X(z)P(z) \tag{15.7}$$

where $P(z)$ *is an arbitrary polynomial of degree less than* κ .

3°. *If* $\kappa < 0$ *then problem (15.1) is solvable if and only if*

$$\int_\Gamma \frac{g(t)t^l}{G_+(t)}\, dt = 0 \quad (0 \le l \le -\kappa - 1)\,. \tag{15.8}$$

If this condition is satisfied then the solution to problem (15.1) is unique and is given by (15.6).

Notice that the above formulated theorem remains true when G is an arbitrary $L_\infty(\Gamma)$–function admitting a q–factorization. The proof of this is identical to that of Theorem 5.1 in [GK 1, 3.15].

Moreover, Theorem 15.1 remains valid for some non–closed curves as well. If, for instance, Γ consists of simple open arcs then one completes these arcs to the closed curve $\widetilde{\Gamma}$ under preservation of the orientation of Γ , defines the function $X(z)$ by means of the factors in the q–factorization of the function

$$\widetilde{G}(t) = \begin{cases} G(t) & \text{if} \quad t \in \Gamma \\ 1 & \text{if} \quad t \in \widetilde{\Gamma}\backslash\Gamma\,, \end{cases}$$

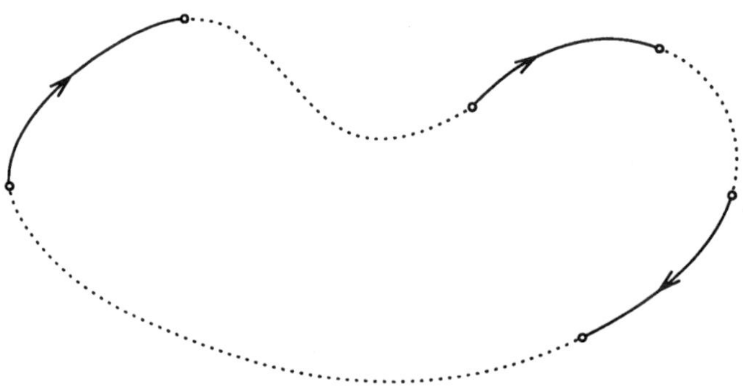

Figure 9.5

and solves the boundary value problem with the right–hand side

$$\widetilde{g}(t) = \begin{cases} g(t) & \text{if} \quad t \in \Gamma \\ 0 & \text{if} \quad t \in \widetilde{\Gamma}\backslash\Gamma\,. \end{cases}$$

That is, the solution $\Phi(z)$ to problem (15.1) with the boundary condition $\Phi_+(t) = \widetilde{G}(t)\ \Phi_-(t) + \widetilde{g}(t)$ is forced to satisfy the condition $\Phi(t) = \Phi_+(t) - \Phi_-(t) = 0$ almost everywhere on $\widetilde{\Gamma}\backslash\Gamma$. On the other hand, this means that the function

$$\Phi(z) - \frac{1}{2\pi i} \int\limits_{\widetilde{\Gamma}} \frac{\varphi(\tau)\, d\tau}{\tau - z} = \frac{1}{2\pi i} \int\limits_{\Gamma} \frac{\varphi(\tau)\, d\tau}{\tau - z},$$

which means that the solution to this problem is analytic at each point $t \in \widetilde{\Gamma}\backslash\Gamma$.

To illustrate this, let us consider some examples.

Example 15.1. Solve problem (15.1) with

$$G(t) = \begin{cases} 1 & \text{if } \ t \in \mathbf{T},\ \operatorname{Im} t > 0 \\ -1 & \text{if } \ t \in \mathbf{T},\ \operatorname{Im} t < 0 \end{cases} \tag{15.9}$$

and with the right-hand side $g(t) \equiv 1$.

Solution. In case $p = 2$ the function G is not 2-non-singular, and so we limit our considerations to the cases $1 < p < 2$ and $2 < p < \infty$.

If $1 < p < 2$ then the function G permits the q-factorization (3.3)

$$G(t) = (t^2 - 1)^{-1/2}\, t\, (1 - t^{-2})^{1/2}, \tag{15.10}$$

and if $2 < p < \infty$ the q-factorization

$$G(t) = (t^2 - 1)^{1/2}\, t^{-1}\, (1 - t^{-2})^{-1/2}. \tag{15.11}$$

For $1 < p < 2$ we have

$$X(z) = \begin{cases} (z^2 - 1)_1^{-1/2} & \text{if } \ |z| < 1 \\ (z^2 - 1)_2^{-1/2} & \text{if } \ |z| > 1, \end{cases} \tag{15.12}$$

where, in case $|z| < 1$, the function $(z^2 - 1)_1^{-1/2}$ denotes a branch of the function $(z^2 - 1)^{-1/2}$ which is analytic in the complex plane with cut along $(-\infty, -1] \cup [1, \infty)$, and in case $|z| > 1$, the function $(z^2 - 1)_2^{-1/2}$ is taken as a branch of $(z^2 - 1)^{-1/2}$ which is analytic in the plane with cut along $[-1, 1]$. Further, these branches are chosen in such a way that their quotient coincides with the function G on $\mathbf{T}$. By Theorem 15.1, if $1 < p < 2$ then problem (15.1) is solvable, and its general solution is given by

$$\Phi(z) = \frac{X(z)}{2\pi i} \int\limits_{\mathbf{T}} \frac{d\tau}{G_+(\tau)(\tau - z)} + \alpha X(z),$$

with an arbitrary constant α.

Notice that $G_+^{-1} \in L_p^+(\mathbf{T})$ which implies

$$\frac{1}{2\pi i}\int_{\mathbf{T}} \frac{d\tau}{G_+(\tau)(\tau - z)} = \begin{cases} X_+^{-1}(z) & \text{if } |z| < 1 \\ 0 & \text{if } |z| > 1 \end{cases}$$

and

$$\frac{X(z)}{2\pi i}\int_{\mathbf{T}} \frac{d\tau}{G_+(\tau)(\tau - z)} = \begin{cases} 1 & \text{if } |z| < 1 \\ 0 & \text{if } |z| > 1 \,. \end{cases}$$

Consequently,

$$\Phi(z) = \begin{cases} 1 + \alpha X(z) & \text{if } |z| < 1 \\ \alpha X(z) & \text{if } |z| > 1 \,. \end{cases}$$

If $p > 2$ then

$$X(z) = \begin{cases} (1 - 1/z^2)_1^{1/2} & \text{if } |z| < 1 \\ (1 - 1/z^2)_2^{1/2} & \text{if } |z| > 1 \,, \end{cases}$$

where the upper and lower functions are branches of analytic functions in the plane with cut along $(-\infty, -1] \cup [1, \infty)$ and $[-1, 1]$, respectively, the quotient of which coincides with G .

The solvability condition (15.8) takes the form

$$\int_{\mathbf{T}} \frac{dt}{G_+(t)} = 0 \,,$$

which is evidently satisfied. So the problem is uniquely solvable, and

$$\Phi(z) = \begin{cases} 1 & \text{if } |z| < 1 \\ 0 & \text{if } |z| > 1 \,. \end{cases}$$

Example 15.2. Solve problem (15.1) for the functions $G(t) \equiv -1$ and $g(t) \equiv 1$ on the curve $\Gamma = \{z \in \mathbf{C} : |z| = 1,\ \mathrm{Im}\, z \le 0\}$.

Solution. Define $\widetilde{\Gamma} = \mathbf{T}$,

$$\widetilde{G}(t) = \begin{cases} 1 & \text{if } t \in \mathbf{T},\ \mathrm{Im}\, t > 0 \\ -1 & \text{if } t \in \mathbf{T},\ \mathrm{Im}\, t < 0 \end{cases}$$

and

$$\widetilde{g}(t) = \begin{cases} 0 & \text{if } t \in \mathbf{T},\ \mathrm{Im}\, t > 0 \\ 1 & \text{if } t \in \mathbf{T}, \mathrm{Im}\, t < 0 \,. \end{cases}$$

So this problem leads to the preceding one with a different right–hand side.

If $p > 2$ then the solvability condition

$$\int_{\Gamma} \frac{dt}{\sqrt{t^2 - 1}} = 0$$

is violated, and so the problem has no solution.

In case $1 < p < 2$ the problem proves to be solvable, and the general solution is

$$\Phi(z) = \frac{X(z)}{2\pi i} \int_{\Gamma} \sqrt{t^2 - 1} \frac{dt}{t - z} + \alpha X(z) , \tag{15.13}$$

where α is an arbitrary constant and the function $X(z)$ is defined in (15.12). When defining this function we note to the equality $G_+(z)/G_-(z) = \widetilde{G}(z)$. It implies that $G_+(t) = G_-(t)$ along the upper semi-circle and thus, we may think of the function $X(z)$ in (15.13) as a function which is analytic in $\mathbf{C}\backslash\Gamma$.

Example 15.3. Find a solution to problem (15.1) in $L_2(\alpha, \beta)$ when $G(t) = \exp(2\pi i\delta)$ with $-1/2 < \delta < 1/2$.

Solution. The function

$$\widetilde{G}(t) = \begin{cases} \exp(2\pi i\delta) & \text{if } \ t \in [\alpha, \beta] \\ 1 & \text{if } \ t \in \widetilde{\Gamma}\backslash[\alpha, \beta] \end{cases}$$

permits the factorization

$$\widetilde{G}(t) = \widetilde{G}_-(t)\widetilde{G}_+(t)$$

on the curve $\widetilde{\Gamma}$ defined in the proof of Theorem 5.5 with the factors

$$G_+(z) = \frac{(\beta - z)^\delta}{(z - \alpha)^\delta}, \quad G_-(z) = \left(\frac{z - i}{\beta - z}\right)^\delta \left(\frac{z - \alpha}{z - i}\right)^\delta .$$

Put

$$X(z) = \begin{cases} G_+(z) & \text{if } \ z \in F_\Gamma^+ \\ G_-^{-1}(z) & \text{if } \ z \in F_\Gamma^- . \end{cases}$$

Since $G_+(z)/G_-(z) = 1$ on $\widetilde{\Gamma}\backslash[\alpha, \beta]$, the functions $G_+(z)$ and $G_-^{-1}(z)$ are branches of the function

$$X(z) = \left(\frac{\beta - z}{z - \alpha}\right)^\delta ,$$

which is analytic in $\mathbf{C}\backslash[\alpha, \beta]$.

Now Theorem 15.1 yields that our problem is uniquely solvable for each right–hand side, and its solution is given by

$$\Phi(z) = \frac{1}{2\pi i} \left(\frac{\beta - z}{z - \alpha}\right)^\delta \int_\alpha^\beta \left(\frac{\tau - \alpha}{\beta - \tau}\right)^\delta \frac{g(\tau)\, d\tau}{\tau - z}$$

where $\left(\frac{\beta - z}{z - \alpha}\right)^\delta$ stand for a $\mathbf{C}\backslash[\alpha, \beta]$ branch of this function which is analytic.

2. The Dirichlet problem. Find a holomorphic in $\mathbf{C}\backslash[\alpha,\beta]$ function which admits a representation by the Cauchy type integral

$$\Phi(z) = \int_{\alpha}^{\beta} \frac{\varphi(\tau)\,d\tau}{\tau - z}$$

with the density function $\varphi \in L_p(\alpha,\beta)$, and which satisfies the boundary condition

$$\operatorname{Re} \Phi(t) = f_1(t)\,, \quad \operatorname{Im} \Phi(t) = f_2(t) \tag{15.14}$$

for $\alpha \leq t \leq \beta$, where f_1 and f_2 are given real–valued functions in $L_p(\Gamma)$.

Solution. Define

$$\Psi(z) := \frac{\Phi(z) + \overline{\Phi(\overline{z})}}{2}, \quad \Omega(z) := \frac{\Phi(z) - \overline{\Phi(\overline{z})}}{2}\,.$$

Then the boundary conditions (15.14) can be rewritten as

$$\Psi^+(t) + \Psi^-(t) = 2f(t) \tag{15.15}$$

$$\Omega^+(t) - \Omega^-(t) = 2g(t)\,, \tag{15.16}$$

where $2f(t) = f_1(t) + f_2(t),\ 2g(t) = f_1(t) - f_2(t)$. Moreover the functions Ψ and Ω should be subject to the identities

$$\overline{\Omega(\overline{z})} = -\Omega(z), \quad \overline{\Psi(\overline{z})} = \Psi(z)\,. \tag{15.17}$$

The function Ω in (15.16) is uniquely determined by

$$\Omega(z) = \frac{1}{\pi i} \int_{\alpha}^{\beta} \frac{g(\tau)\,d\tau}{\tau - z}\,,$$

and thus it clearly satisfies the first condition in (15.17). In order to solve the boundary value problem (15.15) we need a factorization of the function

$$\widetilde{G}(t) = \begin{cases} -1 & \text{if} \quad t \in [\alpha,\beta] \\ 1 & \text{if} \quad t \in \widetilde{\Gamma}\backslash[\alpha,\beta]\,. \end{cases}$$

As in the preceding example, we employ the methods from the proof of Theorem 5.5. For $1 < p < 2$ the q–factorization of the function $\widetilde{G}$ is

$$\widetilde{G}(t) = G_-(t)(t - i)G_+(t)$$

with the factors

$$G_-(z) = \left(\frac{z-i}{\beta - z}\right)^{-1/2} \left(\frac{z-\alpha}{z-i}\right)^{1/2}, \quad G_+(z) = \frac{(\beta - z)^{-1/2}}{(z-\alpha)^{1/2}} .$$

In this case,

$$X(z) = \left\{ \begin{array}{lll} G_+(z) & \text{if} & z \in F_\Gamma^+ \\ (z-i)^{-1} G_-^{-1}(z) & \text{if} & z \in F_\Gamma^- \end{array} \right\} = \frac{1}{\sqrt{(\beta - z)(z-\alpha)}} \qquad (z \in \mathbf{C} \backslash [\alpha, \beta]) .$$

Thus, for these values of p, the problem (15.15) is solvable for each right-hand side, and its general solution is

$$\begin{aligned} \Psi(z) &= \frac{1}{\pi i \sqrt{(\beta - z)(z-\alpha)}} \int_\alpha^\beta \frac{\sqrt{(\beta-\tau)(\tau-\alpha)} f(\tau)\, d\tau}{\tau - z} + \\ &+ \frac{c}{\sqrt{(\beta - z)(z-\alpha)}} \qquad (c = const.) . \end{aligned}$$

Herein, the branch of the function $X(z) = [(\beta - z)(z-\alpha)]^{-1/2}$ has to be chosen in such a way that $\overline{X(\overline{z})} = X(z)$ and $\sqrt{(\beta-\tau)(\tau-\alpha)} = X_+(\tau)$. In this situation, $(\overline{X(\overline{z})})_+ = \overline{X_-(z)} = -\overline{X_+(z)} = -X_+(z)$ (notice that $X_-(t) = -X_+(t)$ along the interval). From this we obtain that the second identity in (15.17) is satisfied if and only if $c \in \mathbf{R}$. So, the final solution in case $1 < p < 2$ is

$$\begin{aligned} \Phi(z) &= \frac{1}{\pi i} \int_\alpha^\beta \frac{g(\tau)\, d\tau}{\tau - z} + \frac{c}{\sqrt{(\beta - z)(z-\alpha)}} + \\ &+ \frac{1}{\pi i \sqrt{(\beta - z)(z-\alpha)}} \int_\alpha^\beta \frac{\sqrt{(\beta-\tau)(\tau-\alpha)}\, f(\tau)\, d\tau}{\tau - z} \end{aligned}$$

with an arbitrary real constant c .

Now let $p > 2$. Then the q-factorization of the function $\widetilde{G}(t)$ is given by

$$\widetilde{G}(t) = G_-(t)(t-i)^{-1}\, G_+(t)$$

with

$$G_-(t) = \left(\frac{z-i}{\beta - z}\right)^{1/2} \left(\frac{z-i}{z-\alpha}\right)^{1/2}, \quad G_+(z) = (z-\alpha)^{1/2}\, (z-\beta)^{1/2} ,$$

and furthermore we obtain in this case

$$X(z) = \frac{\sqrt{(\beta - z)(z-\alpha)}}{z - i} .$$

The problem (15.5) is solvable if and only if

$$\int_{\alpha}^{\beta} \frac{f(t)\,dt}{\sqrt{(t-\alpha)(\beta-t)}} = 0 \tag{15.18}$$

If this condition is satisfied, then

$$\begin{aligned}
\Psi(z) &= \frac{\sqrt{(\beta-z)(z-\alpha)}}{\pi i(z-i)} \int_{\alpha}^{\beta} \frac{f(\tau)(\tau-i)\,d\tau}{\sqrt{(\beta-\tau)(\tau-\alpha)}(\tau-z)} \\
&= \frac{\sqrt{(\beta-z)(z-\alpha)}}{\pi i(z-i)} \int_{\alpha}^{\beta} \frac{f(\tau)\,d\tau}{\sqrt{(\beta-\tau)(\tau-\alpha)}} + \\
&\qquad \frac{\sqrt{(\beta-z)(z-\alpha)}}{\pi i} \int_{\alpha}^{\beta} \frac{f(\tau)\,d\tau}{\sqrt{(\beta-\tau)(\tau-\alpha)}(\tau-z)} \\
&= \frac{\sqrt{(\beta-z)(z-\alpha)}}{\pi i} \int_{\alpha}^{\beta} \frac{f(\tau)\,d\tau}{\sqrt{(\beta-\tau)(\tau-\alpha)}(\tau-z)}\,.
\end{aligned}$$

Finally, in case $p > 2$, the condition (15.18) is necessary and sufficient for the solvability of our problem. In this situation the unique solution is given by

$$\Phi(z) = \frac{1}{\pi i} \int_{\alpha}^{\beta} \frac{g(\tau)\,d\tau}{\tau - z} + \frac{\sqrt{(\beta-z)(z-\alpha)}}{\pi i} \int_{\alpha}^{\beta} \frac{f(\tau)\,d\tau}{\sqrt{(\beta-\tau)(\tau-\alpha)}(\tau-z)}\,. \tag{15.19}$$

Here the branches of the root function have been chosen as in the preceding case.

3. The Dirichlet problem for the plane with cut and its generalization. In [GK 1, 3.15] we explained the application of the theory of singular integral equations with continuous coefficients to the solution of the aforementioned problems. Now we continue these considerations by looking at problems which involve the theory of singular integral equations with piecewise continuous coefficients. For convenience we repeat the problem and the general part of its solution. On the plane **C** with a cut along the positive real semiaxis $(y = 0, x \geq 0)$ find the solution to the differential equation

$$\frac{\partial^2 u}{\partial y^2} - \nu\left(i\frac{\partial}{\partial x}\right) u = 0 \tag{15.20}$$

satisfying the boundary condition

$$\begin{aligned}
u(x,+0) &= a(x) \quad (x > 0) \\
u(x,-0) &= b(x) \quad (x > 0)
\end{aligned} \tag{15.21}$$

where a and b are given functions and ν is a given polynomial not vanishing on $\mathbf{R}$.

In particular, if $\nu(x) = x^2 + 1$ then equation (15.20) takes the form

$$\frac{\partial^2 u}{\partial x^2} + \frac{\partial^2 u}{\partial y^2} - u = 0 .$$

In this case, problem (15.20)–(15.21) is referred to as the *Dirichlet problem.*

The function $u(x, y)$ and all its partial derivatives up through the second order are required to satisfy the conditions

$$\left. \frac{\partial^m u(x,y)}{\partial^{m_1} x \partial^{m_2} y} \right|_{y=y_0} \in L_2(\mathbf{R}) \tag{15.22}$$

for any fixed $y_0 \neq 0$. Furthermore, we suppose the functions

$$\frac{\partial^m u(x,y)}{\partial^{m_1} x \partial^{m_2} y}$$

converge uniformly to zero as $y \to \infty$.

The conditions imposed on the functions $u(x, \pm 0)$, $a(x)$ and $b(x)$ will be formulated below.

For every fixed $y \in \mathbf{R}$, we define

$$\widehat{u}(x, y) = \int_{-\infty}^{\infty} u(t, y) e^{itx}\, dt , \tag{15.23}$$

with the Fourier transformation being understood in the sense of the space $L_2(\mathbf{R})$.

Applying the Fourier transform to both sides of equation (15.20) we obtain the ordinary linear differential equation

$$\widehat{u}''_{yy} - \nu(x)\widehat{u} = 0 , \tag{15.24}$$

in which x is regarded as a parameter.

By the radical

$$\sqrt{\nu(x)} ,$$

we shall always understand its branch satisfying the condition $\mathrm{Re}\sqrt{\nu(x)} \geq 0$.

In searching a solution to the equation $\widehat{u}'' - \nu(x)\widehat{u} = 0$, we separately consider the cases $y > 0$ and $y < 0$. Afterwards the solution functions are combined along the negative real semi-axis $(y = 0,\ x < 0)$ by requiring the equalities

$$u(x, +0) = u(x, -0), \quad u'_y(x, +0) = u'_y(x, -0) \quad (x < 0) \tag{15.25}$$

The solution to equation (15.24) can be written in the form

$$\widehat{u}(x, y) = \begin{cases} \alpha(x) e^{-y\sqrt{\nu(x)}} + \alpha_1(x) e^{y\sqrt{\nu(x)}} & \text{if} \quad y > 0 \\ \beta(x) e^{y\sqrt{\nu(x)}} + \beta_1(x) e^{-y\sqrt{\nu(x)}} & \text{if} \quad y < 0 . \end{cases} \tag{15.26}$$

Since the function $u(x,y)$ converges uniformly to zero as $y \to \infty$ we have $\widehat{u}(x,\infty) = 0$ for each $x \in \mathbf{R}$. This implies that $\alpha_1(x) = \beta_1(x) \equiv 0$ and

$$\widehat{u}(x) = \begin{cases} \alpha(x)\, e^{-y\sqrt{\nu(x)}} & \text{if} \quad y > 0 \\ \beta(x)\, e^{y\sqrt{\nu(x)}} & \text{if} \quad y < 0 \end{cases} . \tag{15.27}$$

With regard to the equations (15.25) the functions $u(x,\pm 0)$ can be represented in the form

$$u(x,+0) = f_-(x) + a_+(x), \quad u(x,-0) = f_-(x) + b_+(x)\,, \tag{15.28}$$

where

$$a_+(x) = \begin{cases} a(x) & \text{if} \quad x > 0 \\ 0 & \text{if} \quad x < 0 \end{cases} , b_+(x) = \begin{cases} b(x) & \text{if} \quad x > 0 \\ 0 & \text{if} \quad x < 0 \end{cases} ,$$

and $f_-(x)$ is, for the time being, an unknown function which is equal to zero if $x > 0$.

Set

$$u'_y(x,+0) - u'_y(x,-0) = f_+(x)\,. \tag{15.29}$$

Due to conditions (15.25), we have $f_+(x) = 0$ in case $x < 0$. Now (for the time being formally) apply the Fourier transformation (15.23) to either part of the equations (15.28) and (15.29) to obtain the equations

$$\widehat{u}(x,+0) = \Phi_-(x) + \widehat{a}_+(x), \quad \widehat{u}(x,-0) = \Phi_-(x) + \widehat{b}_+(x) \tag{15.30}$$

and

$$\widehat{u}'_y(x,+0) - \widehat{u}'_y(x,-0) = \Phi_+(x)\,. \tag{15.31}$$

Up to this point we have not formulated conditions which the functions $a(t), b(t)$ and $u(x,\pm 0)$ have to satisfy, since it is more convenient to describe these conditions in the language of their Fourier transforms.

For some $p \in (1,\infty)$ the given functions a and b, as well as the function u, are required to satisfy the conditions

$$\widehat{a}_+ \in L_p(\mathbf{R}),\ \widehat{b}_+ \in L_p(\mathbf{R}),\ \widehat{u}(x,\pm 0) \in L_p(\mathbf{R})\,. \tag{15.32}$$

Under these conditions we can assert that $\Phi_-(x) \in L_p(\mathbf{R})$ and, since $f_-(x) = 0$ for $x > 0$, the function Φ_- in fact belongs to $QL_p(\mathbf{R})$.

The equalities (15.30) and (15.27) imply

$$\alpha(x) = \Phi_-(x) + \widehat{a}_+(x), \quad \beta(x) = \Phi_-(x) + \widehat{b}_+(x)\,. \tag{15.33}$$

Moreover, we conclude from equality (15.27) that

$$\widehat{u}'_y(x,+0) - \widehat{u}'_y(x,-0) = -\sqrt{\nu(x)}\,(\alpha(x) + \beta(x))\,. \tag{15.34}$$

This identity means that the function $\Phi_+(x)$ is subject to the condition

$$\frac{\Phi_+(x)}{\sqrt{\nu(x)}} \in L_p(\mathbf{R})$$

which is equivalent to

$$\frac{\Phi_+(x)}{(x+i)^{n/2}} \in L_p(\mathbf{R}) \,, \tag{15.35}$$

with n referring to the degree of the polynomial $\nu(x)$.

So far we have not made any demands on $\widehat{u}'_y(x,\pm 0)$. If the function $u(x,y)$ has the property that if $\widehat{u}'_y(x,+0) - \widehat{u}'_y(x,-0)$ belongs to $L_r(\mathbf{R})$ for some $r \in (1,\infty)$, then $\Phi_+(x) \in L_r(\mathbf{R})$. Since $f_+(x) = 0$ whenever $x < 0$ we even have $\Phi_+(x) \in PL_r(\mathbf{R})$. If, in addition, $n \geq 2$, then the function

$$\Psi_+(x) = \frac{\Phi_+(x)}{(x+i)^{n/2}}$$

generally belongs to $PL_s(\mathbf{R})$ for any $s \in (1,\infty)$. However, when solving the problem under consideration we are satisfied with membership of $\Psi_+(x)$ in $PL_p(\mathbf{R})$ for the one value of p occuring in conditions (15.32), and this is exactly what we shall demand.

Thus, the given functions a , b , and $u(x,y)$ which are to be determined are required to fulfill the conditions (15.32) as well as the condition

$$\frac{\widehat{u}'_y(x,+0) - \widehat{u}'_y(x,-0)}{(x+i)^{n/2}} \in PL_p(\mathbf{R}) \tag{15.36}$$

for some number $p \in (1,\infty)$.

From equalities (15.31), (15.34) and (15.33) we get

$$\frac{\Phi_+(x)}{\sqrt{\nu(x)}} + 2\Phi_-(x) = -\widehat{a}_+(x) - \widehat{b}_+(x) \,. \tag{15.37}$$

After solving the boundary value problem, we put Φ_- into (15.33), and there put α and β into (15.27). Ultimately we find the desired function $u(x,y)$:

$$u(x,y) = \begin{cases} \frac{1}{2\pi} \int\limits_{-\infty}^{\infty} (\Phi_-(t) + \widehat{a}_+(t)) e^{-y\sqrt{\nu(t)} - itx}\, dt & \text{if} \quad y > 0 \\ \frac{1}{2\pi} \int\limits_{-\infty}^{\infty} (\Phi_-(t) + \widehat{b}_+(t)) e^{y\sqrt{\nu(t)} - itx}\, dt & \text{if} \quad y < 0 \end{cases} \,. \tag{15.38}$$

Let us return to the boundary value problem (15.37). If the degree of the polynomial $\nu(x)$ is positive, then the coefficient $1/\sqrt{\nu(x)}$ of the boundary value problem (15.37) is degenerate at infinity, and for the solution of this problem one can use the method for solving singular equations with degenerate coefficients explained in [GK 1,3.12]. With this aim, $1/\sqrt{\nu(x)}$ is represented in the form of a product

$$\frac{1}{\sqrt{\nu(x)}} = M(x)\frac{1}{(x+i)^{n/2}}, \tag{15.39}$$

where $M(x)$ is a bounded function which does not vanish on $\overline{\mathbf{R}}$.

In problem (15.37), setting

$$\Psi_+(x) = \frac{\Phi_+(x)}{(x+i)^{n/2}}, \qquad \Psi_-(x) = 2\Phi_-(x)$$

and

$$\Psi(x) = \Psi_+(x) + \Psi_-(x)\,,$$

and using the identity (15.39) we arrive at the singular integral equation

$$MP_{\mathbf{R}}\Psi + Q_{\mathbf{R}}\Psi = g, \tag{15.40}$$

where $g(x) = -\widehat{a}_+(x) - \widehat{b}_+(x)$.

Thus, in order to solve the problem (15.20)–(15.21) it suffices to solve the singular integral equation (15.40), to set $\Psi_- = Q_{\mathbf{R}}\Psi$, and to rewrite the solution (15.38) (with $\Phi_-(x)$ replaced by $\frac{1}{2}\Psi_-(x)$) in the form

$$u(x,y) = \begin{cases} \frac{1}{4\pi}\int\limits_{-\infty}^{\infty} (\Psi_-(t) + 2\widehat{a}_+(t))e^{-y\sqrt{\nu(x)}-itx}\,dt & \text{if} \quad y>0 \\ \frac{1}{4\pi}\int\limits_{-\infty}^{\infty} (\Psi_-(t) + 2\widehat{b}_+(t))e^{y\sqrt{\nu(x)}-itx}\,dt & \text{if} \quad y<0 \end{cases} \tag{15.41}$$

Let us consider some concrete examples.

Example 1°. For the Dirichlet problem

$$\begin{cases} \frac{\partial^2 u}{\partial y^2} + \frac{\partial^2 u}{\partial x^2} - u = 0 & \\ u(x,+0) = a(x),\ u(x,-0) = b(x) & \text{if} \quad x>0 \\ u(x,-0) = u(x,+0),\ u'_y(x,-0) = u'_y(x,+0) & \text{if} \quad x<0 \end{cases}, \tag{15.42}$$

the function $M(x)$ is given by

$$M(x) = \left(\frac{x+i}{x-i}\right)^{1/2}.$$

In order to determine the function $\Psi_- = Q_{\mathbf{R}}\Psi$, we consider the singular integral equation

$$(MP+Q)\Psi = g\,, \tag{15.43}$$

where $P = P_{\mathbf{R}}$ and $Q = Q_{\mathbf{R}}$.

If $1 < p < 2$ then the function M admits a factorization in $L_p(\mathbf{R})$ with zero index,

$$M(x) = \frac{1}{\sqrt{x-i}}\sqrt{x+i}\,,$$

and if $p > 2$ the factorization of M is

$$M(x) = \sqrt{x-i}\left(\frac{x+i}{x-i}\right)\frac{1}{\sqrt{x+i}}$$

with index -1. This shows that, for $1 < p < 2$, the equation (15.43) is uniquely solvable for each right-hand side and that

$$\Psi_-(x) = \frac{1}{\sqrt{x-i}}Qg\sqrt{x-i}\,,$$

whereas in case $p > 2$ the equation (15.43) is also solvable for each right-hand side, but the homogenous equation now possesses a non-trivial solution. The general solution of the equation (15.43) in case $p > 2$ is

$$\Psi(x) = \frac{x-i}{\sqrt{x+i}}\,P\frac{g}{\sqrt{x-i}} + \sqrt{x-i}\,Q\frac{g}{\sqrt{x-i}} + \frac{c_1}{\sqrt{1+x^2}}\,,$$

with an arbitrary constant c_1. Hence, it follows that

$$\Psi_-(x) = \sqrt{x-i}\,Q\frac{g}{\sqrt{x-i}} + \frac{c}{\sqrt{x-i}}$$

for some arbitrary constant c. Now it remains to put the function $\Psi_-(x)$ into (15.41) in order to obtain the general solution of our problem for $p < 2$ as well as for $p > 2$.

Let, for instance, $a(x) = e^{-x}$ and $b(x) \equiv 0$. Then $\widehat{b}_+(x) \equiv 0$ and $\widehat{a}_+(x) = i/(x+i)$, and these functions belong to $L_p(\mathbf{R})$ for every $p \in (1,\infty)$. The choice of the space $L_p(\mathbf{R})$ depends in which space the solution to the equation (15.43) is required as well as the space the functions

$$\widehat{u}(x,+0) = \widehat{a}_+(x) + \Psi_-(x)/2$$

and

$$\widehat{u}(x,-0) = \widehat{b}_+(x) + \Psi_-(x)/2$$

are to be members of.

We distinguish between the two possibilities $1 < p < 2$ and $p > 2$. In the first case (if $1 < p < 2$),

$$\begin{aligned}
\Psi_-(x) &= \frac{-i}{\sqrt{x-i}}Q\frac{\sqrt{x-i}}{x+i} \\
&= \frac{-i}{\sqrt{x-i}}Q\frac{\sqrt{x-i}-\sqrt{-2i}+\sqrt{-2i}}{x+i} \\
&= \frac{-i}{\sqrt{x-i}}Q\frac{1}{\sqrt{x-i}+1-i} = \frac{-i}{x-i+(1-i)\sqrt{x-i}}\,,
\end{aligned} \qquad (15.44)$$

and in the second version $(p > 2)$,

$$\begin{aligned}
\Psi_-(x) &= -i\sqrt{x-i}\,Q\frac{1}{\sqrt{x-i}\,(x+i)} + \frac{c}{\sqrt{x-i}} \\
&= -\frac{\sqrt{x-i}}{2}Q\frac{(x+i)-(x-i)}{\sqrt{x-i}\,(x+i)} + \frac{c}{\sqrt{x-i}} \\
&= -\frac{\sqrt{x-i}}{2}Q\left(\frac{1}{\sqrt{x-i}} - \frac{\sqrt{x-i}}{x+i}\right) + \frac{c}{\sqrt{x-i}} \\
&= \frac{-1+i}{2(\sqrt{x-i}+1-i)} + \frac{c}{\sqrt{x-i}}\,.
\end{aligned} \tag{15.45}$$

It is interesting to notice that the function $\Psi_-(x)$ defined by (15.44) belongs to the collection of all solutions of the form (15.45) as one can easily check by setting $c = (1-i)/2$ in (15.45).

Thus, the solution to problem (15.42) with the functions $a(x) = e^{-x}$ and $b(x) = 0$ is given by

$$u(x,y) = \begin{cases} \frac{1}{4\pi}\int\limits_{-\infty}^{\infty}\left(\frac{(i-1)\sqrt{t-i}}{2(t+i)} + \frac{i}{t+i} + \frac{c}{\sqrt{t-i}}\right)\exp(-y\sqrt{1+t^2}-itx)\,dt & \text{if} \quad y>0 \\ \frac{1}{4\pi}\int\limits_{-\infty}^{\infty}\left(\frac{(i-1)\sqrt{t-i}}{2(t+i)} - \frac{i}{t+i} + \frac{c}{\sqrt{t-i}}\right)\exp(y\sqrt{1+t^2}-itx)\,dt & \text{if} \quad y<0 \end{cases}.$$

The same function $g(x) = \frac{i}{x+i}$ appears if the functions a and b are equal to $a(x) = b(x) = \frac{1}{2}e^{-x}$. In this case, the solution is continuous in the whole plane except for the point $z = 0$, and the solution is given by

$$u(x,y) = \frac{1}{8\pi}\int\limits_{\infty}^{\infty}\left(\frac{(i-1)\sqrt{t-i}}{t+i} + \frac{c}{\sqrt{t-i}}\right)\exp(-|y|\sqrt{1+t^2}-itx)\,dt\,.$$

Example 2°. For the problem

$$\begin{cases} \frac{\partial^2 u}{\partial y^2} + \frac{\partial^2 u}{\partial x^2} + 2\frac{\partial u}{\partial x} - u = 0 \\ u(x,+0) = a(x),\ u(x,-0) = b(x) \quad (x>0) \\ u(x,+0) = u(x,-0),\ u'_y(x,+0) = u'_y(x,-0) \quad (x<0) \end{cases}$$

the polynomial ν equals $\nu(x) = (x+i)^2$, and due to the special choice of the branch of the root function $\sqrt{z}$ we have

$$\sqrt{\nu(x)} = \sqrt{(x+i)^2} = \begin{cases} x+i & \text{if} \quad x>0 \\ -x-i & \text{if} \quad x<0\,. \end{cases} \tag{15.46}$$

Consequently,

$$M(x) = \frac{x+i}{\sqrt{(x+i)^2}} = \begin{cases} 1 & \text{if} \quad x>0 \\ -1 & \text{if} \quad x<0 \end{cases}.$$

To solve the equation

$$(MP+Q)\Psi = g$$

we factor the function $M(x)$ in the space $L_p(\mathbf{R})$.

To this end we denote by $M_-(z)$ the function $\sqrt{z}$ which is analytic in the plane $\mathbf{C}$ with a cut along the "positive" imaginary axis ($x=0, y\geq 0)$, and by $M_+(z)$ the function $\sqrt{z}$ analytic in the plane $\mathbf{C}$ with a cut along the "negative"imaginary axis $(x=0, y\leq 0)$. The branches of these functions are chosen such that $M_+(1) = M_-(1) = 1$. In this situation, the following equalities hold for all points $x \in \mathbf{R}\ (x \neq 0)$:

$$M(x) = M_-(x)M_+^{-1}(x) \tag{15.47}$$

and

$$M(x) = M_-^{-1}(x)M_+(x)\,. \tag{15.48}$$

Next we are going to clarify to which spaces $L_p(\mathbf{R})$ the factorizations (15.47) and (15.48) correspond.

The operator

$$(Bf)(t) = \frac{1}{1-t} f\left(i\,\frac{t+1}{1-t}\right)$$

provides us with an isomorphism between the spaces $L_p(\mathbf{R})$ and $L_p(\mathbf{T}, |t-1|^{p-2})$. In Section 9.3 we obtained the factorization of the function

$$a(t) = \begin{cases} 1 & \text{if } \ t\in\mathbf{T},\ \operatorname{Im} t > 0 \\ -1 & \text{if } \ t\in\mathbf{T},\ \operatorname{Im} t < 0 \end{cases}$$

in the space $L_p(\mathbf{T}, |t-1|^{\alpha}|t+1|^{\beta})$. This factorization depends on in which of the intervals $(-1, p/2-1)$ and $(p/2-1, p-1)$ the numbers α and β are located. We have $\alpha = p-2$ and $\beta = 0$ in our example, and this means that

$$-1 < \beta < \frac{p}{2} - 1 < \alpha < p-1 \quad \text{in case} \quad p > 2 \tag{15.49}$$

and

$$-1 < \alpha < \frac{p}{2} - 1 < \beta < p-1 \quad \text{in case} \quad 1 < p < 2\,. \tag{15.50}$$

If the conditions (15.49) are satisfied then the factorization of the function a is

$$a(t) = \left(1-\frac{1}{t}\right)^{-\frac{1}{2}} \left(1+\frac{1}{t}\right)^{\frac{1}{2}} (t-1)^{\frac{1}{2}}\,(t+1)^{-\frac{1}{2}}\,,$$

and if (15.50) is in force then

$$a(t) = \left(1-\frac{1}{t}\right)^{\frac{1}{2}} \left(1+\frac{1}{t}\right)^{-\frac{1}{2}} (t-1)^{-\frac{1}{2}}(t+1)^{\frac{1}{2}}\,.$$

This yields finally that the equality (15.47) is just the factorization of the function $M(x)$ in spaces $L_p(\mathbf{R})$ if $p > 2$, and equation (15.48) if $1 < p < 2$. Hence, for all $p \neq 2$, the operator $MP + Q$ is invertible on $L_p(\mathbf{R})$, the equation $(MP + Q)\varphi = \psi$ is solvable for each right–hand side ψ , and

$$\psi_-(x) = \begin{cases} M_-(x)QM_-^{-1}(x)g(x) & \text{if} \quad p > 2 \\ M_-^{-1}(x)QM_-(x)g(x) & \text{if} \quad 1 < p < 2\,. \end{cases}$$

Let, for instance, $g(x) = \frac{\delta}{x+i}$ with δ standing for a fixed constant. Then, if $1 < p < 2$,

$$\begin{aligned} \psi_-(x) &= M_-^{-1}(x)QM_-(x)\frac{\delta}{x+i} \\ &= \frac{\delta}{M_-(x)}Q\frac{M_-(x) - M_-(-i)}{x+i} \\ &= \frac{\delta}{x+i}\left(1 - \frac{M_-(-i)}{M_-(x)}\right) = \frac{\delta}{x+i}\left(1 - \frac{1-i}{\sqrt{2x}}\right), \end{aligned} \tag{15.51}$$

and analogous, for $p > 2$,

$$\psi_-(x) = \frac{\delta}{x+i}\left(1 - \frac{\sqrt{2x}}{1-i}\right). \tag{15.52}$$

As a special consequence of this we obtain that the solution to the Dirichlet problem

$$\begin{cases} \frac{\partial^2 u}{\partial y^2} + \frac{\partial^2 u}{\partial x^2} + 2\frac{\partial u}{\partial x} - u = 0 & \\ u(x,+0) = e^{-x},\ u(x,-0) = 0 & (x > 0) \\ u(x,+0) = u(x,-0),\ u'_y(x,+0) = u'_y(x,-0) & (x < 0) \end{cases} \tag{15.53}$$

depends on the space $L_p(\mathbf{R})$ to which the function $\widehat{u}(x,0)$ is required to belong for $x < 0$. For the values $p \in (1,2)$, the solution $u(x,y)$ has the form

$$u(x,y) = \begin{cases} \frac{1}{4\pi}\int\limits_{-\infty}^{\infty}\left(\frac{1+i}{(t+i)\sqrt{2t}} + \frac{i}{t+i}\right)\exp\left(-y\sqrt{(t+i)^2} - itx\right)\,dt & (y > 0) \quad (15.54) \\ \frac{1}{4\pi}\int\limits_{-\infty}^{\infty}\left(\frac{1+i}{(t+i)\sqrt{2t}} - \frac{i}{t+i}\right)\exp\left(y\sqrt{(t+i)^2} - itx\right)\,dt & (y < 0) \quad (15.55) \end{cases}$$

whereas for $p > 2$

$$u(x,y) = \begin{cases} \frac{1}{8\pi}\int\limits_{-\infty}^{\infty}\frac{(i-1)\sqrt{2t}+2i}{t+i}\exp\left(-y\sqrt{(t+i)^2} - itx\right)\,dt & (y > 0) \quad (15.56) \\ \frac{1}{8\pi}\int\limits_{-\infty}^{\infty}\frac{(i-1)\sqrt{2t}-2i}{t+i}\exp\left(y\sqrt{t+i)^2} - itx\right)\,dt & (y < 0) \quad (15.57) \end{cases}$$

Herein, $\sqrt{(x+i)^2}$ is defined by the equality (15.46), and $\sqrt{2z}$ is assumed to be continuous in the plane with a cut along the “positive” imaginary axis.

4. Wiener-Hopf equations with discontinuous symbols. On the space $L_2(0,\infty)$ we consider the integral equation

$$\varphi(t) - \frac{1-i}{\pi}\int_0^\infty \frac{\sin(t-s)}{t-s}\varphi(s)\,ds = e^{-t}\,, \tag{15.58}$$

which is a Wiener–Hopf equation. The kernel function $k(t) = \frac{\sin t}{t}$ of this integral operator is not contained in $L_1(\mathbf{R})$, and its Fourier transform is discontinuous:

$$\widehat{\left(\frac{\sin t}{t}\right)} = \pi\chi_{[-1,1]}(x)\,.$$

As in [GK 1, 3.14] we continue the functions $\varphi(t)$ and e^{-t} onto $(-\infty,0)$ by zero, and after applying the Fourier transformation we arrive at the singular integral equation

$$(cP+Q)\Phi(x) = \frac{i}{x+i} \tag{15.59}$$

on the space $L_2(\mathbf{R})$, where

$$c(x) = \begin{cases} i & \text{if} \quad x\in[-1,1] \\ 1 & \text{if} \quad x\in\mathbf{R}\backslash[-1,1]\,, \end{cases}$$

and φ is the inverse Fourier transform of the function $\Phi_+ = P\Phi$. In order to solve the equation (15.57) it is more convenient to consider the equivalent equation

$$(\widetilde{c}P_{\mathbf{T}} + Q_{\mathbf{T}})\widetilde{\varphi} = 1/2 \tag{15.60}$$

on the space $L_2(\mathbf{T})$ where now

$$\widetilde{c}(t) = \begin{cases} i & \text{if} \quad t\in\mathbf{T}, \operatorname{Re} t<0 \\ 1 & \text{if} \quad t\in\mathbf{T}, \operatorname{Re} t>0 \end{cases} \tag{15.61}$$

and

$$\widetilde{\varphi}(t) = \frac{1}{1-t}\,\Phi\left(i\frac{t+1}{1-t}\right)\,.$$

The function $\widetilde{c}(t)$ is 2-non-singular, and its 2-index is equal to zero. Let $\widetilde{c}(t) = \widetilde{c}_-(t)\widetilde{c}_+(t)$ refer to the 2-factorization of the function $\widetilde{c}$. Then the unique solution to the equation (15.60) has the form

$$\widetilde{\varphi}(t) = \frac{1}{2}(\widetilde{c}_+^{-1}P_{\mathbf{T}}\widetilde{c}_-^{-1} + \widetilde{c}_-Q_{\mathbf{T}}\widetilde{c}_-^{-1})\,,$$

from which it follows that

$$(P_{\mathbf{T}}\widetilde{\varphi})(t) = \frac{1}{2\widetilde{c}_+(t)\widetilde{c}_-(\infty)}\,. \tag{15.62}$$

Next we are going to factorize the function $\widetilde{c}(t)$. Define

$$g_1(z) = R^{\frac{1}{4}} e^{\frac{i\theta}{4}} \quad \left(-\frac{\pi}{2} \leq \theta < \frac{3\pi}{2}\right) ,$$

$$g_2(z) = R^{-\frac{1}{4}} e^{-\frac{i\theta}{2}} \quad \left(-\frac{3\pi}{2} \leq \theta < \frac{\pi}{2}\right) .$$

Then one checks without difficulty that

$$g_1(z)g_2(z) = \begin{cases} 1 & \text{if} \quad \mathrm{Re} z > 0 \\ i & \text{if} \quad \mathrm{Re} z < 0 . \end{cases}$$

Set

$$\widetilde{c}_-(t) = g_2\left(\frac{1-it}{t-i}\right)$$

and

$$\widetilde{c}_+(t) = g_1\left(\frac{1-it}{t-i}\right) .$$

The function $\widetilde{c}_-(t)$ is analytic in the plane $\mathbf{C}$ with a cut along the interval $[-i,i]$, and the function $\widetilde{c}_+(t)$ is analytic in the plane $\mathbf{C}$ with a cut along the pair of rays $(-i\infty,-i] \cup [i,+i\infty)$. It is immediate that the equality $\widetilde{c}(t) = \widetilde{c}_-(t)\widetilde{c}_+(t)$ yields a 2–factorization of the function $\widetilde{c}$. Since $\widetilde{c}_-(\infty) = g_2(-i) = \exp(i\pi/8)$ one has

$$(P_{\mathbf{T}}\widetilde{\varphi})(t) = \frac{1}{2\exp(i\pi/8) g_1\left(\frac{1-it}{t-i}\right)} ,$$

and it follows that

$$\begin{aligned} \Phi_+(x) &= \frac{2i}{x+i}(P_{\mathbf{T}}\widetilde{\varphi})\left(\frac{x-i}{x+i}\right) \\ &= \frac{i\exp(-i\pi/8)}{x+i} g_1^{-1}\left(\frac{x-1}{x+1}\right) . \end{aligned}$$

Further, the definition of the function g_1 implies that

$$g_1^{-1}\left(\frac{x-1}{x+1}\right) = \left|\frac{x+1}{x-1}\right|^{\frac{1}{4}} \alpha(x) \tag{15.63}$$

with

$$\alpha(x) = \begin{cases} 1 & \text{if} \quad |x| > 1 \\ \exp\left(-\frac{i\pi}{4}\right) & \text{if} \quad |x| < 1 . \end{cases}$$

Thus,

$$\Phi_+(x) = \begin{cases} \frac{\exp(i\pi/8)}{x+i}\left(\frac{1+x}{1-x}\right)^{1/4} & \text{if} \quad |x| < 1 \\ \frac{\exp(3\pi i/8)}{x+i}\left(\frac{x+1}{x-1}\right)^{1/4} & \text{if} \quad |x| > 1 . \end{cases} \tag{15.64}$$

This shows that the solution φ to the equation (15.58) is given by the integral

$$\varphi(x) = \frac{\exp(i\pi/8)}{2\pi} \int_{-\infty}^{\infty} \left(\frac{1+s}{1-s}\right)^{1/4} \frac{\exp(-isx)}{s+i}\, ds .$$

Here, the choice of the branch of the function under the integral sign is in accordance with (15.63) and (15.64).

5. Integral equations with Hilbert kernel. We consider the integral equation

$$\left(e^{\frac{ix}{3}} + 1\right)\varphi(x) + \frac{e^{\frac{ix}{3}} - 1}{2\pi i} \int_0^{2\pi} \cot\frac{s-x}{2}\varphi(s)\, ds = e^{\frac{ix}{3}} + 1 \tag{15.65}$$

in the complex Hilbert space $L_2(0, 2\pi)$. For its solution we substitute $t = e^{ix}$, $s = e^{i\tau}$ and so come to the equivalent equation in the space $L_2(\mathbf{T})$

$$(t^{\frac{1}{3}} P_{\mathbf{T}} + Q_{\mathbf{T}} + K)u = \frac{1}{2}(t^{\frac{1}{3}} + 1) ,$$

where $u(e^{ix}) = \varphi(x)$ and

$$(Ku)(t) = \frac{1 - t^{\frac{1}{3}}}{4\pi i} \int_{\mathbf{T}} \frac{u(\tau)\, d\tau}{\tau} .$$

The operator $A_0 = t^{\frac{1}{3}} P_{\mathbf{T}} + Q_{\mathbf{T}}$ is a singular integral operator with the piecewise continuous coefficient $a(t) = t^{1/3}$. This function is 2–non–singular, and its 2–index is equal to zero. Due to Theorem 4.2, the operator A_0 is invertible, and

$$A_0^{-1} = (t-1)^{\frac{1}{3}}(P_{\mathbf{T}} + t^{-\frac{1}{3}} Q_{\mathbf{T}}) \left(1 - \frac{1}{t}\right)^{\frac{1}{3}} I .$$

Since the operator K has rank one, the operator $A = t^{1/3}P_{\mathbf{T}} + Q_{\mathbf{T}} + K$ proves to be a Φ–operator on the space $L_2(\mathbf{T})$ with $\operatorname{Ind} A = 0$ and $\dim \ker A \leq 1$. We show that actually $\dim \ker A = 0$ and, hence, the operator A is even invertible. For let $u_0 \in \ker A$. If $Ku_0 = 0$ then $A_0 u_0 = 0$ and so $u_0 = 0$.

Now suppose $Ku_0 \neq 0$. Then we can assume that

$$Ku_0 = t^{1/3} - 1 ,$$

and the equation $Au_0 = 0$ can be rewritten as

$$\begin{aligned} t^{1/3} P_{\mathbf{T}} u_0 + Q_{\mathbf{T}} u_0 + t^{1/3} - 1 &= 0 , \\ t^{1/3}(P_{\mathbf{T}} u_0 + 1) &= 1 - Q_{\mathbf{T}} u_0 , \end{aligned}$$

or

$$(t-1)^{1/3}(P_{\mathbf{T}}u_0+1) = (1-Q_{\mathbf{T}}u_0)\left(1-\frac{1}{t}\right)^{1/3} .$$

The left-hand side of the latter equality belongs to $L_2^+(\mathbf{T})$, and its right-hand side to $L_2^-(\mathbf{T})$. So there must exist a number $\beta \in \mathbf{C}$ $(\beta \neq 0)$ such that

$$P_{\mathbf{T}}u_0 = \beta(t-1)^{-1/3} - 1, \quad Q_{\mathbf{T}}u_0 = 1 - \beta\left(1-\frac{1}{t}\right)^{-1/3}$$

and, consequently, $u_0(t) = \beta\left((t-1)^{-1/3} - (1-\frac{1}{t})^{-1/3}\right)$. A straightforward computation yields

$$Au_0 = \frac{\beta(t^{1/3}-1)(1+i\sqrt{3})}{4}$$

which is a contradiction to our assumption $u_0 \in \ker A$. Thus we have shown that $\dim\ker A = 0$, the operator A is invertible, and the equation (15.65) is uniquely solvable. One easily checks that the function $\varphi(x) \equiv 1$ solves the equation (15.65). In this example, the rank one perturbation K of the operator A_0 preserves the invertibility, that is, both operators A_0 and $A_0 + K$ are invertible. An analogous situation was seen in equation (15.26) in [GK 1, Chapter 3] when $a(x) = \cos\alpha \neq 0$ and $b(x) = \sin\alpha$. However, if $\cos\alpha = 0$ this equation then the operator A_0 is invertible but the operator $A_0 + K$ fails to be invertible.

The following theorem provides us with a criterion for the invertibility of the operator A .

Theorem 15.2. *Let $a, b \in L_\infty(0, 2\pi)$,*

$$(A\varphi)(x) = a(x)\varphi(x) + \frac{b(x)}{2\pi i}\int_0^{2\pi} \cot\frac{s-x}{2}\varphi(s)\,ds ,$$

and

$$(A_0u)(t) = a_0(t)u(t) + \frac{b_0(t)}{\pi i}\int_{\mathbf{T}} \frac{\varphi(\tau)\,d\tau}{\tau - t} ,$$

where $a_0(e^{ix}) = a(x)$ and $b_0(e^{ix}) = b(x)$.

Then the operator A is invertible in $L_p(0, 2\pi)$ $(1 < p < \infty)$ if and only if the following conditions are fulfilled

1)

$$\operatorname{ess}\inf_{0\le x\le 2\pi} |a^2(x) - b^2(x)| \neq 0 , \tag{15.66}$$

2) the function

$$c(t) = \frac{a_0(t)+b_0(t)}{a_0(t)-b_0(t)} \tag{15.67}$$

admits a factorization $c(t) = c_-(t)c_+(t)$ *in the space* $L_p(\mathbf{T})$ *and*

3)

$$c_+(0)c_-(\infty) \neq -1 . \tag{15.68}$$

The two conditions 1) and 2) yield a criterion for the invertibility of the operator A_0 on $L_p(\mathbf{T})$, and the additional condition (15.68) ensures the invertibility of the operator A. Notice further that if $a, b \in \mathbf{C}$ and $a^2 - b^2 \neq 0$, then condition (15.68) only means that $a \neq 0$.

Proof of the Theorem. Define $t = \exp ix$, $\tau = \exp is$, $u(e^{ix}) = \varphi(x)$ and $(Bu)(x) = u(e^{ix})$. Then $B^{-1}AB = A_0 + K$ where the operator A_0 has been introduced in the formulation of the theorem, and the rank one operator K is defined by

$$(Ku)(t) = -\frac{b_0(t)}{2\pi i} \int_{\mathbf{T}} \frac{u(\tau)\, d\tau}{\tau} .$$

If the operator A is invertible then A_0 is a Φ-operator on $L_p(\mathbf{T})$ with index $\operatorname{Ind} A_0 = 0$, and from Theorem 3.1 of Chapter 8 and Theorem 4.1 of Chapter 7 we conclude that the conditions (15.66) and (15.67) are satisfied.

We claim that the condition (15.68) is satisfied as well. Assume to the contrary, that $c_+(0)c_-(\infty) = -1$. We shall show that in this situation the function

$$u_0(t) = c_-(t) - c_+^{-1}(t) \tag{15.69}$$

belongs to the kernel of the operator $B^{-1}AB$, which contradicts the invertibility of A. A straightforward computation yields that

$$B^{-1}AB = (a_0 - b_0)[cP + Q - dK_0] ,$$

with $d = (c-1)/2$, $K_0 u = \frac{1}{2\pi i} \int_{\mathbf{T}} \frac{u(\tau)\, d\tau}{\tau}$, and

$$B^{-1}ABu_0 = b_0(c_-(\infty) + c_+^{-1}(0)) = 0 .$$

This proves the necessity of the conditions of the theorem. We are nowgoing to verify their sufficiency.

Again by Theorems 3.1 of Chapter 8 and 4.1 of Chapter 7, the conditions (15.66) and (15.67) guarantee the invertibility of the operator A_0. Thus, A is a Φ-operator with index $\operatorname{Ind} A = 0$. To finish the proof of the theorem, it suffices to check whether $\dim \ker A = 0$ or, equivalently, $\dim \ker (cP + Q - dK_0) = 0$. If $b_0 = 0$ then $d = 0$, and the assertion is evident. So we assume that $b_0 \neq 0$. Let $f \in \ker (cP + Q - dK_0)$. Our first goal is to prove the equality $K_0 f = 0$. Indeed, if $K_0 f \neq 0$, then we may assume that $K_0 f = 2$ and so rewrite the equation $(cP + Q - dK_0)f = 0$ in the form $c(Pf - 1) = -1 - Qf$, or

$$c_+(1 - Pf) = c_-^{-1}(1 + Qf) . \tag{15.70}$$

Since $c_+ \in L_q^+(\mathbf{T})$, $1 - Pf \in L_p^+(\mathbf{T})$, $c_-^{-1} \in L_q^-(\mathbf{T})$ and $1 + Qf \in L_p^-(\mathbf{T})$, both sides of the identity (15.70) belong to $L_1^+(\mathbf{T}) \cap L_1^-(\mathbf{T})$ and, consequently, they must be equal to a certain constant α. Thus, $Pf = 1 - \alpha c_+^{-1}$, $Qf = \alpha c_- - 1$, and $f = \alpha(c_- - c_+^{-1})$ $(\alpha \neq 0)$. This implies that

$$0 = (cP + Q - dK_0)f = \alpha b_0(c_-(\infty) + c_+^{-1}(0)) .$$

The numbers α and $c_-(\infty) + c_+^{-1}(0)$ are different from zero. Since the function b_0 is non-zero, too, this leads to a contradiction. Hence $K_0 f = 0$, but then $A_0 f = 0$ and so $f = 0$. ■

We look for the solution to the equation

$$a(x)\varphi(x) + \frac{b(x)}{2\pi i} \int_0^{2\pi} \cot \frac{s - x}{2} \varphi(s)\, ds = \psi(x) \tag{15.71}$$

under the assumptions that $a^2(x) - b^2(x) \neq 0$ and that the function $G(t) = (a_0(t) + b_0(t))(a_0(t) - b_0(t))^{-1}$ admits a factorization, $G(t) = G_-(t)G_+(t)$, in the space $L_2(\mathbf{T})$. Here $a_0(e^{ix}) = a(x)$ and $b_0(e^{ix}) = b(x)$.

As in the proof of Theorem 15.2 we go over to the equivalent equation in $L_2(\mathbf{T})$: $(A_0 + K)\varphi_0 = \psi_0$. Since the function G is factorizable in $L_2(\mathbf{T})$ with index zero, the operator A_0 is invertible in $L_2(\mathbf{T})$. Setting $f_0 = A_0^{-1}\psi_0$ we arrive at the integral equation

$$\varphi_0 - A_0^{-1} b_0 K_0 \varphi_0 = f_0 , \tag{15.72}$$

where

$$K_0\varphi_0 = \frac{1}{2\pi i} \int_{\mathbf{T}} \frac{\varphi_0(\tau)\, d\tau}{\tau} .$$

One immediately checks that

$$A_0^{-1} b_0 = \frac{c_-(t) - c_+^{-1}(t)}{2c_-(\infty)} \qquad (= v(t))$$

and

$$K_0 A_0^{-1} b_0 = \frac{1}{2}(1 - c_-^{-1}(\infty)c_+^{-1}(0)) .$$

The equation (15.72) implies that $\varphi_0 = f_0 + \lambda v$ where the number λ is determined by the equation

$$\lambda(1 + c_-^{-1}(\infty)c_+^{-1}(0)) = 2K_0 f_0 . \tag{15.73}$$

We distinguish two cases.

1. $c_+(0)c_-(\infty) \neq -1$. In this case, the operator A is invertible. Using the above derived formula

$$\varphi_0 = f_0 + \frac{c_- - c_+^{-1}}{c_-(\infty) + c_+^{-1}(0)} K_0 f_0$$

we find the unique solution of (15.71) as

$$\varphi(x) = f(x) + \frac{c_-(e^{ix}) - c_+^{-1}(e^{ix})}{(c_-(\infty) + c_+^{-1}(0))2\pi} \int_0^{2\pi} f(s)\, ds, \tag{15.74}$$

where

$$f(x) = \frac{a(x)\psi(x)}{a^2(x) - b^2(x)} - \frac{b(x)c_-(e^{ix})}{\pi(a(x)+b(x))} \int_0^{2\pi} \frac{(1 - \cot\frac{x-s}{2})\psi(s)\, ds}{c_-(e^{is})(a(s)-b(s))} . \tag{15.75}$$

2. $c_+(0)c_-(\infty) = -1$. In this situation, the equation (15.71) is not invertible for any right side. The equation (15.73) implies that the condition $K_0 f_0 = 0$ is just the solvability criterion for (15.71). Since

$$f_0 = A_0^{-1}\psi_0 = (c_+^{-1}P_{\mathbf{T}} + c_- Q_{\mathbf{T}})c_-^{-1}(a_0 - b_0)^{-1}\psi_0 ,$$

we find

$$K_0 f_0 = c_+^{-1}(0) K_0 c_-^{-1}(a_0 - b_0)^{-1}\psi_0 .$$

Thus, the equation (15.71) is solvable if and only if

$$\int_0^{2\pi} \frac{\psi(x)\, dx}{c_-(e^{ix})(a(x) - b(x))} = 0 . \tag{15.76}$$

If this condition is satisfied then the λ in equation (15.73) can be replaced by an arbitrary number and, consequently, the general solution to equation (15.71) has the form

$$\varphi(x) = f(x) + \lambda \frac{c_-(e^{ix})b(x)}{a(x)+b(x)} ,$$

where f is defined by (15.75) and λ is an arbitrary constant.

9.16 Exercises

16.1. Determine all pairs (α, β) of complex numbers such that the function

$$a(t) = \begin{cases} t^\alpha & \text{if} \quad t \in \mathbf{T}\,, \operatorname{Im} t \geq 0 \\ t^\beta & \text{if} \quad t \in \mathbf{T}\,, \operatorname{Im} t < 0\,, \end{cases}$$

which is continuous at all points $t \in \mathbf{T} \setminus \{\pm 1\}$, is p-non-singular, and compute the p-index of this function.

16.2. Find all pairs (α, β) of complex numbers such that the functions

$$a(t) = t^\alpha, \quad b(t) = t^\beta, \quad c(t) = a(t)b(t) ,$$

which are continuous on $\mathbf{T} \setminus \{1\}$, are p-non-singular, and so that

$$\text{ind } c^{p,1} = \text{ind } a^{p,1} + \text{ind } b^{p,1} .$$

16.3 Solve the equation

$$(aP_\mathbf{T} + Q_\mathbf{T})\varphi = \psi$$

with

$$a(t) = \begin{cases} 1 & \text{if} \quad t \in \mathbf{T}, \ \text{Re}\, t > 0 \\ -1 & \text{if} \quad t \in \mathbf{T}, \ \text{Re}\, t < 0 \end{cases}$$

in the spaces $L_4(\mathbf{T})$ and $L_{4/3}(\mathbf{T})$.

16.4. Let Γ stand for the boundary of the annulus $1 \le |z| \le 2$. Determine all triples (α, β, γ) of complex numbers such that the function

$$a(t) = \begin{cases} \alpha t^\gamma & \text{if} \quad |t| = 1 \\ \beta t^\gamma & \text{if} \quad |t| = 2 , \end{cases}$$

which is continuous at each point $t \in \Gamma \setminus \{1, 2\}$, is p-non-singular, and describe the p-factorization of the function a in this case.

Hint. In case $\alpha\beta \ne 0$ one can choose

$$a_-(z) = \begin{cases} \alpha & \text{if} \quad |z| \le 1 \\ \beta & \text{if} \quad |z| \ge 2 . \end{cases}$$

16.5. Solve the equation $(aP + Q)\varphi = \psi$ under the conditions of the previous exercise.

16.6. Put $\Gamma = \{z \in \mathbf{C} : \ \text{Im}\, z \le 0, |z| = 1\}$. Solve the equation

$$\varphi(t) - \frac{1}{2\pi} \int_\Gamma \frac{\varphi(\tau)\, d\tau}{\tau - t} = \psi(t)$$

in $L_2(\Gamma)$.

16.7. Let $\alpha \in \mathbf{R}, 1 < p < \infty$, and $\alpha \pm 1/p \notin \mathbf{Z}$. Determine the general solution to the equation

$$\cos(\pi\alpha)\varphi(t) + \frac{\sin(\pi\alpha)}{\pi} \int_0^1 \frac{\varphi(s)\, ds}{s - t} = \psi(t)$$

in $L_p(0,1)$.

Hint. Since replacing α by $\alpha+1$ does not influence the equation we can assume the values of α to belong to a certain interval of length 1. In case $p \geq 2$ the interval $\frac{1}{p} < \alpha < \frac{1}{q}$ proves to be advantageous, and in case $p \leq 2$, the interval $-\frac{1}{q} < \alpha < \frac{1}{p}$. In both situations, the number γ appearing in Theorem 5.5 coincides with α .

16.8. Prove that if the singular integral operator $aI + bS_\Gamma$ with constant coefficients is one–sided invertible on the space $L_2(\alpha,\beta)$ (i.e. $\Gamma = [\alpha,\beta]$) then it is in fact two–sided invertible.

16.9. Determine all numbers $\lambda \in \mathbf{C}$ such that the equation

$$\lambda\varphi(t) + \frac{\operatorname{sgn} t}{\pi i} \int_{-\infty}^{\infty} \frac{\varphi(s)\,ds}{s-t} = \psi(t)$$

is normally solvable in $L_2(\mathbf{R})$. For these λ's, establish a solvability criterion of the given equation, and solve it.

16.10. Let $\alpha \in \mathrm{R}$. Establish solvability criteria of the equation

$$\cos(\pi\alpha)\varphi(t) + \frac{\sin(\pi\alpha)}{\pi} \int_0^{\infty} \frac{\varphi(s)\,ds}{s-t} = \psi(t)\ ,$$

in $L_p(0,\infty)$ and describe the general solution if these criteria are satisfied.

16.11. Solve the equation

$$\varphi(t) + \frac{1}{\pi} \int_0^{\infty} \frac{\varphi(s)\,ds}{s-t} = \frac{1}{t+1}$$

in $L_2(0,\infty)$.

16.12. Show that Theorem 13.1 remains true if the function $a(t) = (1-t)^{-\alpha}$ is replaced by $(t-t_0)^{-\alpha}$ where t_0 is an arbitrarily chosen point on $\mathbf{T}$ (confer Corollary 13.1).

16.13. Solve the equation $Ax = y$ in l_4 where

$$A = (a_{j-k})_{j,k=0}^{\infty}, \quad a_m = \frac{1}{2m-1}\ .$$

16.14. Solve the equation $T_a x = y$ in l_2 where

$$a(t) = \begin{cases} 1 & \text{if} \quad t \in \mathbf{T},\ \operatorname{Im} t < 0 \\ \lambda & \text{if} \quad t \in \mathbf{T},\ \operatorname{Im} t > 0\ , \end{cases}$$

and λ is a fixed complex number.

Hint. Use the factorization of this function obtained in Section 9.4.

16.15. Solve the equations given by the following systems in the space $\widetilde{l}_2$:

a)

$$\begin{cases} \sum_{k=-\infty}^{\infty} \frac{\xi_k}{3j-3k+1} = \eta_j & (j \geq 0) \\ \sum_{k=-\infty}^{\infty} \frac{\xi_k}{3k-3j-1} = \eta_j & (j < 0) , \end{cases}$$

b)

$$\begin{cases} \sum_{k=-\infty}^{\infty} \frac{\xi_k}{3j-3k+1} = \eta_j & (j \geq 0) \\ \sum_{k=-\infty}^{\infty} \frac{\xi_k}{3k-3j+1} = \eta_j & (j < 0) , \end{cases}$$

c)

$$\sum_{k=0}^{\infty} \frac{\xi_k}{3j-3k+1} + \sum_{k=-\infty}^{-1} \frac{\xi_k}{3k-3j-1} = \eta_j \qquad (j = 0, \pm 1, \ldots) ,$$

d)

$$\sum_{k=0}^{\infty} \frac{\xi_k}{3j-3k+1} + \sum_{k=-\infty}^{-1} \frac{\xi_k}{3k-3j+1} = \eta_j \qquad (j = 0, \pm 1, \ldots) .$$

Comments and references

For the theory of singular integral equations with piecewise Hölder continuous coefficients in special classes of piecewise Hölder continuous functions we refer to the monographs by MUSKHELISHVILI [1] and GAKHOV [1].

9.1.–9.4. The investigation of singular integral equations with piecewise continuous coefficients on spaces L_p with weight was initiated by KHVEDELIDZE [1]. In particular he obtained some sufficient conditions for the Fredholmness of such operators. A theorem similar to the Theorem 3.1 was proved by SHAMIR [1] in case $\rho \equiv 1$ and $\Gamma = \mathbf{R}$, and by WIDOM [1] for a measurable set $\Gamma \subset \mathbf{R}$.

The results in Section 9.3 are also related to WIDOM's paper [2]. In the general case, the basic results of sections 9.1–9.4 were proved by GOHBERG and KRUPNIK [1,2,4]. In KHAIKIN [2,3] there is proposed a method to study and solve singular integral equations in the case when their coefficients do not vanish and the function $c = ab^{-1}$ is not $\{p, \rho\}$–non–singular.

9.5. The Theorem 5.1 is taken from GOHBERG/KRUPNIK [9] where also certain generalizations of Theorem 5.1 were obtained. We also mention the results of SCHWARTZ [1] and SCHLEIFF[1] which reveal close connections to this theorem as well. Sufficient conditions for the one–sided invertibility of the operator S_Γ on $L_p(\Gamma, \rho)$ were derived by many authors (see KHVEDELIDZE [1], p. 37). Theorem 5.3 goes back in its general form TO WIDOM [1] and to GOHBERG/ KRUPNIK [4]; and Theorem 5.4 is a generalization of a well–known result (see the monographs by MUSKHELISHVILI [1, p. 451] and GAKHOV [1, p. 195]).

9.7. The presented method for the inversion of singular integral operators was communicated to the authors by DYNIN. In ESKIN [1] an analogous approach is applied to some classes of pseudodifferential operators on L_2.

9.9 The estimates of the norms $\|S_\Gamma\|, \|P_\Gamma\|$ and $\|Q_\Gamma\|$ on the spaces $L_p(\Gamma)$ go back to GOHBERG/KRUPNIK [2,3].

9.10. The results in this section belong to DUDUCHAVA [1-4].

9.11. This section is an explanation of GOHBERG/KRUPNIK 's paper [2]. For the proof of Theorem 11.1 see SEMENOV[1].

9.12. These results are due to SPITKOVSKI[1].

9.13–9.14. DUDUCHAVA [5, 6] proved the basic results of these sections under stronger restrictions on the coefficients a, b of the operators $T_a, \Pi(a,b)$ and $\Pi'(a,b)$, and by another method. The presented generalizations of DUDUCHAVA's results were established in GOHBERG/KRUPNIK [8] by means of the local principle (see [GK 1, Chapter 5]). Theorem 13.1 was derived by DUDUCHAVA [5.6], Proposition 13.1 has been taken from the monograph HARDY/LITTLEWOOD/POLYA [1], and Proposition 14.1 concerning the interpolation property of compactness was obtained by KRASNOSELSKII [1]. Finally, DUDUCHAVA [5, 6] was the first to succeed in deriving Fredholm criteria for Toeplitz operators with piecewise continuous coefficients on spaces l_p.

Chapter 10

Singular integral operators on non–simple curves

In this chapter we generalize the results of the preceding chapter to the case of non–simple curves. The chapter is divided into four sections. The first two are in preparation the third one which contains the main theorem. The last small section contains exercises.

10.1 Technical lemmas

Throughout this section we let Γ denote a closed non–simple curve which consists of n simple closed curves $\Gamma_1, \ldots, \Gamma_n$ having exactly one point, t_0, in common. Further we suppose the regions [1] $F^+_{\Gamma_j}$ $(=: F^+_j)$ to be pairwise disjoint and that, running counterclockwise around the point t_0, the region F^+_j is followed by F^+_{j+1} $(j = 1, \ldots, n-1)$. Finally, let the curve Γ be oriented such that the region F^+_j is located on the left–hand side of Γ_j.

By $PC_0(\Gamma)$ we denote the collection of all functions f which are continuous at each point $t \in \Gamma$ different from t_0 and which possess finite limits $f_j(t_0+0)$ and $f_j(t_0-0)$ $(j = 1, \ldots, n)$ if t tends to t_0 along Γ_j on the arc which is directed away from t_0 or to t_0, respectively. Further we let $PC_+(\Gamma)$ stand for the subset of $PC_0(\Gamma)$ consisting of all functions f so that $f_j(t_0+0) = f_j(t_0-0)(j = 1, \ldots, n)$, whereas $PC_-(\Gamma)$ refers to the set of all functions $f \in PC_0(\Gamma)$ satisfying $f_j(t_0-0) = f_{j+1}(t_0+0)$ $(j = 1, \ldots, n-1)$ and $f_n(t_0-0) = f_1(t_0+0)$. The intersection $PC_+(\Gamma) \cap PC_-(\Gamma)$ just coincides with the class $C(\Gamma)$ of all continuous functions on Γ.

[1] Remember that F^+_Γ stands for the region which is bounded by the curve Γ.

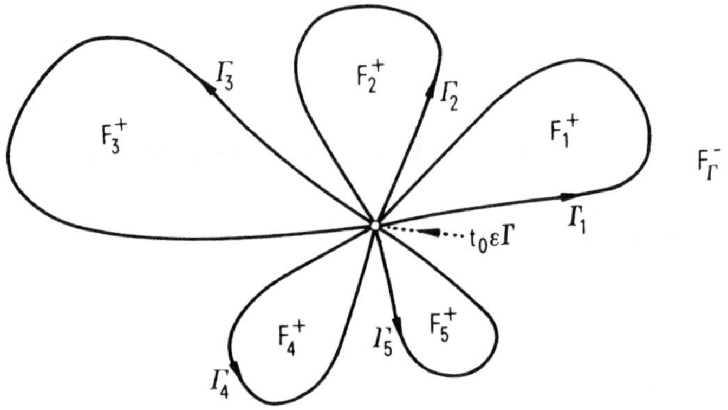

Figure 10.1

Let $\rho(t) = \prod_{k=0}^{m} |t - t_k|^{\beta_k}$, where $t_0, t_1, \ldots, t_m$ are different points on Γ and β_k are real numbers fulfilling the inequalities $-1 < \beta_k < p - 1 \quad (k = 0, \ldots, m)$. By Theorem 4.3 in Chapter 1, [GK 1] the operators $aP_\Gamma - P_\Gamma aI$ and $aQ_\Gamma - Q_\Gamma aI$ are compact on $L_p(\Gamma, \rho)$ for each function $a \in C(\Gamma)$. Similar assertions for functions in $PC_\pm(\Gamma)$ read as follows:

Proposition 1.1. *If $a \in PC_+(\Gamma)$ then the operator $P_\Gamma a P_\Gamma - aP_\Gamma$ is compact on $L_p(\Gamma, \rho)$.*

Proof. Write the function a as $a = bg$ where $b \in C(\Gamma)$ and g is a function which is constant on each curve Γ_j . We claim that $P_\Gamma g P_\Gamma = gP_\Gamma$. Indeed, if r is a rational function in $R(\Gamma)$ then $P_\Gamma r =: r_+ \in R_+(\Gamma)$ by Theorem 1.1 of Chapter 1, [GK 1](see also Section 1 of Chapter 6). Since the functions gr_+ and r_+ differ on Γ_j by at most a constant factor, we have $S_{\Gamma_j} g r_+ = g r_+$ for $t \in \Gamma_j$ and $S_{\Gamma_j} g r_+ = 0$ at all other points of Γ . Thus, $S_\Gamma g P_\Gamma r = g P_\Gamma r$, from which it follows that $S_\Gamma g P_\Gamma = g P_\Gamma$ and $P_\Gamma g P_\Gamma = g P_\Gamma$. Since, moreover, the operator $P_\Gamma bI - bP_\Gamma$ is compact and $a = bg$, we arrive at the compactness of $P_\Gamma a P_\Gamma - aP_\Gamma$. ∎

Proposition 1.2. *If $a \in PC_-(\Gamma)$ then the operator $Q_\Gamma a Q_\Gamma - aQ_\Gamma$ is compact on $L_p(\Gamma, \rho)$.*

Proof. Let $s_j \in F_j^+ \quad (j = 1, \ldots, n)$ be arbitrary points and let $g_j \; (j = 1, \ldots, n)$ refer to the unique branch of the function

$$g_j(t) := \alpha_j \left(\frac{t - s_j}{t - s_{j+1}} \right)^{\delta_j}$$

(with $s_{n+1} = s_1$ and $a_j, \delta_j \in \mathbf{C}$) , which is holomorphic in the extended complex plane with a cut along a curve connecting the points s_j and s_{j+1} in $F_i^+ \cup F_{i+1}^+ \cup \{t_0\}$. Without loss of generality we assume that $\inf_{t\in\Gamma} |a(t)| \neq 0$. Then we can choose the numbers α_j and δ_j in such a way that the quotient $a(t)/g_j(t)$ tends to one as $t \to t_0$ along the subarc of Γ_j which is directed to t_0 and that the function $g_j(t)$ tends to one as $t \to t_0$ along the subarc of Γ_j which is directed away from t_0 . The special choice of the cuts ensures that $g_j \in PC_-(\Gamma)$. Put $g := g_1 \cdot \ldots \cdot g_n$. It is easy to see that the function $b := a/g$ is continuous on Γ .

Next we are going to show that $Q_\Gamma g Q_\Gamma = g Q_\Gamma$. For let r be an arbitrary function in $R(\Gamma)$. By Theorem 1.1 of Chapter 1, [GK 1], $Q_\Gamma r = r_-$. The function gr_- is holomorphic in F_Γ^- , continuous in $\overline{F_\Gamma^-} \backslash \{t_0\}$, bounded in a neighbourhood of the point t_0 , and equal to zero at infinity. Theorem 4.8 of Chapter 2, [GK 1] implies that $Q_\Gamma g r_- = g r_-$, and hence, $Q_\Gamma g Q_\Gamma = g Q_\Gamma$. ■

The Propositions 1.1 and 1.2 show that, whenever $a \in PC_+(\Gamma)$ and $b \in PC_-(\Gamma)$,

$$Q_\Gamma a Q_\Gamma - Q_\Gamma a I \in \mathcal{T}(L_p(\Gamma,\rho)), \; P_\Gamma b P_\Gamma - P_\Gamma b I \in \mathcal{T}(L_p(\Gamma,\rho)), \tag{1.1}$$

where $\mathcal{T}(L_p(\Gamma,\rho))$ denotes the collection of compact operators on $L_p(\Gamma,\rho)$. This implies the following

Proposition 1.3. *Let a and b be arbitrary functions which simultaneously belong to one of the classes $PC_+(\Gamma)$ or $PC_-(\Gamma)$. Then the operators $P_\Gamma a P_\Gamma b P_\Gamma - P_\Gamma ab P_\Gamma$ and $Q_\Gamma a Q_\Gamma b Q_\Gamma - Q_\Gamma ab Q_\Gamma$ are compact on $L_p(\Gamma,\rho)$.*

Theorem 1.1. *Let a and b be functions which simultaneously belong to $PC_+(\Gamma)$ or $PC_-(\Gamma)$. Then the operator $A = aP_\Gamma + bQ_\Gamma$ is a Φ-operator on $L_p(\Gamma,\rho)$ if and only if*

$$\operatorname{ess}\inf_{t\in\Gamma} |a(t)| > 0 \quad \text{and} \quad \operatorname{ess}\inf_{t\in\Gamma} |b(t)| > 0 \, . \tag{1.2}$$

If these conditions are satisfied then

$$\operatorname{Ind} A = \frac{1}{2\pi} \sum_{j=1}^{n} [\arg b(t)/a(t)]_{\Gamma_j} \, . \tag{1.3}$$

If, on the other hand, the condition (1.2) is violated then A is neither a Φ_+- nor a Φ_--operator.

Proof. To start with, assume that $a, b \in PC_+(\Gamma)$ and that the conditions (1.2) are fulfilled. Define $c := a/b$. Obviously, $c \in PC_+(\Gamma)$ and, by Proposition 1.1, the operator A can be written in the form

$$A = b(P_\Gamma c P_\Gamma + Q_\Gamma) + T \tag{1.4}$$

for some operator $T \in \mathcal{T}(L_p(\Gamma, \rho))$. By means of Proposition 1.3 one checks that the operator

$$M := (P_\Gamma c^{-1} P_\Gamma + Q_\Gamma) b^{-1} I$$

is a regularization of A . Thus, by Theorem 7.1' of Chapter 4, [GK 1], the operator A proves to be a Φ–operator in the space $L_p(\Gamma, \rho)$.

Introduce functions $c^{(k)}$ via

$$c^{(k)}(t) := \begin{cases} c(t) & \text{if} \quad t \in \Gamma_k , \\ 1 & \text{if} \quad t \in \Gamma \backslash \Gamma_k . \end{cases}$$

Obviously, $c^{(k)} \in PC_+(\Gamma)$ and $c = c^{(1)} c^{(2)} \cdot \ldots \cdot c^{(n)}$. Using Proposition 1.1 again, we can write the operator A as

$$A = b \prod_{k=1}^{n} (c^{(k)} P_\Gamma + Q_\Gamma) + T_1 \quad (T_1 \in \mathcal{T}(L_p(\Gamma, \rho))) .$$

Due to the continuity of the functions $c^{(k)}$ on Γ_k (see Theorem 7.1 Chapter 3, [GK 1] [1]) the operators $c^{(k)} P_{\Gamma_k} + Q_{\Gamma_k}$ are Φ–operators on $L_p(\Gamma_k, \rho)$ with the index

$$\text{Ind } (c^{(k)} P_{\Gamma_k} + Q_{\Gamma_k}) = -\frac{1}{2\pi} [\arg c^{(k)}(t)]_{\Gamma_k} .$$

Now, by Theorem 1.1 of Chapter 7, [GK 1], $\text{Ind } (c^{(k)} P_\Gamma + Q_\Gamma) = \text{Ind } (c^{(k)} P_{\Gamma_k} + Q_{\Gamma_k})$, and Theorem 6.1 in Chapter 4 [GK 1] implies

$$\text{Ind } A = \sum_{k=1}^{n} \text{Ind } (c^{(k)} P_\Gamma + Q_\Gamma) .$$

This formula, in combination with the previously obtained equalities, yields (1.3).

Next we verify that if A is a Φ– or a $\Phi_\pm$–operator and if a and b are functions in $PC_+(\Gamma)$ then condition (1.2) is satisfied. To this end we approximate the function b by a function $d \in PC_+(\Gamma)$ such that $\inf_{t \in \Gamma} |d(t)| > 0$ and that the operator $A_1 := aP_\Gamma + dQ_\Gamma$ is a Φ– or $\Phi_\pm$–operator again. We represent this operator as

$$A = d \prod_{k=1}^{n} (g^{(k)} P_\Gamma + Q_\Gamma) + T_2 , \text{ [2]} \tag{1.5}$$

where $g := ad^{-1}$ and T_2 is a compact operator. If we interchange the factors in the first term on the right of (1.5), then we have only to replace the operator T_2 by another

[1] The original Theorem 7.1 of Chapter 3, [GK 1] deals with curves consisting of Lyapunov curves, but evidently it also holds for piecewise Lyapunov curves.

[2] $g^{(k)}$ is defined analogously to $c^{(k)}$.

compact operator. Thus, by Theorem 7.1' in Chapter 4, [GK 1], all operators $g^{(k)}P_\Gamma + Q_\Gamma$ are simultaneously Φ- or $\Phi_\pm$-operators in $L_p(\Gamma,\rho)$ and consequently, following Theorem 1.1 of Chapter 7, [GK 1], the operators $g^{(k)}P_{\Gamma_k} + Q_{\Gamma_k}$ $(k = 1,\dots,n)$ are Φ- or $\Phi_\pm$-operators in $L_p(\Gamma_k,\rho)$.

Now, taking the continuity of the functions $g^{(k)}$ on Γ_k into account, Theorem 7.3 of Chapter 4, [GK 1] entails that $g^{(k)} \in GC(\Gamma_k)$ and hence

$$\operatorname{ess}\inf_{t\in\Gamma} |g(t)| > 0 ,$$

from which the first of the conditions (1.2) follows.

For the second condition we represent the operator A in the form

$$A = a(P_\Gamma + Q_\Gamma f Q_\Gamma)(I + P_\Gamma f Q_\Gamma)$$

with $f = a^{-1}b$. Since the operators aI and $I + P_\Gamma f Q_\Gamma$ are invertible, the operator $A_2 := P_\Gamma + Q_\Gamma f Q_\Gamma$ is a Φ- or $\Phi_\pm$-operator. Considering the equality

$$A_2 = \prod_{k=1}^{n}(P_\Gamma + Q_\Gamma f^{(k)} I) + T_3 \; {}^{1}$$

with a certain $T_3 \in \mathcal{T}(L_p(\Gamma,\rho))$ we conclude that all operators $P_\Gamma + Q_\Gamma f^{(k)}I$ are simultaneously Φ-, Φ_+- or Φ_--operators in $L_p(\Gamma,\rho)$. Moreover, together with these operators, the operators $P_{\Gamma_k} + Q_{\Gamma_k} f^{(k)} I$ $(k = 1,\dots,n)$ are Φ-, Φ_+- or Φ_-- operators in the corresponding spaces $L_p(\Gamma_k,\rho)$. Since $f^{(k)}|\Gamma_k \in C(\Gamma_k)$, we even have $f^{(k)}|\Gamma_k \in GC(\Gamma)$ by Theorem 7.3 in Chapter 4, [GK 1]. Thus,

$$\operatorname{ess}\inf_{t\in\Gamma} |f(t)| > 0 ,$$

which yields the second of the conditions (1.2). This proves the theorem in case $a, b \in PC_+(\Gamma)$.

Now let $a, b \in PC_-(\Gamma)$. Then the operator A can be rewritten as $A = a(P_\Gamma + Q_\Gamma c Q_\Gamma) + T$ with $c := b/a$ and $T \in \mathcal{T}(L_p(\Gamma,\rho))$, and the operator $M := (P_\Gamma + Q_\Gamma c^{-1} Q_\Gamma)a^{-1}I$ is a regularization for A . The proof of the necessity of the conditions of the theorem as well as the derivation of the index formula proceed analogously to the case $PC_+(\Gamma)$ with the only difference being that another factorization $c = c^{(0)}c^{(1)}\dots c^{(n)}$ of the function c is needed. For let $\tau_1,\dots,\tau_n$ be points on $\Gamma_1,\Gamma_2,\dots,\Gamma_n$,respectively, which are different from t_0 , and dividing the curve Γ into n non-closed arcs $\gamma_1,\dots,\gamma_n$ (which contain t_0). Furthermore we pick a function $c^{(0)}$ which is continuous on Γ and possesses the

[1] $f^{(k)}$ is defined as $c^{(k)}$ at p.154

following properties: (1) $c^{(0)}(t) \neq 0 \quad (t \in \Gamma)$, (2) $c^{(0)}(\tau_k) = c(\tau_k) \quad (k = 1, \ldots, n)$, and (3) $\operatorname{ind} c^{(0)} = 0$. On defining

$$c^{(k)}(t) := \begin{cases} c(t)/c^{(0)}(t) & \text{if} \quad t \in \gamma_k \\ 1 & \text{if} \quad t \in \Gamma \backslash \gamma_k \end{cases} \quad (k = 1, \ldots, n) ,$$

one obtains immediately

$$\operatorname{Ind} A = \sum_{k=1}^{n} \operatorname{Ind} (P_{\gamma_k} + c^{(k)} Q_{\gamma_k}) \tag{1.6}$$

and

$$\sum_{k=1}^{n} [\arg b(t)/a(t)]_{\Gamma_k} = \sum_{k=1}^{n} [\arg c^{(k)}(t)]_{\gamma_k} . \tag{1.7}$$

Since the functions $c^{(k)}(k = 1, \ldots, n)$ are continuous on γ_k and equal to one at the endpoints of the arc γ_k , one has (see Theorem 4.1 of Chapter 9)

$$\operatorname{Ind} (P_{\gamma_k} + c^{(k)} Q_{\gamma_k}) = \frac{1}{2\pi} [\arg c^{(k)}(t)]_{\gamma_k} .$$

This in combination with (1.6) and (1.7) gives formula (1.3). ■

10.2 A preliminary theorem

In this section we consider singular integral equations with coefficients in $PC_0(\Gamma)$ in the space $L_p(\Gamma, \rho)$. The curve Γ and the weight ρ are exactly the same as in the preceding section.

Theorem 2.1. *Let* $a, b \in PC_0(\Gamma)$. *Then the operator* $A = aP_\Gamma + bQ_\Gamma$ *is a* Φ- *or* $\Phi_\pm$*-operator on* $L_p(\Gamma, \rho)$ *if and only if*

$$\operatorname{ess} \inf_{t \in \Gamma} |a(t)| > 0, \qquad \operatorname{ess} \inf_{t \in \Gamma} |b(t)| > 0 \tag{2.1}$$

and

$$c_1(t_0 + 0) \cdot \ldots \cdot c_n(t_0 + 0) f_\delta(\mu) + c_1(t_0 - 0) \cdot \ldots \cdot c_n(t_0 - 0)(1 - f_\delta(\mu)) \neq 0 \tag{2.2}$$

where $\delta := 2\pi(1 + \beta_0)/p$, $c := ab^{-1}$, $0 \leq \mu \leq 1$, $c_j(t_0 \pm 0) := \lim_{t \to t_0 \pm 0, \, t \in \Gamma_j} c(t)$, *and*

$$f_\delta(\mu) = \begin{cases} \frac{\sin \theta\mu}{\sin \theta} e^{i\theta(\mu - 1)} & (\theta = \pi - \delta) \quad \text{if} \quad \delta \neq \pi \\ \mu & \text{if} \quad \delta = \pi . \end{cases}$$

If the conditions (2.1) and (2.2) are satisfied then the operator A is a Φ-operator with the index

$$\begin{aligned}\operatorname{Ind} A &= -\frac{1}{2\pi}\sum_{j=1}^{n}[\arg c(t)]_{t\in\Gamma_j} - \frac{1}{2\pi}[\arg\{c_1(t_0+0)\cdot\ldots\cdot c_n(t_0+0)f_\delta(\mu)\\ &+ c_1(t_0-0)\cdot\ldots\cdot c_n(t_0-0)(1-f_\delta(\mu))\}]_{\mu=0}^{1}\,. \end{aligned} \tag{2.3}$$

For the proof we need the following.

Lemma 2.1. *Each function $c \in PC_0(\Gamma)$ satisfying the conditions $c_j(t_0 \pm 0) \neq 0$ can be represented in the form $c = fxg$ where $g \in PC_+(\Gamma)$, $f \in PC_-(\Gamma)$, and x is a function which is subject to the following conditions: (1) $x(t) = 1$ if $t \in \Gamma\backslash\Gamma_1$, (2) $x_1(t_0 - 0) = 1$, and (3) $x_1(t_0 + 0) = c_1(t_0 + 0)\cdot\ldots\cdot c_n(t_0 + 0)/c_1(t_0 - 0)\cdot\ldots\cdot c_n(t_0 - 0)$.*

Proof. Let x be an arbitrary function in $PC_0(\Gamma)$ which satisfies the conditions (1)-(3) and which does not vanish on Γ. By g we denote a function in $PC_+(\Gamma)$ such that $g_k(t_0+0) = c_1(t_0+0)\cdot\ldots\cdot c_k(t_0+0)/c_1(t_0-0)\cdot\ldots\cdot c_{k-1}(t_0-0)$. Further we suppose that $\operatorname{ess\,inf}|g(t)| > 0$ $(t \in \Gamma)$. Define $f = cg^{-1}x^{-1}$. Then, because of

$$f_k(t_0-0) = f_{k+1}(t_0+0) = c_1(t_0-0)\cdot\ldots\cdot c_k(t_0-0)/c_1(t_0+0)\cdot\ldots\cdot c_k(t_0+0) \quad (k = 1,\ldots,n-1)$$

and

$$f_n(t_0-0) = f_1(t_0+0) = c_1(t_0-0)\cdot\ldots\cdot c_n(t_0-0)/c_1(t_0+0)\cdot\ldots\cdot c_n(t_0+0)\,,$$

one has $f \in PC_-(\Gamma)$. ∎

Proof of Theorem 2.1. Let the conditions (2.1) and (2.2) be satisfied. We rewrite the function $c = ab^{-1}$ in the form $c = fxg$ with f, x, and g being functions appearing in Lemma 2.1. Then, by Propositions 1.1 and 1.2,

$$cP_\Gamma + Q_\Gamma = f(xP_\Gamma + Q_\Gamma)(gP_\Gamma + f^{-1}Q_\Gamma) + T \tag{2.4}$$

with $T \in \mathcal{T}(L_p(\Gamma,\rho))$. The operator $gP_\Gamma + f^{-1}Q_\Gamma$ is a Φ–operator since the operator $g^{-1}P_\Gamma + fQ_\Gamma$ is one of its regularizations. As a consequence of Theorem 1.1 of Chapter 7, the operator $xP_\Gamma + Q_\Gamma$ is a Φ–, Φ_+– or Φ_-–operator in $L_p(\Gamma,\rho)$. But the conditions (2.1) and (2.2) imply that

$$\operatorname*{ess\,inf}_{t\in\Gamma_1} |x(t)| \neq 0 \quad\text{and}\quad x_1(t_0+0)f_\delta(\mu) + x_1(t_0-0)(1-f_\delta(\mu)) \neq 0 \quad (0 \le \mu \le 1)\,,$$

and so, following Theorem 3.1 of Chapter 9, the operator $xP_{\Gamma_1} + Q_{\Gamma_1}$ is a Φ–operator in $L_p(\Gamma_1,\rho)$. Hence the operator $xP_\Gamma + Q_\Gamma$ is actually a Φ–operator on $L_p(\Gamma,\rho)$. This proves the sufficiency of the conditions (2.1) and (2.2).

For their necessity we first claim that the operator $A = aP_\Gamma + bQ_\Gamma$ cannot be a $\Phi_\pm$–operator. Indeed, each pair of functions $a, b \in PC_0(\Gamma)$ can be uniformly approximated by functions $a', b' \in PC_0(\Gamma)$ satisfying the conditions (2.1) and (2.2) (indeed, the functions a' and b' may even be chosen from the piecewise rational functions). By what we have already shown, the operator $A' = a'P_\Gamma + b'Q_\Gamma$ is a Φ–operator, and since the operator A' is sufficiently close to A (in the norm) the operator A is also a Φ–operator by Theorem 6.4 of Chapter 4, [GK 1].

We are going to show that $\text{ess inf}_{t\in\Gamma} |a(t)| > 0$. For assume the contrary, that is, the operator A is a Φ–operator, and $\text{ess inf}_{t\in\Gamma} |a(t)| = 0$. Then, evidently, one can choose an operator $D = dP_\Gamma + bQ_\Gamma$ which is sufficiently close to A with respect to the operator norm such that the function d satisfies the conditions: $d(\tau) = 0$ at a certain point τ of Γ different from t_0 but $d(t') \neq 0$ and $d(t'') \neq 0$ when t' and t'' refer to the end points of a certain arc l containing the point τ but not the point t_0 . Let g be an arbitrary function which is continuous on Γ , different from zero on $\Gamma\backslash l$, and equal to d on l ; and let h denote a function which is equal to one on l and to dg^{-1} on $\Gamma\backslash l$. Then $d = gh$, and hence $D = (hP_\Gamma + bQ_\Gamma)(gP_\Gamma + Q_\Gamma) + T$ with a compact operator T . Since D is a Φ–operator, the operator $gP_\Gamma + Q_\Gamma$ is a Φ– or Φ_+operator by Lemma 15.2 of Chapter 4. But this is impossible since g is a continuous function which vanishes at the point τ . Thus, $\text{ess inf}_{t\in\Gamma} |a(t)| > 0$. The estimate $\text{ess inf}_{t\in\Gamma} |b(t)| > 0$ can be proved analogously.

Next we show the necessity of condition (2.2). For we write the operator $cP_\Gamma + Q_\Gamma$ with $c = a/b$ in the form (2.4). Since $cP_\Gamma + Q_\Gamma$ is a Φ–operator, the operator $xP_\Gamma + Q_\Gamma$ proves to be a Φ– or Φ_-–operator (by Proposition 15.1 of Chapter 4 [GK 1]). As we have already remarked above, this implies that the operator $xP_{\Gamma_1} + Q_{\Gamma_1}$ is a Φ– or Φ_-–operator in $L_p(\Gamma_1, \rho)$. But now Theorem 3.1 of Chapter 9 tells us that $x_1(t_0+0)f_\delta(\mu) + x_1(t_0-0)(1 - f_\delta(\mu)) \neq 0$ $(0 \leq \mu \leq 1)$; that is, the validity of (2.2).

So it remains to verify condition (2.3). The combination of the formula $\text{Ind } A = \text{Ind }(cP_\Gamma + Q_\Gamma)$ with equation (2.4) and the identity $gP_\Gamma + f^{-1}Q_\Gamma = (gP_\Gamma + Q_\Gamma)f^{-1}(fP_\Gamma + Q_\Gamma) + T$ $(T \in \mathcal{T}(L_p(\Gamma, \rho)))$ yields immediately

$$\text{Ind } A = \text{Ind }(xP_\Gamma + Q_\Gamma) + \text{Ind }(gP_\Gamma + Q_\Gamma) + \text{Ind }(fP_\Gamma + Q_\Gamma) \, . \tag{2.5}$$

Now, Theorem 3.1 of Chapter 9 and Theorem 1.1 of Chapter 7 imply that

$$\begin{aligned} \text{Ind }(xP_\Gamma + Q_\Gamma) &= \text{Ind }(xP_{\Gamma_1} + Q_{\Gamma_1}) \mid L_p(\Gamma_1, \rho) \\ &= -\frac{1}{2\pi}[\arg x(t)]_{\Gamma_1} - \frac{1}{2\pi}[\arg(x_1(t_0+0)f_\delta(\mu) \\ &+ x_1(t_0-0)(1 - f_\delta(\mu)))]_{\mu=0}^{1} = 0 \, . \end{aligned} \tag{2.6}$$

Further, by Theorem 1.1,

$$\operatorname{Ind}(gP_\Gamma + Q_\Gamma) = -\frac{1}{2\pi}\sum_{j=1}^{n}[\arg g(t)]_{\Gamma_j} \tag{2.7}$$

and

$$\operatorname{Ind}(fP_\Gamma + Q_\Gamma) = -\frac{1}{2\pi}\sum_{j=1}^{n}[\arg f(t)]_{\Gamma_j}\ . \tag{2.8}$$

Combining the identities (2.5)-(2.8) one arrives at (2.3). ■

An analogous theorem holds for the operator $A = P_\Gamma aI + Q_\Gamma bI\ (a,b \in PC_0(\Gamma))$. This is a consequence of Theorem 6.1 of Chapter 7. However notice that the application of this requires that it be shown that

$$\operatorname{ess}\inf_{t\in\Gamma}|a(t)| > 0, \qquad \operatorname{ess}\inf_{t\in\Gamma}|b(t)| > 0 \tag{2.9}$$

whenever the operator $P_\Gamma aI + Q_\Gamma bI$ is a $\Phi-$ or $\Phi_\pm$–operator. But this can be verified in the same way as was done for the operator $aP_\Gamma + bQ_\Gamma$ in Theorem 2.1.

10.3 The main theorem

Let Γ be a non–simple curve, and denote by $PC(\Gamma)$ the class of all functions possessing finitely many discontinuities of the first kind on Γ . To each function $a \in PC(\Gamma)$ and to each point $t_0 \in \Gamma$ we associate two numbers $a(t_0{+}0)$ and $a(t_0{-}0)$ in the following way: Let t_0 be a point belonging to each of the n non–closed arcs $\gamma_1,\ldots,\gamma_n\ (n = 1,2,\ldots)$ but not being an interior point of any of these arcs. Further we suppose the arcs $\gamma_1,\ldots,\gamma_m\ (m \le n)$ to be directed to t_0 , and the arcs $\gamma_{m+1},\ldots,\gamma_n$ to be directed away from t_0 . For $0 < m < n$ we define

$$\begin{aligned} a(t_0 - 0) &:= a_1(t_0 - 0)\cdot\ldots\cdot a_m(t_0 - 0)\ , \\ a(t_0 + 0) &:= a_{m+1}(t_0 + 0)\cdot\ldots\cdot a_n(t_0 + 0)\ , \end{aligned} \tag{3.1}$$

with

$$a_j(t_0 - 0) = \lim_{t\to t_0, t\in\gamma_j} a(t) \quad (j = 1,\ldots,m)\ ,$$

$$a_j(t_0 + 0) = \lim_{t\to t_0, t\in\gamma_j} a(t) \quad (j = m+1,\ldots,n)\ .$$

If $m = 0$ then we put $a(t_0 - 0) = 1, a(t_0 + 0) = a_1(t_0 + 0)\cdot\ldots\cdot a_n(t_0 + 0)$, and in case $m = n$ we define $a(t_0 + 0) = 1$ and $a(t_0 - 0) = a_1(t_0 - 0)\cdot\ldots\cdot a_n(t_0 - 0)$. If, in particular, the point t_0 is the starting or end point of a non–closed subarc of the curve Γ then either $m = 0, n = 1$ or $m = n = 1$, and if t_0 is a non–singular point of the curve Γ

then $a(t_0+0)$ and $a(t_0-0)$ are nothing other than the left–sided and right–sided limits of the function a as $t \to t_0$. Let us agree upon the following notation. If γ is a simple arc which is divided into the subarcs $\gamma_1, \ldots, \gamma_q$ by the discontinuity points of the function $a(\in PC(\Gamma))$, then we denote the sum $\sum_{k=1}^{q} [\arg a(t)]_{\gamma_k}$ by $[\arg a]_\gamma$.

Theorem 3.1. *Let the non–simple curve Γ be the union of r simple arcs $\Gamma_1, \ldots, \Gamma_r$, let $t_1, \ldots, t_m$ denote all discontinuity points of the functions $a, b(\in PC(\Gamma))$, $t_{m+1}, \ldots, t_s$ all end points (or starting points) of the non–closed arcs of the curve Γ, and let $\rho(t) = \prod_{k=1}^{s} |t-t_k|^{\beta_k}$ $(-1 < \beta_k < p-1,\ k = 1, \ldots, s)$ and $\delta_k = 2\pi(1+\beta_k)/p$ $(1 < p < \infty)$.*

1. The operator $A = aP_\Gamma + bQ_\Gamma$ is a Φ– operator on $L_p(\Gamma, \rho)$ if and only if

$$\operatorname{ess\,inf}_{t\in\Gamma} |a(t)| > 0, \quad \operatorname{ess\,inf}_{t\in\Gamma} |b(t)| > 0, \tag{3.2}$$

and

$$c(t_k+0) f_{\delta_k}(\mu) + c(t_k-0)(1 - f_{\delta_k}(\mu)) \neq 0 \quad (0 \le \mu \le 1, k = 1, \ldots, s) \tag{3.3}$$

with $c := ab^{-1}$.

2. If the conditions (3.2) and (3.3) are satisfied and if

$$\kappa := \frac{1}{2\pi} \sum_{j=1}^{r} [\arg c(t)]_{\Gamma_j} + \frac{1}{2\pi} \sum_{k=1}^{s} [\arg(c(t_k+0) f_{\delta_k}(\mu) + c(t_k-0)(1 - f_{\delta_k}(\mu)))]_{\mu=0}^{1}, \tag{3.4}$$

then the operator A is invertible, invertible from the left or invertible from the right depending on whether the number κ is equal to zero, positive or negative, respectively. If $\kappa > 0$ then $\dim \operatorname{coker} A = \kappa$, *and in case $\kappa < 0$ one has* $\dim \ker A = -\kappa$.

3. If one of the conditions (3.2) and (3.3) is violated then A is neither a Φ_+– nor a Φ_-–operator.

Proof. The proof consists of five steps.

1. We prove the first part of the theorem in the special case when Γ is the composition of n simple non–closed arcs possessing exactly one common point, t_0, which is not an inner point of any of the arcs. Further we suppose the functions a and b to be equal to one at those end points (or starting points) of the non–closed arcs which are different from t_0. In this situation we complete the curve Γ to a curve $\widetilde{\Gamma}$ satisfying all assumptions of Section 10.1, and we extend the functions a and b onto $\widetilde{\Gamma} \backslash \Gamma$ by setting them to one there. One readily checks that, for the operator $\widetilde{a} P_{\widetilde{\Gamma}} + \widetilde{b} Q_{\widetilde{\Gamma}}$ acting on $L_p(\widetilde{\Gamma}, \rho)$, the conditions (3.2) and (3.3) as well as the formula (3.4) coincide with the conditions (2.1),(2.2), and with formula (2.3), respectively. This fact in combination with Theorem 1.1 of Chapter 7 yields the validity of the first part of Theorem 3.2 in the special case under consideration and proves, moreover, the formula Ind $A = -\kappa$.

2. Next we consider the case when Γ is a non–simple closed curve and the functions a and b are continuous at each point of the curve Γ except for the point t_0. Furthermore,

we let $\inf_{t\in\Gamma} |a(t)b(t)| > 0$. Define $c := a/b$, and let γ denote a part of the curve Γ which contains the point t_0 and which satisfies all assertions made in the previous step. The function c can be represented as $c = dg$ with $g \in C(\Gamma)$ and $d|(\Gamma\backslash\gamma) = 1$. Then, $A = b(dP_\Gamma + Q_\Gamma)(gP_\Gamma + Q_\Gamma) + T$, with T a compact operator. The operator $gP_\Gamma + Q_\Gamma$ is a Φ–operator since $g \in GC(\Gamma)$ and, by Theorem 1.1 of Chapter 7, the operator $dP_\Gamma + Q_\Gamma$ is a Φ– or $\Phi_\pm$–operator if and only if the operator $dP_\gamma + Q_\gamma$ has the same property. By the previous step, the condition (3.3) (which coincides for both operators $aP_\Gamma + bQ_\Gamma$ and $dP_\gamma + Q_\gamma$) is necessary and sufficient for the operator $aP_\Gamma + bQ_\Gamma$ to be a Φ–operator in the space $L_p(\Gamma, \rho)$. Since formula (3.4) holds for the operator $dP_\Gamma + Q_\Gamma$ and since $\operatorname{Ind}(gP_\Gamma + Q_\Gamma) = -(1/2\pi)[\arg g(t)]_\Gamma$ we conclude that

$$\operatorname{Ind} A = -\frac{1}{2\pi}[\arg(d|\gamma)^{p,\rho}(t_0, \mu)]_{\mu=0}^{1} - \frac{1}{2\pi}[\arg d(t)]_\gamma - \frac{1}{2\pi}[\arg g(t)]_\Gamma = -\kappa\,. \tag{3.5}$$

3. Our next goal is the proof of the first part of the theorem as well as of the formula $\operatorname{Ind} A = -\kappa$ in the general situation. Invoking Theorem 1.1 of Chapter 7 it is sufficient to confine ourselves to the case when Γ is a closed curve. Then one can show, as in the proof of Theorem 2.1, that the condition (3.2) is satisfied whenever the operator $aP_\Gamma + bQ_\Gamma$ is a Φ–operator.

Let $c = a/b$. By Proposition 3.1 of Chapter 7 we can rewrite the operator $cP_\Gamma + Q_\Gamma$ in the form

$$cP_\Gamma + Q_\Gamma = \prod_{j=1}^{s}(c^{(j)}P_\Gamma + Q_\Gamma) + T\,,$$

where T is a compact operator. Furthermore $c^{(0)} \in C(\Gamma)$, each of the functions $c^{(j)}$ $(j = 1, \ldots, s)$ is continuous on $\Gamma\backslash\{t_j\}$, and $c := c^{(0)}c^{(1)} \cdot \ldots \cdot c^{(s)}$. Then the preceding step yields both the validity of the theorem, as well as the formula $\operatorname{Ind} A = -\kappa$.

4. If the condition (3.2) is satisfied then one can show by repeating the arguments of the proof of Theorem 5.3 of Chapter 7 that at least one of the numbers $\dim\ker A$ or $\dim\ker A^*$ is equal to zero. Now the previous step implies the proof of the second assertion of the theorem.

5. As in Theorem 2.1 one can see that the operator A is never a Φ_+– or Φ_-–operator, which finishes the proof. ∎

An analogous result can be established for the operator $P_\Gamma aI + Q_\Gamma bI$ (see the remark at the end of the preceding section).

Let us finally mention that if the functions a and b in $PC(\Gamma)$ are subject to the conditions (3.2) and (3.3) and if Γ is a closed non–simple curve then the function $c := ab^{-1}$ admits a factorization $c := c_- t^\kappa c_+$ in $L_p(\Gamma, \rho)$. By means of the factors $c_\pm$ appearing in this factorization, one can derive formulas for the (possibly one–sided) inverse of the operator A and for a basis of the kernel or cokernel of A , and moreover, one also obtains solvability criteria for the equation $A\varphi = f$. The same arguments as in Section 9.4 show that these results carry over to general (not necessarily closed) non–simple curves, too.

We conclude this section with some examples.

1°. Let Γ denote Bernoulli's lemniscate [1] which is defined in polar coordinates by $\rho^2 = \cos 2\theta$. Define

$$a(t) = \begin{cases} \alpha & \text{if} \quad t \in \Gamma, \ \mathrm{Re}\, t > 0 \\ \beta & \text{if} \quad t \in \Gamma, \ \mathrm{Re}\, t < 0 \end{cases} \tag{3.6}$$

with complex numbers α and β , and let $A = aP_\Gamma + Q_\Gamma$ be defined on $L_p(\Gamma, \rho)$ with $1 < p < \infty$ and $\rho(t) = |t|^\gamma$ $(-1 < \gamma < p-1)$. The problem is to study the equation $A\varphi = f$ in $L_p(\Gamma, \rho)$.

Solution. The conditions $\alpha \neq 0,\ \beta \neq 0$ are necessary for the one–sided invertibility of the operator A (Theorem 1.1). In the course of the proof of Proposition 1.1 we have seen that $P_\Gamma g P_\Gamma = g P_\Gamma$ for each function g which is constant on every Γ_j . Thus, if $\alpha, \beta \neq 0$, then

$$(aP_\Gamma + Q_\Gamma)(a^{-1}P_\Gamma + Q_\Gamma) = (a^{-1}P_\Gamma + Q_\Gamma)(aP_\Gamma + Q_\Gamma) = I ,$$

and this shows that A is invertible and that $A^{-1} = a^{-1}P_\Gamma + Q_\Gamma$ in this case.

2°. Consider the same curve and the function

$$a(t) = \left(\frac{2t-1}{2t+1}\right)^\gamma \quad (\gamma \in \mathbf{C}) , \tag{3.7}$$

thought of as the restriction onto Γ of the function $a(z)$ which is analytic in the extended complex plane with a cut along the interval [-1/2, 1/2].

In Proposition 1.2 we have shown that the function a defined by (3.7) fulfills the identities $Q_\Gamma a Q_\Gamma = a Q_\Gamma$ and $Q_\Gamma a^{-1} Q_\Gamma = a^{-1} Q_\Gamma$. Thus, the operator $A = aP_\Gamma + Q_\Gamma$ is invertible on $L_p(\Gamma, \rho)$, and $A^{-1} = (\ P_\Gamma + aQ_\Gamma)a^{-1}I$.

3°. Consider the same curve once more, but now with the function

$$a(t) = \begin{cases} \alpha \frac{(2t-1)^\gamma}{(2t+1)^{\gamma+1}} & \text{if} \quad t \in \Gamma, \ \mathrm{Re}\, t > 0 \\ \beta \frac{(2t-1)^\gamma}{(2t+1)^{\gamma+1}} & \text{if} \quad t \in \Gamma, \ \mathrm{Re}\, t < 0 . \end{cases} \tag{3.8}$$

It is necessary for the at least one–sided invertibility of the operator $A = aP_\Gamma + Q_\Gamma$ that both α and β be nonzero. If these conditions are satisfied then the function $a(t)$ admits the factorization

$$a(t) = a_-(t)(2t+1)^{-1}a_+(t) \tag{3.9}$$

with

$$a_-(t) = \left(\frac{2t-1}{2t+1}\right)^\gamma \tag{3.10}$$

[1]The orientation of the curve is in accordance with the increase of the parameter θ from $-\frac{\pi}{4}$ to $\frac{\pi}{4}$ and from $\frac{3\pi}{4}$ to $\frac{5\pi}{4}$.

and

$$a_+(t) = \begin{cases} \alpha & \text{if} \quad t \in \Gamma,\ \mathrm{Re}\, t > 0 \\ \beta & \text{if} \quad t \in \Gamma,\ \mathrm{Re}\, t < 0\,. \end{cases} \tag{3.11}$$

Since $\operatorname{ind}(2t+1) = -1$, the operator $A = aP_\Gamma + Q_\Gamma$ is invertible from the right, $\ker A = \operatorname{span}\{a_+^{-1}(1-a)\}$, and the operator

$$R = a_-(a^{-1}P_\Gamma + Q_\Gamma)a_-^{-1}I$$

is one of the right inverses of A.

4°. Finally, we consider the function

$$a(t) = \begin{cases} \alpha\frac{(2t-1)^{\gamma+1}}{(2t+1)^\gamma} & \text{if} \quad t \in \Gamma,\ \mathrm{Re}\, t > 0 \\ \beta\frac{(2t-1)^{\gamma+1}}{(2t+1)^\gamma} & \text{if} \quad t \in \Gamma,\ \mathrm{Re}\, t < 0\,. \end{cases} \tag{3.12}$$

on the curve Γ introduced in the first example.

In case $\alpha, \beta \neq 0$ the function a is factorizable into

$$a(t) = a_-(t)(2t-1)a_+(t)$$

with the functions $a_-(t)$ and $a_+(t)$ being defined by (3.10) and (3.11). Since $\operatorname{ind}(2t-1) = 1$, the operator $A = aP_\Gamma + Q_\Gamma$ is left invertible,

$$R = a_-(a^{-1}P_\Gamma + Q_\Gamma)a_-^{-1}I$$

is one of its left inverses, and the equation $A\varphi = f$ is solvable if and only if

$$\int_\Gamma t^{-1}f(t)a_-^{-1}(t)\, dt = 0 \tag{3.13}$$

The latter fact is a consequence of condition (4.5) of Chapter 7.

10.4 Exercises

10.1. Let Γ be the closed curve which is defined in polar coordinates by $\rho = 2\cos 3\theta - 1$. Study the equation $(t^\gamma P_\Gamma + Q_\Gamma)\varphi = f \quad (\gamma \in \mathbf{C})$ in the space $L_p(\Gamma)$ $(1 < p < \infty)$, that is, characterize the invertibility of the operator $A = t^\gamma P_\Gamma + Q_\Gamma$, determine $\dim\ker A$, $\dim\operatorname{coker} A$ and $\ker A$, and establish criteria for the solvability of the equation $A\varphi = f$.

10.2. Let $\Gamma = [-1,1] \cup [-i,i]$ and

$$a(t) = \begin{cases} \alpha & \text{if} \quad t \in [-1,0) \\ \beta & \text{if} \quad t \in (0,1] \\ \gamma & \text{if} \quad t \in [-i,0) \\ \delta & \text{if} \quad t \in (0,i]\,. \end{cases}$$

Describe conditions for the one-sided invertibility of the operator $A = aP_\Gamma + Q_\Gamma$ on the space $L_p(\Gamma)$. Do these conditions depend on the orientation of Γ ?

10.3. Solve the equation $(aP_\Gamma + Q_\Gamma)\varphi = 1$ for the functions a defined by (3.6), (3.7), (3.8) and (3.12) with γ replaced by 1 .

Comments and references

For the theory of singular integral equations with piecewise Hölder continuous coefficients we refer to the monograph MUSKKHELISHVILI [1]. The results of Chapter 10 are taken from GOHBERG and KRUPNIK'S paper [7].

Chapter 11

Singular integral operators with coefficients having discontinuities of almost periodic type

This chapter is devoted to the inversion of singular integral operators with bounded coefficients possessing discontinuities of the second kind of a special type.

The chapter is divided into five sections the first two of which are in preparation for the third one, which contains the main result. In the fourth section, which can be thought of as a completion of Section 3.12, we study singular integral operators with continuous coefficients vanishing at some points. Section 11.5 contains some exercises.

11.1 Almost periodic functions and their factorization

An *almost periodic polynomial* is a function $p(\lambda)(-\infty < \lambda < \infty)$ which is representable as a linear combination of functions of the form $\exp(i\delta\lambda)$ with $\delta \in \mathbf{R}$.

By Π_c we denote the closure in $L_\infty(\mathbf{R})$ of the set of all almost periodic polynomials. The class Π_c is a subalgebra of $L_\infty(\mathbf{R})$, and its elements are just the functions which are almost periodic in the sense of Bohr (see FAVARD [1]). Let Π_w stand for the subset of Π_c consisting of all functions $a \in \Pi_c$ of the form

$$a(\lambda) = \sum_{j=1}^{\infty} a_j e^{i\delta_j\lambda} \quad (-\infty < \lambda < \infty) \tag{1.1}$$

which are subject to the condition

$$\|a\|_w := \sum_{j=1}^{\infty} |a_j| < \infty .$$

Clearly, Π_w is a Banach algebra under the norm $\|.\|_w$.

A function $a \in \Pi_c$ is called *non-degenerated* if

$$\operatorname{ess} \inf_{-\infty<\lambda<\infty} |a(\lambda)| > 0 .$$

To each non-degenerated function $a \in \Pi_c$ we associate the number $\operatorname{ind}_\Pi a$ defined by

$$\operatorname{ind}_\Pi a := \lim_{l\to\infty} \frac{1}{2l} [\arg a(\lambda)]_{-l}^{l} .$$

The existence of the limit in this equality is a standard attribute of the almost periodic functions (see LEVITAN [1], p. 129 Theorem 2.7.1). The real number $\operatorname{ind}_\Pi a$ is referred to as the *almost periodic index of the function* a. The almost periodic index $\operatorname{ind}_\Pi a$ is stable under small perturbations of the function a in the algebra Π_c. Furthermore it is easy to see that $\operatorname{ind}_\Pi \exp(i\delta\lambda) = \delta$.

It is well known (consult GELFAND, RAIKOV, SHILOV [1]) that if $a \in \Pi_w$ is a non-degenerated function, then the function a^{-1} is also contained in the algebra Π_w. An analogous assertion holds for the algebra Π_c.

Let Π_w^+ (Π_w^-) stand for the subalgebra of the algebra Π_w which consists of all functions of the form (1.1) with all numbers δ_j being non-negative (non-positive). Evidently, the functions in Π_w^+ (Π_w^-) admit analytic extensions into the upper (lower) half plane. Finally we mention that if $a \in \Pi_w^+$ (Π_w^-) and if

$$\inf_{\operatorname{Im}\lambda\geq 0} |a(\lambda)| > 0 \quad \left(\inf_{\operatorname{Im}\lambda\leq 0} |a(\lambda)| > 0\right)$$

then the function a^{-1} belongs to Π_w^+ (Π_w^-).

Theorem 1.1. *Each non-degenerated function $a \in \Pi_w$ is factorizable into the product*

$$a(\lambda) = a_-(\lambda) e^{i\lambda\nu} a_+(\lambda) \quad (-\infty < \lambda < \infty) ,$$

where $a_+ \in \Pi_w^+$, $a_- \in \Pi_w^-$, $\nu = \operatorname{ind}_\Pi a$, *and*

$$\inf_{\operatorname{Im}\lambda\geq 0} |a_+(\lambda)| > 0 , \qquad \inf_{\operatorname{Im}\lambda\leq 0} |a_-(\lambda)| > 0 .$$

Proof. Let $a \in \Pi_w$ and $\nu = \operatorname{ind}_\Pi a$. Then the almost periodic index of the function $b := a\exp(-i\nu\lambda)$ is equal to zero and, by the well known generalization of the WIENER-LEVI theorem (see GELFAND, RAIKOV, SHILOV[1] and SHILOV [1]), the function $c = \log b$ belongs to the algebra Π_w. Thus, the function c can be written as

$$c(\lambda) = \sum_{j=1}^{\infty} c_j \exp(i\gamma_j\lambda) \quad \left(\sum_{j=1}^{\infty} |c_j| < \infty\right) .$$

Define

$$c_+(\lambda) := \sum_{\gamma_j \geq 0} c_j \exp(i\gamma_j \lambda), \quad c_-(\lambda) := \sum_{\gamma_j < 0} c_j \exp(i\gamma_j \lambda) .$$

Then

$$a(\lambda) = a_-(\lambda) e^{i\nu\lambda} a_+(\lambda) ,$$

where $a_\pm(\lambda) = \exp c_\pm(\lambda)$, and the conditions (1.3) are obviously satisfied.

It follows from inequality (1.2) that $\nu = \mathrm{ind}_\pi a$. ■

11.2 Lemmas on functions with discontinuities of almost periodic type

Let Γ be a closed composed curve consisting of the simple closed curves $\Gamma_1, \ldots, \Gamma_m$, and let t_0 be an arbitrary point on Γ . For definiteness we suppose that $t_0 \in \Gamma_k$.

Furthermore, let ω_0 denote a continuous function mapping the curve $\Gamma_k \backslash \{t_0\}$ one-to-one onto the real axis in such a manner that

$$\lim_{t \succ t_0, t \to t_0} \omega_0(t) = -\infty \quad \text{and} \quad \lim_{t \prec t_0, t \to t_0} \omega_0(t) = \infty ,$$

and let p refer to a certain almost periodic function in Π_c .

We say that t_0 is a *discontinuity point of almost periodic type of a function* $a \in L_\infty(\Gamma)$ *with characteristics* $\{p, \omega_0\}$ if [1]

$$\lim_{t \to t_0} (a(t) - p(\omega_0(t))) = 0 . \tag{2.1}$$

Notice that this definition implies that

$$\sup_{-\infty < \lambda < \infty} |p(\lambda)| \leq \sup_{t \in \Gamma_k} |a(t)| \tag{2.2}$$

and

$$\inf_{-\infty < \lambda < \infty} |p(\lambda)| \geq \inf_{t \in \Gamma_k} |a(t)|. \tag{2.3}$$

Indeed, these estimates are consequences of (2.1) and of the following simple property of the almost periodic function $p(\lambda)$: Given $\varepsilon > 0$ then, in each neighbourhood of the point at infinity, one can find numbers λ_1 and λ_2 satisfying

$$|p(\lambda_1)| \geq \sup_{-\infty < \lambda < \infty} |p(\lambda)| - \varepsilon ,$$

[1] Observe that this definition only relates to the values of the function ω_0 on a sufficiently small neighbourhood of the point t_0 .

$$|p(\lambda_2)| \le \inf_{-\infty<\lambda<\infty} |p(\lambda)| + \varepsilon .$$

Also note that the estimates (2.2) and (2.3) remain valid if the $\sup|a(t)|$ and the $\inf|a(t)|$ are merely taken over an arbitrary neighbourhood of the point t_0.

A function $a \in L_\infty(\Gamma)$ is said to be *non–degenerate* if

$$\text{ess} \inf_{t\in\Gamma} |a(t)| > 0 . \tag{2.4}$$

Lemma 2.1. *Let $a \in L_\infty(\Gamma)$ be a non–degenerate function which is continuous on $\Gamma\backslash\{t_1,\dots,t_n\}$, and let $t_1,\dots,t_n$ be its discontinuity points of almost periodic type with the associated characteristics $\{p_j,\omega_j\}$ $(j = 1,\dots,n)$. Then the function a can be represented in the form*

$$a(t) = p_1(\omega_1(t))p_2(\omega_2(t))\cdot\ldots\cdot p_n(\omega_n(t))a_c(t) \tag{2.5}$$

with $a_c \in C(\Gamma)$.

Proof. The functions p_j $(j = 1,\dots,n)$ are non–degenerate due to estimate (2.3). We define the function a_c by

$$a_c(t) := \frac{a(t)}{p_1(\omega_1(t))p_2(\omega_2(t))\cdot\ldots\cdot p_n(\omega_n(t))} ,$$

and show that the function a_c is continuous. Indeed, since the functions p_j $(j = 1,\dots,n)$ are non–degenerate we have $p_j^{-1} \in \Pi_c$, and the equality $a(t)p_j^{-1}(\omega_j(t)) = 1 + p_j^{-1}(\omega_j(t))(a(t) - p_j(\omega_j(t)))$ implies that the function $a(t)p_j^{-1}(\omega_j(t))$ is continuous at the point t_j. ■

To each function a satisfying the conditions of Lemma 2.1 we associate the $(n+1)$ numbers

$$\text{ind}\,(a,t_j) := \text{ind}_\Pi\, p_j \quad (j = 1,2,\dots,n)$$

and

$$\text{ind}\, a := \text{ind}\, a_c ,$$

which will be called the *indices of the function* a. All numbers $\text{ind}\,(a,t_j)$ are real whereas $\text{ind}\, a$ is an integer. The numbers $\text{ind}\,(a,t_j)$ can be computed via the formula

$$\text{ind}\,(a,t_j) = \lim_{\left(\substack{t'\to t_j,\, t''\to t_j \\ t''\prec t_j \prec t'}\right)} \frac{1}{\omega_j(t'') - \omega_j(t')}[\arg a(t)]_{t'}^{t''} . \tag{2.6}$$

Indeed,

$$\frac{1}{\omega_j(t'') - \omega_j(t')}\,[\arg p_j(\omega_j(t))]_{t'}^{t''} = \frac{1}{l_2 - l_1}\,[\arg p_j(\lambda)]_{l_1}^{l_2}$$

with $l_1 = \omega_j(t') < 0$ and $l_2 = \omega_j(t'') > 0$, and consequently,

$$\lim_{\substack{t'\to t_j, t''\to t_j \\ t''\prec t_j \prec t'}} \frac{1}{\omega_j(t'') - \omega_j(t')} [\arg p_j(\omega_j(t))]_{t'}^{t''} = \mathrm{ind}_\Pi\, p_j \, . \tag{2.7}$$

By equality (2.5), the function a can be written as $a = p_j b_j$ where b_j is a function which is continuous on a neighbourhood of the point t_j and which does not vanish on Γ . This implies that the left hand side of (2.6) coincides with the right hand side of (2.7).

Let $t_1, t_2, \ldots, t_m$ be different points on the curve Γ and ω_j $(j = 1, 2, \ldots, m)$ be a function satisfying the condition (2.0) at the point t_j . By $\mathcal{A}(t_1, \ldots, t_m)$ we denote the collection of all continuous on $\Gamma \backslash \{t_1, \ldots, t_m\}$ functions f having at $t_1, \ldots, t_m$ discontinuities of almost periodic type with characteristics of the form $\{p_j, \omega_j\}$ where p_j belongs to Π_c .

Lemma 2.2. *The set $\mathcal{A}(t_1, t_2, \ldots, t_m)$ is a subalgebra of the algebra $L_\infty(\Gamma)$. Each function $a \in \mathcal{A}(t_1, t_2, \ldots, t_m)$ admits a representation of the form*

$$a(t) = \sum_{j=1}^{m} p_j(\omega_j(t)) + b(t) \quad (t \in \Gamma) \tag{2.8}$$

with $b \in C(\Gamma)$ and $p_j \in \Pi_c$.

The linear hull of all rational functions in $R(\Gamma)$ and of all functions of the form $\exp(\nu\omega_j(t))$ $(j = 1, 2, \ldots, m; -\infty < \nu < \infty)$ is a dense subset of $\mathcal{A}(t_1, t_2, \ldots, t_m)$.

Proof. Clearly, the set $\mathcal{A} = \mathcal{A}(t_1, t_2, \ldots, t_m)$ forms an algebra under the usual operations with functions. We claim that this algebra is closed. For let $\{a_j\}_{j=1}^\infty$ be a Cauchy sequence of functions in $\mathcal{A}$. This sequence converges in the norm of $L_\infty(\Gamma)$ to a function $a \in L_\infty(\Gamma)$ which can only be discontinuous at the points $t_1, t_2, \ldots, t_m$. Let t_j be a discontinuity point of the function a , and denote by $\{p_j^{(k)}, \omega_j\}$ the characteristics of the function a_k at the point $t_j (k = 1, 2, \ldots, \ j = 1, \ldots, m)$. The estimate (2.2) shows that $\{p_j^{(k)}\}_{k=1}^\infty$ $(j = 1, \ldots, m)$ is a Cauchy sequence. Denote the limit in $L_\infty(\Gamma)$ by p_j $(j = 1, \ldots, m)$. Now it is easy to see that the function a admits a discontinuity of almost periodic type with characteristics $\{p_j, \omega_j\}$ at the point t_j $(j = 1, \ldots, m)$; that is, $a \in \mathcal{A}$. Furthermore, one immediately checks that $a - p_1 - p_2 - \ldots - p_m \in C(\Gamma)$. Thus, equality (2.8) holds, which implies the last assertion of the lemma. ■

Let us agree upon to write $\mathcal{A}_0(t_1, t_2, \ldots, t_m)$ instead of $\mathcal{A}(t_1, t_2, \ldots, t_m)$ in case the curve Γ is the unit circle and the function ω_j is defined by

$$\omega_j(t) = -i\frac{t + t_j}{t - t_j} \quad (j = 1, 2, \ldots, m) \, .$$

One readily verifies that

$$\text{ind}\,(a, t_j) = \frac{1}{4} \lim_{\theta \searrow 0} \theta[\arg a(e^{i\varphi})]_{\varphi=\theta_j+\theta}^{\theta_j+2\pi-\theta} \quad (t_j = e^{i\theta_j})$$

whenever $a \in \mathcal{A}_0(t_1, t_2, \ldots, t_m)$.

11.3 The main theorem

Throughout this and the following section we assume the functions ω_j appearing in the definition [1] of the algebra $\mathcal{A}(t_1, t_2, \ldots, t_m)$ map the curve Γ one–to–one onto the real axis, are differentiable on Γ , possess derivatives $d\omega_j/dt$ which are non–vanishing on Γ , and belong to $H_\mu(\Gamma)$ for some $0 < \mu < 1$.

Theorem 3.1. *Let* $t_1, t_2, \ldots, t_m$ *be different points on the closed composed curve* Γ *and* $a, b \in \mathcal{A}(t_1, t_2, \ldots, t_m)$. *Then the operator* $aP_\Gamma + bQ_\Gamma$ *is a* Φ*– or* $\Phi_\pm$*–operator on the space* $L_p(\Gamma, \rho)$ [2] *if and only if*

1) $\inf_{t\in\Gamma} |a(t)| > 0$ *and* $\inf_{t\in\Gamma} |b(t)| > 0$,

2) the numbers $\text{ind}\,(ab^{-1}, t_j)$ $(j = 1, \ldots, m)$ *are all non–negative or all non–positive.*

If the condition 1) is satisfied and if the numbers $\text{ind}\,(ab^{-1}, t_j)$ $(j = 1, 2, \ldots, m)$ *are non–negative and not all equal to zero, then the operator* $aP_\Gamma + bQ_\Gamma$ *is left–invertible, and*

$$\dim \text{coker}\,(aP_\Gamma + bQ_\Gamma) = \infty \ .$$

If the condition 1) is satisfied and if the numbers $\text{ind}\,(ab^{-1}, t_j)$ $(j = 1, 2, \ldots, m)$ *are non–positive and not all equal to zero then the operator* $aP_\Gamma + bQ_\Gamma$ *is right–invertible, and*

$$\dim \ker\,(aP_\Gamma + bQ_\Gamma) = \infty \ .$$

Finally, if the condition 1) is satisfied and $\text{ind}\,(ab^{-1}, t_1) = \text{ind}\,(ab^{-1}, t_2) = \ldots = \text{ind}\,(ab^{-1}, t_m) = 0$, *then the operator* $aP_\Gamma + bQ_\Gamma$ *is left invertible, right invertible, or two–sided invertible depending on whether the number* $\text{ind}\,(ab^{-1})$ *is positive, negative, or equal to zero, respectively. In this situation,*

$$\dim \text{coker}\,(aP_\Gamma + bQ_\Gamma) = \text{ind}\,(ab^{-1}) \quad \textit{if} \quad \text{ind}\,(ab^{-1}) > 0$$

and

$$\dim \ker\,(aP_\Gamma + bQ_\Gamma) = -\text{ind}\,(ab^{-1}) \quad \textit{if} \quad \text{ind}\,(ab^{-1}) < 0 \ .$$

In order to prepare for the proof of this theorem, we formulate two lemmas.

---

[1] See the preceding section for this definition.

[2] In this and the following section we suppose all weights $\rho(t)$ are of the form $\rho(t) := \prod_{j=1}^{r} |t - t_j|^{\beta_j}$ $(r \geq m)$ and that $-1 < \beta_j < p - 1$ $(1 < p < \infty, j = 1, 2, \ldots, r)$.

Lemma 3.1. *Let* a *be a function of the form*

$$a(t) = t^\kappa \exp\left(\nu \frac{t+t_0}{t-t_0}\right) \qquad (|t| = 1)\ ,$$

where $|t_0| = 1$, ν *is a real number, and* κ *is an integer. Then, in case* $\nu > 0$, *the operator* $aP_{\mathbf{T}}+Q_{\mathbf{T}}$ *is left-invertible on the space* $L_p(\mathbf{T},\rho)$ *and* $\dim \operatorname{coker}(aP_{\mathbf{T}}+Q_{\mathbf{T}}) = \infty$, *whereas in case* $\nu < 0$, *the operator* $aP_{\mathbf{T}}+Q_{\mathbf{T}}$ *is right invertible and* $\dim \ker(aP_{\mathbf{T}}+Q_{\mathbf{T}}) = \infty$.

Proof. The operator $aP_{\mathbf{T}}+Q_{\mathbf{T}}$ can be written as $aP_{\mathbf{T}}+Q_{\mathbf{T}} = (P_{\mathbf{T}}aP_{\mathbf{T}}+Q_{\mathbf{T}})(I+Q_{\mathbf{T}}aP_{\mathbf{T}})$. Since the operator $I+Q_{\mathbf{T}}aP_{\mathbf{T}}$ is invertible with

$$(I+Q_{\mathbf{T}}aP_{\mathbf{T}})^{-1} = I - Q_{\mathbf{T}}aP_{\mathbf{T}}\ ,$$

it suffices to prove the theorem for the operator $B_a := P_{\mathbf{T}}aI|L_p^+(\mathbf{T},\rho)$.

First we show that if $\nu < 0$, then the operator $X := B_aB_{a^{-1}}$ is invertible in $L_p^+(\mathbf{T},\rho)$. In case $\kappa \le 0$ we have $a^{-1} \in L_\infty^+(\mathbf{T})$. Thus, the operator $B_{a^{-1}}$ is a right inverse of B_a , that is, $X = I$. Let us now consider the case $\kappa > 0$. In this situation the operator B_a can be factorized into $B_a = D_\nu U_\kappa$ with

$$U_\kappa := P_{\mathbf{T}}t^\kappa I|L_p^+(\mathbf{T},\rho) \quad \text{and} \quad D_\nu := P_{\mathbf{T}}\exp\left(\nu\frac{t+t_0}{t-t_0}\right) I|L_p^+(\mathbf{T},\rho)\,.$$

Hence, $X = D_\nu U_\kappa U_{-\kappa} D_{-\nu}$, and taking into account that $U_\kappa U_{-\kappa} = I - R_\kappa$, where R_κ refers to the projection operator defined by

$$R_\kappa\left(\sum_{j=0}^{\infty} a_j t^j\right) := \sum_{j=0}^{\kappa-1} a_j t^j\ ,$$

one immediately gets

$$X = I - D_\nu R_\kappa D_{-\nu}\ .$$

The intersection of the spaces $\operatorname{im} D_{-\nu}$, and $\operatorname{im} R_\kappa$ consists of the zero vector only. Indeed, let $D_{-\nu}\varphi = \eta$ with $\eta = R_\kappa \eta$. Then, obviously, $\varphi = D_\nu R_\kappa \eta$, and the equality

$$D_\nu R_\kappa = R_\kappa D_\nu R_\kappa\ , \tag{3.1}$$

which is easily verified, implies that $\varphi = R_\kappa D_\nu R_\kappa \eta = R_\kappa \varphi$. Hence, the functions φ and η are polynomials in t of degree less than or equal to $\kappa - 1$. On the other hand, the equality $D_\nu \varphi = \eta$ means that

$$\exp\left(-\nu\frac{t+t_0}{t-t_0}\right)\varphi(t) = \eta(t)\ ,$$

which is only possible if $\varphi = \eta = 0$.

Because of $\|D_\nu\| = \|D_{-\nu}\| = 1$, and since the finite–dimensional projection R_κ is orthogonal in $L_2^+(\mathbf{T})$ the preceding assertion implies that $\|D_\nu R_\kappa D_{-\nu}\| \le \|R_\kappa D_{-\nu}\| < 1$. Thus, the operator $X = I - D_\nu R_\kappa D_{-\nu}$ is invertible in both $L_2^+(\mathbf{T})$ and $L_p^+(\mathbf{T},\rho)$. The latter can be explained as follows: In the space $L_p^+(\mathbf{T},\rho)$ one has $\ker X \subseteq \operatorname{im} R_\kappa$. Indeed, if $X\varphi = 0$ then $\varphi = D_\nu R_\kappa D_{-\nu}\varphi$, and so via (3.1) it follows that $\varphi = R_\kappa\varphi$. Thus, the subspace $\ker X$ is identical for each of the spaces $L_p^+(\mathbf{T},\rho)$, and since $\ker X = \{0\}$ in $L_2^+(\mathbf{T})$ we conclude that $\ker X = \{0\}$ for all spaces $L_p^+(\mathbf{T},\rho)$. This shows that the operator $I - D_\nu R_\kappa D_{-\nu}$ is invertible in all spaces $L_p^+(\mathbf{T},\rho)$ since the projection R_κ is finite dimensional.

In case $\nu < 0$ the operator B_a is right invertible, and one of its right inverses is $B_{a^{-1}}(B_a B_{a^{-1}})^{-1}$. The kernel $\ker D_\nu$ is infinite dimensional since it contains all functions of the form

$$\exp\left(\mu\frac{t+t_0}{t-t_0}\right) + \alpha_\mu \exp\left(-\nu\frac{t+t_0}{t-t_0}\right)$$

with $0 \le \mu \le -\nu$ and $\alpha_\mu := -\exp(\mu+\nu)$. This observation, in combination with the equality $B_a = U_\kappa D_\nu$, yields that $\dim\ker B_a = \infty$ if $\kappa \le 0$.

If $\kappa > 0$ then $\dim\operatorname{coker} U_\kappa < \infty$, and hence $\dim(\ker D_\nu \cap \operatorname{im} U_\kappa) = \infty$. By the equality $B_a = D_\nu U_\kappa$, we have $\dim\ker B_a = 0$ in case $\kappa > 0$, too.

Finally, if $\nu > 0$ then the operator B_a is left invertible on the space $L_p(\mathbf{T},\rho)$, and one of its left–inverses is given by $(B_{a^{-1}}B_a)^{-1}B_{a^{-1}}$. Since $\dim\ker B_{a^{-1}} = \infty$ we find that $\dim\operatorname{coker} B_a = \infty$. ∎

Let us present another, shorter proof of Lemma 3.1 which is based on the results of the preceding section.

Second proof. To start with let $\kappa = 0$. If $\nu > 0$ then $a \in L_\infty^+$ and, thus,

$$(a^{-1}P_{\mathbf{T}} + Q_{\mathbf{T}})(aP_{\mathbf{T}} + Q_{\mathbf{T}}) = I\,.$$

Define

$$\psi_\mu(t) := \exp\left(\mu\frac{t+t_0}{t-t_0}\right) - [\exp(\nu-\mu)]\exp\left(\nu\frac{t+t_0}{t-t_0}\right),$$

with $0 \le \mu \le \nu$. It is immediate that $(P_{\mathbf{T}}\overline{a} + Q_{\mathbf{T}})\psi_\mu = 0$ and consequently $\psi_\mu \in \ker(aP_{\mathbf{T}} + Q_{\mathbf{T}})^*$. Thus, the operator $A_0 = aP_{\mathbf{T}} + Q_{\mathbf{T}}$ is left–invertible if $\nu > 0$ and $\kappa = 0$, and $\dim\ker A_0^* = \infty$.

If $\nu < 0$ and $\kappa = 0$ then

$$(aP_{\mathbf{T}} + Q_{\mathbf{T}})(a^{-1}P_{\mathbf{T}} + Q_{\mathbf{T}}) = 0\,,$$

and the functions φ_μ defined by

$$\varphi_\mu(t) := \exp\left(\mu\frac{t+t_0}{t-t_0}\right) - \exp\left((\mu+\nu)\frac{t+t_0}{t-t_0}\right) + \exp(-\mu-\nu)$$

with $0 < \mu < -\nu$ belong to the kernel of A_0. Thus, the operator A_0 is right invertible, and $\dim\ker A_0 = \infty$ in this case.

Now suppose $\kappa \neq 0$. Then one can write the operator A_0 in the form

$$A_0 = (at^{-\kappa}P_{\mathbf{T}} + Q_{\mathbf{T}})(t^{\kappa}P_{\mathbf{T}} + Q_{\mathbf{T}}) + T$$

with T a compact operator. The operator $t^{\kappa}P_{\mathbf{T}} + Q_{\mathbf{T}}$ is a Φ–operator, and by what we have shown above, the operator A_0 admits a left regularization (right regularization) with $\dim\operatorname{coker} A_0 = \infty$ $(\dim\ker A_0 = \infty)$ in case $\nu > 0$ $(\nu < 0)$. Furthermore, Theorem 5.1 in Chapter 7 entails that $\dim\ker A_0 = 0$ (respectively $\dim\operatorname{coker} A_0 = 0$), and from Theorem 5.1 in [GK 1], Chapter 2 (Theorem 5.2 in [GK 1], Chapter 2) we conclude that the operator A_0 is left invertible (respectively right invertible).

Lemma 3.2. *Let $a \in \mathcal{A}_0(t_1, t_2, \ldots, t_m)$ satisfy the condition*

$$\inf_{|t|=1} |a(t)| > 0\,.$$

If the numbers $\operatorname{ind}(a, t_j)$ $(j = 1, 2, \ldots, m)$ *are non–negative and not all equal to zero then the operator $aP_{\mathbf{T}} + Q_{\mathbf{T}}$ is left invertible, and* $\dim\operatorname{coker}(aP_{\mathbf{T}} + Q_{\mathbf{T}}) = \infty$. *If the numbers* $\operatorname{ind}(a, t_j)$ $(j = 1, 2, \ldots, m)$ *are non–positive and not all equal to zero then the operator $aP_{\mathbf{T}} + Q_{\mathbf{T}}$ is right–invertible, and* $\dim\ker(aP_{\mathbf{T}} + Q_{\mathbf{T}}) = \infty$. *If* $\operatorname{ind}(a, t_1) = \operatorname{ind}(a, t_2) = \ldots = \operatorname{ind}(a, t_m) = 0$ *then the operator $aP_{\mathbf{T}} + Q_{\mathbf{T}}$ is left invertible, right invertible, or two–sided invertible depending on whether the number* $\operatorname{ind} a$ *is positive, negative, or equal to zero. Moreover, in this situation,* $\operatorname{Ind}(aP_{\mathbf{T}} + Q_{\mathbf{T}}) = -\operatorname{ind} a$.

Proof. Due to Lemma 2.1 we can represent the function a in the form

$$a(t) = p_1\left(-i\frac{t+t_1}{t-t_1}\right)p_2\left(-i\frac{t+t_2}{t-t_2}\right)\cdot\ldots\cdot p_m\left(-i\frac{t+t_m}{t-t_m}\right)b(t),$$

where the p_j are non–degenerated functions in Π_c and $b \in GC(\Gamma)$. Approximating sufficiently closely each of the functions p_j $(j = 1, \ldots, m)$ by a function $q_j \in \Pi_w$ and the function b by a function $b_0 \in W$, we arrive at

$$a(t) = q_1\left(-i\frac{t+t_1}{t-t_1}\right)q_2\left(-i\frac{t+t_2}{t-t_2}\right)\cdot\ldots\cdot q_m\left(-i\frac{t+t_m}{t-t_m}\right)b_0(t)(1 + m(t))\,,$$

with $m \in L_\infty(\mathbf{T})$ a function satisfying $\sup|m(t)| < 1$. By Theorem 1.1, the function q_j $(j = 1, \ldots, m)$ is factorizable into

$$q_j(\lambda) = q_j^-(\lambda)e^{i\nu_j\lambda}q_j^+(\lambda) \qquad (-\infty < \lambda < \infty)$$

with $q_j^\pm \in L_\infty^\pm(\mathbf{T})$ and $1/q_j^\pm \in L_\infty^\pm(\mathbf{T})$ and, by Corollary 3.2 of Chapter 3, the function b_0 factors into $b_0 = b_- t^{\kappa} b_+$. Thus, the function a admits the following factorization with respect to the curve $\mathbf{T}$:

$$a(t) = a_-(t)t^\kappa \prod_{j=1}^{m} \exp\left(\nu_j \frac{t+t_j}{t-t_j}\right)(1+n(t))a_+(t) ,$$

where

$$a_\pm(t) := \prod_{j=1}^{m} q_j^\pm \left(-i\frac{t+t_j}{t-t_j}\right) b_\pm(t) ,$$

and $n \in L_\infty(\mathbf{T})$ is a function with $\sup |n(t)|$ $(t \in \mathbf{T})$ sufficiently small. Obviously, $a_\pm \in L_\infty^\pm(\mathbf{T})$ and $1/a_\pm \in L_\infty^\pm(\mathbf{T})$, and moreover, $\nu_j = \text{ind}\,(a,t_j)$ $(j = 1,\ldots,m)$ and $\kappa = \text{ind}\, a$. Define

$$\psi(t) := t^\kappa \prod_{j=1}^{m} \exp\left(\nu_j \frac{t+t_j}{t-t_j}\right) .$$

Then the operator $aP_\mathbf{T} + Q_\mathbf{T}$ can be written as

$$aP_\mathbf{T} + Q_\mathbf{T} = a_-(\psi(1+n)P_\mathbf{T} + Q_\mathbf{T})(a_+P_\mathbf{T} + a_-^{-1}Q_\mathbf{T}) .$$

Here the operators a_-I and $a_+P_\mathbf{T} + a_-^{-1}Q_\mathbf{T}$ are invertible:

$$(a_-I)^{-1} = a_-^{-1}I \quad \text{and} \quad (a_+P_\mathbf{T} + a_-^{-1}Q_\mathbf{T})^{-1} = a_+^{-1}P_\mathbf{T} + a_-Q_\mathbf{T} .$$

Now suppose all numbers ν_j $(j = 1,2,\ldots,m)$ to be non-negative and $\nu_1 > 0$. Then

$$\begin{aligned}\psi(1+n)P_\mathbf{T} + Q_\mathbf{T} &= \left(t^\kappa \exp\left(\nu_1\frac{t+t_1}{t-t_1}\right)P_\mathbf{T} + Q_\mathbf{T}\right)\left(\exp\left(\nu_2\frac{t+t_2}{t-t_2}\right)P_\mathbf{T} + Q_\mathbf{T}\right)\cdot \\ &\quad \ldots\cdot\left(\exp\left(\nu_m\frac{t+t_m}{t-t_m}\right)P_\mathbf{T} + Q_\mathbf{T}\right) + \psi n P_\mathbf{T} .\end{aligned}$$

Following Lemma 3.1, the first term on the right–hand side of this equation is a left invertible operator with an infinite–dimensional cokernel. Since $\sup|n(t)|$ is sufficiently small, this assertion remains valid for the operator $\psi(1+n)P_\mathbf{T} + Q_\mathbf{T}$.

Now let all numbers ν_j be non–positive and $\nu_1 < 0$. Then one easily checks that the operator

$$\prod_{j=2}^{m}\left(\exp\left(\nu_j\frac{t+t_j}{t-t_j}\right)P_\mathbf{T} + Q_\mathbf{T}\right)^{-1}\left(t^\kappa\exp\left(\nu_1\frac{t+t_1}{t-t_1}\right)P_\mathbf{T} + Q_\mathbf{T}\right)^{-1} \tag{3.2}$$

is a right inverse for the operator $\psi P_\mathbf{T} + Q_\mathbf{T}$. By Lemma 3.1, the operator (3.2) possesses an infinite–dimensional cokernel, and thus $\dim\ker\,(\psi P_\mathbf{T}+Q_\mathbf{T}) = \infty$. This implies that the operator $\psi(1+n)P_\mathbf{T}+Q_\mathbf{T}$ is also right invertible and that $\dim\ker\,(\psi(1+n)P_\mathbf{T}+Q_\mathbf{T}) = \infty$.

In case $\nu_1 = \nu_2 = \ldots = \nu_m = 0$ the assertion of the lemma is evident, finishing the proof of the lemma. ■

Proof of Theorem 3.1. The necessity of the conditions in the first assertion of the theorem follows from Theorem 4.1 of Chapter 7, and their sufficiency will follow once the other assertions of the theorem have been established.

Because of Theorems 1.1 and 1.2 of Chapter 7 we may assume without loss of generality that Γ is a simple closed curve. Invoking Theorem 3.1 of Chapter 7 concerning the separation of singularities, we can further limit ourselves to the case when the system $t_1, \ldots, t_m$ consists of a single point, say t_1.

Denote the characteristics of the function $c = ab^{-1}$ at the point t_1 by $\{p, \omega\}$, and consider the function $\zeta = g(t)$ with

$$g(t) := \frac{\omega(t) - i}{\omega(t) + i},$$

which maps the curve Γ onto the unit circle $\mathbf{T}$. Define furthermore

$$\rho(t) := \prod_{k=1}^{n} |t - t_k|^{\beta_k}, \quad \rho_0(\zeta) := \prod_{k=1}^{n} |\zeta - \zeta_k|^{\beta_k} \; (\zeta_k = g(t_k)),$$

and introduce the operator $Y \in L(L_p(\mathbf{T}, \rho_0),\ L_p(\Gamma, \rho))$ by $(Y\varphi)(t) := \varphi(g(t))$. Taking into account that the derivative dg/dt is continuous on Γ and does not vanish there, one immediately gets the boundedness of the operator Y. Moreover, the operator Y is invertible, and $(Y^{-1}f)(\zeta) = f(\beta(\zeta))$, where β is the inverse function of g.

Since the derivative dg/dt belongs to H_μ (see Section 7.2) the operator $cP_\Gamma + Q_\Gamma$ can be rewritten as

$$cP_\Gamma + Q_\Gamma = Y(c_0 P_\mathbf{T} + Q_\mathbf{T})Y^{-1} + T, \tag{3.3}$$

where $c_0(\zeta) = c(\beta(\zeta))$ and T stands for a compact operator.

Now suppose condition 1) is satisfied. We claim that if the operator $A_0 = c_0 P_\mathbf{T} + Q_\mathbf{T}$ is at least one-sided invertible then the operator $A_1 = cP_\Gamma + Q_\Gamma$ is invertible from the same side, and

$$\operatorname{Ind} A_0 = \operatorname{Ind} A_1. \tag{3.4}$$

Indeed, if the operator A_0 is at least one-sided invertible then, due to identity (3.3), the operator A_1 admits a regularization from the same side, and equality (3.4) holds. By Theorem 5.1 of Chapter 7, the operator A_1 is also at least one-sided invertible. So equality (3.4) shows that both operators A_0 and A_1 are invertible from the same side. Substituting ζ by $g(t)$ the difference

$$c_0(\zeta) - p\left(i\frac{\zeta + \zeta_1}{\zeta - \zeta_1}\right)$$

becomes

$$c(t) - p(\omega(t)),$$

which yields that $c_0 \in \mathcal{A}_0(t_1)$. It remains to apply Lemma 3.2 to finish the proof of the theorem. ■

11.4 Operators with continuous coefficients – the degenerate case

We are now going to employ the results of the previous section in order to complete our knowledge about operators of the form $aP_\Gamma + bQ_\Gamma$ with continuous coefficients a and b having zeros. For suppose the functions $a, b \in C(\Gamma)$ to be of the form $a = \rho_1 a_0$ and $b = \rho_2 b_1$ with $a_0, b_1 \in \mathcal{A}(t_1, t_2, \ldots, t_m)$ and

$$\rho_1(t) := \prod_{j=1}^{m}(t - t_j)^{n_j}, \quad \rho_2(t) := \prod_{j=1}^{m}(t - t_j)^{k_j}$$

where n_j and k_j stand for certain non-negative integers. We denote the greatest common divisor of the polynomials ρ_1 and ρ_2 by ρ_0 . Furthermore we define $\rho_+ := \rho_1\rho_0^{-1}$ and $\rho_- := t^{-m}\rho_2\rho_0^{-1}$ where m refers to the degree of the polynomial $\rho_2\rho_0^{-1}$. In the sequel we shall need the following representations of the functions a and b :

$$a = \rho_+\rho_0 a_0 \quad \text{and} \quad b = \rho_-\rho_0 b_0$$

with b_0 abbreviating the function $t^m b_1$. Remember that $\widetilde{L}_p(\rho, \rho_\pm; \Gamma)$ stands for the Banach space of all functions of the form

$$g = \rho_+^{-1} P_\Gamma f + \rho_-^{-1} Q_\Gamma f ,$$

$f \in L_p(\Gamma, \rho)$, with norm given by $\|g\|_{\widetilde{L}_p} := \|f\|_{L_p}$. By $\widehat{L}_p(\rho, \rho_0; \Gamma)$ we denote the Banach space consisting of all functions $\rho_0 f$ with $f \in L_p(\Gamma, \rho)$ provided with the norm $\|\rho_0 f\|_{\widehat{L}_p} = \|f\|_{L_p}$.

The operator $A = aP_\Gamma + bQ_\Gamma$ can be factorized into

$$A = \rho_0(a_0 P_\Gamma + b_0 Q_\Gamma)(\rho_+ P_\Gamma + \rho_- Q_\Gamma) . \tag{4.1}$$

By what we have shown in Section 3.12, the operator $\rho_+ P_\Gamma + \rho_- Q_\Gamma$ extends naturally to an invertible operator $\rho_+ \widetilde{P}_\Gamma + \rho_- \widetilde{Q}_\Gamma \in L(\widetilde{L}_p(\rho, \rho_\pm, \Gamma); L_p(\Gamma, \rho))$, the inverse of which is the operator $\rho_+^{-1} P_\Gamma + \rho_-^{-1} Q_\Gamma$. The operator $\rho_0 I \in L(L_p(\Gamma, \rho), \widehat{L}_p(\rho, \rho_0; \Gamma))$ is in fact an isometry. Thus, by means of (4.1), the operator A can be extended to the operator

$$\widetilde{A} := \rho_0(a_0 P_\Gamma + b_0 Q_\Gamma)(\rho_+ \widetilde{P}_\Gamma + \rho_- \widetilde{Q}_\Gamma) \tag{4.2}$$

belonging to $L(\widetilde{L}_p(\rho, \rho_\pm; \Gamma),\ \widehat{L}_p(\rho, \rho_0; \Gamma))$. Using Theorem 3.1 as a guide we arrive at

Theorem 4.1. *The operator* $\widetilde{A}$ *given by (4.2) is a* $\Phi-$ *or* $\Phi_{\pm}$*-operator if and only if*

1) $\inf_{t\in\Gamma}|a_0(t)|>0$ *and* $\inf_{t\in\Gamma}|b_0(t)|>0$ *and*

2) all the numbers $\mathrm{ind}\,(a_0b_0^{-1},t_j)$ $(j=1,\ldots,m)$ *have the same sign.*

If condition 1) is satisfied and if the numbers $\mathrm{ind}\,(a_0b_0^{-1},t_j)$ $(j=1,\ldots,m)$ *are non-positive and not all equal to zero, then the operator* $\widetilde{A}$ *is right-invertible and* $\dim\ker\widetilde{A}=\infty$.

If condition 1) is satisfied and if the numbers $\mathrm{ind}\,(a_0b_0^{-1},t_j)$ $(j=1,\ldots,m)$ *are non-negative and not all equal to zero then the operator* $\widetilde{A}$ *is left-invertible and* $\dim\mathrm{coker}\,\widetilde{A}=\infty$.

If condition 1) is satisfied and $\mathrm{ind}\,(a_0b_0^{-1},t_1)=\mathrm{ind}\,(a_0,b_0^{-1},t_2)=\ldots=\mathrm{ind}\,(a_0b_0^{-1},t_m)=0$ *then the operator* $\widetilde{A}$ *is left, right or two-sided invertible depending on whether the number* $\mathrm{ind}\,(a_0b_0^{-1})$ *is positive, negative, or equal to zero. In this situation,* $\mathrm{Ind}\,\widetilde{A}=-\mathrm{ind}\,(a_0b_0^{-1})$.

Notice in particular that Theorem 4.1 provides us with a wide class of singular integral operators $A=aP_\Gamma+bQ_\Gamma$, the coefficients a and b of which are continuous and have zeros, and for which one of the numbers $\dim\ker A$ or $\dim\ker A^*$ is infinite. The most interesting case is when $\rho_1=\rho_2=\rho_0$. In this situation, the spaces $\widetilde{L}_p(\rho,\rho_\pm;\Gamma)$ and $L_p(\Gamma,\rho)$ coincide, and Theorem 4.1 yields examples of singular integral operators $A=aP_\Gamma+bQ_\Gamma$ acting on $L_p(\Gamma,\rho)$ with continuous coefficients a and b having finitely many zeros and with the property that one of the numbers $\dim\ker A$ and $\dim\ker A^*$ is infinite.

Let us consider an example. Denote by $a\in C(\Gamma)$ a function of the form $a=\rho_-a_0$ with $a_0\in\mathcal{A}(t_1,\ldots,t_m)$, $\rho_-(t):=\prod_{j=1}^n(1-t^{-1}t_j)^{k_j}$ for certain non-negative integers k_j. Further we let $\widehat{L}_p$ stand for the Banach space of all functions

$$g=P_\Gamma\rho_-f+Q_\Gamma f\,,$$

$f\in L_p(\Gamma,\rho)$, provided with the norm $\|g\|_{\widehat{L}_p}:=\|f\|_{L_p}$. The operator $A=aP_\Gamma+Q_\Gamma$ can be written as

$$A=(P_\Gamma\rho_-+Q_\Gamma)(a_0P_\Gamma+Q_\Gamma)(I+Q_\Gamma(a-a_0)P_\Gamma)\,.$$

Here the operator $I+Q_\Gamma(a-a_0)P_\Gamma$ is invertible on $L_p(\Gamma,\rho)$,

$$(I+Q_\Gamma(a-a_0)P_\Gamma)^{-1}=I-Q_\Gamma(a-a_0)P_\Gamma\,,$$

and the operator $P_\Gamma\rho_-+Q_\Gamma$ acts isometrically from $L_p(\Gamma,\rho)$ onto $\widehat{L}_p$.

If we think of the operator A as acting from $L_p(\Gamma,\rho)$ onto $\widehat{L}_p$ then the following assertion holds:

The operator $A=aP_\Gamma+Q_\Gamma$ *is at least one-sided invertible if and only if the operator* $a_0P_\Gamma+Q_\Gamma\in L(L_p(\Gamma,\rho))$ *is invertible from the same side. Moreover,* $\dim\ker A=\dim\ker\,(a_0P_\Gamma+Q_\Gamma)$*, and* $\dim\mathrm{coker}\,A=\dim\mathrm{coker}\,(a_0P_\Gamma+Q_\Gamma)$.

This assertion allows us to derive several examples of operators $aP_\Gamma + Q_\Gamma$ with the continuous coefficient a vanishing at finitely many points for which one of the equalities

$$\dim\ker(aP_\Gamma + Q_\Gamma) = \infty \quad \text{or} \quad \dim\operatorname{coker}(aP_\Gamma + Q_\Gamma) = \infty$$

holds on the space $L_p(\Gamma, \rho)$. Of course, analogous theorems can be formulated for operators of the form $P_\Gamma aI + Q_\Gamma$ (see Section 3.12).

11.5 Exercises

11.1. Let $a_\nu(t) = \exp\left(\nu\frac{t+t_0}{t-t_0}\right)$ $(\nu \in \mathbf{R})$. Prove that

a) $a_\nu \in L_\infty(\mathbf{T})$,

b) $a_\nu \in L_\infty^+(\mathbf{T})$ if and only if $\nu \geq 0$,

c) $a_\nu \in L_\infty^-(\mathbf{T})$ if and only if $\nu \leq 0$,

11.2. Let $\nu > 0$ and $0 < \mu < \nu$. Determine all quadrupels of numbers $\{\alpha, \beta, \gamma, \delta\}$ such that the function

$$\varphi(t) = \alpha a_\nu + \beta a_\mu + \gamma a_{\mu+\nu} + \delta$$

a) lies in the kernel of the operator $A = a_\nu P_\mathbf{T} + Q_\mathbf{T}$, b) lies in the kernel of the operator $B = P_\mathbf{T} a_\nu I + Q_\mathbf{T}$.

11.3. Let $\mu, \nu \in \mathbf{R}$ $(\nu \geq 0)$, $m, n \in \mathbf{Z}$, $a(t) = t^n a_\nu(t)$ and $b(t) = t^m a_\mu(t)$. What can be said about the one-sided invertibility of the operator

$$A = (bP_\mathbf{T} + Q_\mathbf{T})(aP_\mathbf{T} + Q_\mathbf{T})$$

and about the numbers $\dim\ker A$, $\dim\operatorname{coker} A$, and $\operatorname{Ind} A$?

11.4. Determine all real numbers ν such that the equation

$$(ta_\nu P_\mathbf{T} + Q_\mathbf{T})\varphi = 1$$

is solvable in the space $L_2(\mathbf{T})$.

11.5. Show that the operator $A = aP_\mathbf{T} + Q_\mathbf{T}$ with the continuous coefficient $a(t) = (t - t_0)a_\nu(t)$ is not normally solvable on $L_2(\mathbf{T})$. Construct its extension $\widetilde{A}$ (see equality (4.2)) and study the nature of its one-sided invertibility.

Comments and references

The results of this section were derived by GOHBERG/SEMENCUL [1] and SEMENCUL [1] based on earlier results of GOHBERG/FELDMAN[1] and COBURN/DOUGLAS[1].

Theorem 1.1 is GOHBERG and FELDMAN's (see [1]). SARASON [1] developed a theory of Toeplitz operators with semi-almost periodic symbols. This theory generalizes the results of both the present chapter and Chapter 9 in the case of the unit circle and of the space L_2.

For further results related to Section 11.4 we refer to SEMENCUL [1], PRÖSSDORF [1, 2], CHEBOTAREV[1], KHAIKIN [1], DYBIN [1], DYBIN/ KARAPETIANC [1] and [GK 1, 3.12].

Chapter 12

Singular integral operators with bounded measurable coefficients

The present chapter is devoted to the study of singular integral operators with coefficients in $L_\infty(\Gamma)$. In particular we establish necessary and sufficient conditions for their Fredholmness, and give the Helson-Szegö criterion for the boundedness of the operator S_Γ in the spaces $L_2(\Gamma,\rho)$ as well as some of its generalizations for spaces $L_p(\Gamma,\rho)$.

We denote by $W_p(\Gamma)$ the class of all weight functions ρ such that the operator S_Γ is bounded in the space $L_p(\Gamma,\rho)$, and throughout this chapter (unless otherwise stated) we let Γ stand for a closed Lyapunov curve and suppose that $0 \in F_\Gamma^+$.

12.1 Singular operators with measurable coefficients in the space $L_2(\Gamma)$

In this section we employ the local principle to investigate singular operators of the form $aP_\Gamma + Q_\Gamma$ with a bounded measurable coefficient a in the space $L_2(\Gamma)$. Starting with Section 12.3 we generalize these results to the spaces $L_p(\Gamma)$ and $L_p(\Gamma,\rho)$.

We denote by $\mathcal{S}_2(\Gamma)$ the class of all functions $a \in L_\infty(\Gamma)$ which are subject to the following condition: To each point $\tau \in \Gamma$ there exist an open arc $l(\tau)$ containing the point τ and functions $g_\tau^\pm \in L_\infty^\pm(\Gamma)$ with $1/g_\tau^\pm \in L_\infty^\pm(\Gamma)$ such that the range of the restriction of the function $ag_\tau^+g_\tau^-$ to $l(\tau)$ is located in a certain closed half-plane Λ_τ which does not contain the origin.

Theorem 1.1. *If $a \in \mathcal{S}_2(\Gamma)$ then the operator $aP_\Gamma + Q_\Gamma$ is Fredholm.*

Proof. Let τ be an arbitrary point on Γ, $g_\tau^\pm \in L_\infty^\pm(\Gamma)$, and let $l(\tau)$ be a neighbourhood of τ such that the values of the function $ag_\tau^+g_\tau^-$ at $t \in l(\tau)$ belong to the half plane Λ_τ . Define

$$a_\tau(t) := \begin{cases} a(t) & \text{if} \quad t \in l(\tau) \\ \alpha(g_\tau^+(t)g_\tau^-(t))^{-1} & \text{if} \quad t \in \Gamma \setminus l(\tau) \end{cases}$$

for some fixed complex number $\alpha \in \Lambda_\tau$. We are going to show that the operator $a_\tau P_\Gamma + Q_\Gamma$ is Fredholm in $L_2(\Gamma)$.

Set $b := g_\tau^- a_\tau g_\tau^+$. The range of the function b is a bounded subset of the half–plane Λ_τ . Thus, there is a number $\gamma \in \mathbf{C}$ such that

$$\sup_{t\in\Gamma} |\gamma b(t) - 1| < 1 . \tag{1.1}$$

Now write the operator $A_\tau = a_\tau P_\Gamma + Q_\Gamma$ as

$$A_\tau = (g_\tau^-)^{-1} \left((\gamma b - 1)\, P_\Gamma + I\right) \left((\gamma g_\tau^+)^{-1}\, P_\Gamma + g_\tau^- Q_\Gamma\right) . \tag{1.2}$$

The outer factors on the right–hand side of equality (1.2) are invertible, and their inverses are $g_\tau^- I$ and $\gamma g_\tau^+ P_\Gamma + (g_\tau^-)^{-1} Q_\Gamma$. Because of $\|\widehat{P_\Gamma}\| = 1$ (see Theorem 2.1 of Chapter 7) we further have

$$\inf_{P_\Gamma \in \widehat{P_\Gamma}} \|(\gamma b - 1) P_\Gamma\| < 1 . \tag{1.3}$$

Hence, due to Theorem 7.2 of [GK 1], Chapter 4, the operator A_τ is Fredholm with zero index. Since the function a is equivalent to the function a_τ at each point $\tau \in \Gamma$ and since the operators $a_\tau P_\Gamma + Q_\Gamma$ are Fredholm, example 1.1 in [GK 1], Chapter 5, implies that the operator A is Fredholm, too. ■

Notice that the condition $a \in \mathcal{S}_2(\Gamma)$ is not only sufficient but also necessary for the Fredholmness of the operator $aP_\Gamma + Q_\Gamma$ in the space $L_2(\Gamma)$. This will be shown in Section 12.2.

As a consequence of Theorem 1.1 and Theorem 3.1 of Chapter 8 we obtain

Corollary 1.1. *Each function* $a \in \mathcal{S}_2(\Gamma)$ *admits a factorization* $a = a_- t^\kappa a_+$ *in* $L_2(\Gamma)$ *where* $\kappa = \text{ind}\, a\, |L_2(\Gamma) = -\text{ind}\,(aP_\Gamma + Q_\Gamma)$.

Let $\mathcal{S}_2^\circ(\Gamma)$ stand for the set of all functions in $\mathcal{S}_2(\Gamma)$ which fulfill the following condition: To each point $\tau \in \Gamma$ there exists an open arc $l(\tau)$ containing the point τ such that the range of the restriction of the function a to $l(\tau)$ is located in a certain closed half–plane Λ_τ which does not contain the origin.

The index $\text{ind}\, a|L_2(\Gamma)$ of a function $a \in \mathcal{S}_2^\circ(\Gamma)$ admits a geometrical interpretation. It is obviously sufficient to consider the case of a simple closed curve. Let $l(\tau_1), l(\tau_2), \ldots, l(\tau_n)$ be a covering of the curve Γ , and let t_0 be an arbitrary point on Γ . We cut off the curve Γ at t_0 and replace the point t_0 by two points: the starting point t^- and the end point t^+ . Then we associate a value at t^- of the function $\arg a$ and, following the natural orientation of the curve Γ , we define the function $\varphi_a(t) := \arg a(t)$ on the overlapping intervals in such a manner that $|\arg a(t) - \arg a(t')| < \pi$ whenever t and t' belong to the same interval. In this way we arrive at a certain value $\varphi_a(t^+) = \arg a(t^{\prime})$, and it turns out that $\operatorname{ind} a|L_2(\Gamma) = (1/2\pi)(\varphi_a(t^+) - \varphi_a(t^-))$. Indeed, we shall see below (Lemma 4.1) that each function $a \in \mathcal{S}_2^\circ(\Gamma)$ can be factorized into the product $a = bg$, with $g \in C(\Gamma)$, $g(t) \neq 0$ $(t \in \Gamma)$ and $\operatorname{ind} g = (1/2\pi)(\varphi_a(t^+) - \varphi_a(t^-))$, and with $b \in \mathcal{S}_2^\circ(\Gamma)$ being a function whose range is located in a half-plane $\operatorname{Re} z \geq \delta > 0$. Now one obtains as in the proof of Theorem 1.1, that $bP_\Gamma + Q_\Gamma$ is a Fredholm operator with index zero. Thus, $\operatorname{Ind}(aP_\Gamma + Q_\Gamma) = -\operatorname{ind} g$ and it follows that

$$\operatorname{ind} a|L_2(\Gamma) = \frac{1}{2\pi}(\varphi_a(t^+) - \varphi_a(t^-)) .$$

Notice further that the set $\mathcal{S}_2(\Gamma)$ can also be defined in the case of an arbitrary composed curve Γ . Let $\widetilde{\Gamma}(\supset \Gamma)$ be a closed curve. A function a belongs to $\mathcal{S}_2(\Gamma)$ by definition if $\widetilde{a} \in \mathcal{S}_2(\widetilde{\Gamma})$, where $\widetilde{a}$ refers to the function

$$\widetilde{a}(t) := \begin{cases} a(t) & \text{if } t \in \Gamma \\ 1 & \text{if } t \in \widetilde{\Gamma} \backslash \Gamma . \end{cases}$$

Evidently, the class $\mathcal{S}_2(\Gamma)$ does not depend on the special choice of the curve $\widetilde{\Gamma}$. Theorem 1.1 in Chapter 7 entails that the Theorem 1.1 remains valid in the case of an arbitrary composed curve.

12.2 Necessary conditions in the space $L_2(\Gamma)$

It is the aim of the present section to prove that the condition $a \in \mathcal{S}_2(\Gamma)$ is not only sufficient but also necessary for the Fredholmness of the operator $aP_\Gamma + Q_\Gamma$ in the space $L_2(\Gamma)$. For the proof of the corresponding result we need two lemmas.

Lemma 2.1. *Let $a \in L_\infty(\mathbf{T})$ and $|a(t)| = 1$. Then the operator $A = aP_\mathbf{T} + Q_\mathbf{T}$ is left invertible in $L_2(\mathbf{T})$ if and only if*

$$\rho\,(a, L_\infty^+(\mathbf{T})) < 1 , \tag{2.1}$$

where $\rho(a, L_\infty^+(\mathbf{T}))$ stands for the distance of a from $L_\infty^+(\mathbf{T})$.

Proof. Let (2.1) be satisfied. Then there is a function $h \in L_\infty^+(\mathbf{T})$ such that $\|a - h\|_\infty < 1$, and since $|a(t)| = 1$ we conclude that $\|1 - \bar{a}h\|_\infty < 1$. Define $m := 1 - \bar{a}h$. Then the operator $\bar{a}hP_\mathbf{T} + Q_\mathbf{T} = I + mP_\mathbf{T}$ is invertible on $L_2(\mathbf{T})$. The identity

$\bar{a}hP_{\mathbf{T}} + Q_{\mathbf{T}} = (\bar{a}P_{\mathbf{T}} + Q_{\mathbf{T}})(hP_{\mathbf{T}} + Q_{\mathbf{T}})$ implies that $a^{-1}P_{\mathbf{T}} + Q_{\mathbf{T}}$ is a right invertible operator, and we conclude from Corollary 6.2 of Chapter 7 that the operator A is left invertible.

Now we suppose the operator A to be left invertible. Then the operator $P_{\mathbf{T}}aP_{\mathbf{T}} + Q_{\mathbf{T}}$ is also left invertible. Thus, given an arbitrary function f in $L_2^+(\mathbf{T})$ there is a constant $\delta > 0$ such that $\|(P_{\mathbf{T}}aP_{\mathbf{T}} + Q_{\mathbf{T}})f\| \geq \delta\|f\|$, that is, $\|P_{\mathbf{T}}af\| \geq \delta\|f\|$. Since $\|af\|^2 = \|P_{\mathbf{T}}af\|^2 + \|Q_{\mathbf{T}}af\|^2$, we have $\|Q_{\mathbf{T}}af\|^2 = \|f\|^2 - \|P_{\mathbf{T}}af\|^2 \leq (1 - \delta^2)\|f\|^2$. In other words, $\|Q_{\mathbf{T}}af\| < \varepsilon\|f\|$ $(0 < \varepsilon < 1)$. For each function $g \in \overset{\circ}{L_1^-}(\mathbf{T})$, the function $\bar{g}$ can be represented as $\bar{g} = g_1g_2$ with $g_1, g_2 \in L_2^+(\mathbf{T}),\ g_2(0) = 0$, and $\|g\|_{L_1(\mathbf{T})} = \|g_1\|_{L_2(\mathbf{T})}\|g_2\|_{L_2(\mathbf{T})}$ (cf. HOFFMAN [1,p 107]). This implies that

$$\begin{aligned} |(a,g)| &= |(ag_1, \bar{g}_2)| = |(Q_{\mathbf{T}}ag_1, \bar{g}_2)| \\ &\leq \|Q_{\mathbf{T}}ag_1\|_{L_2(\mathbf{T})}\|g_2\|_{L_2(\mathbf{T})} < \varepsilon\|g_1\|_{L_2(\mathbf{T})}\|g_2\|_{L_2(\mathbf{T})} \\ &= \varepsilon\|g\|_{L_1(\mathbf{T})}. \end{aligned}$$

Thinking of a as a functional on $\overset{\circ}{L_1^-}(\mathbf{T})$ and taking into account that $(\overset{\circ}{L_1^-}(\mathbf{T}))^*$ is isometrically isomorphic to the quotient space $L_\infty(\mathbf{T})/L_\infty^+(\mathbf{T})$ we obtain after identification that $\|a\|_{L_\infty(\mathbf{T})/L_\infty^+(\mathbf{T})} \leq \varepsilon$, from which the estimate (2.1) follows. ■

Lemma 2.2. *Let* $a \in L_\infty(\mathbf{T})$ *and* $|a(t)| = 1$. *Then the operator* $A = aP_{\mathbf{T}} + Q_{\mathbf{T}}$ *is invertible on* $L_2(\mathbf{T})$ *if and only if the function* a *is of the form* $a = gh$ *with* $h^{\pm 1} \in L_\infty^+(\mathbf{T})$ *and* $\operatorname{Re} g(t) \geq \delta > 0$.

Proof. The proof of the sufficiency proceeds as in Lemma 2.1. For the proof of the necessity we remark that the operators A and $\bar{a}P_{\mathbf{T}} + Q_{\mathbf{T}}$ are simultaneously invertible (Theorem 6.2 of Chapter 7). By Lemma 2.1, there exists a function $h_0 \in L_\infty^+(\mathbf{T})$ such that $\|\bar{a} - h_0\|_\infty < 1$, i.e. $\|1 - ah_0\|_\infty < 1$. Define $m := 1 - ah_0$. The operators $aP_{\mathbf{T}} + Q_{\mathbf{T}}$ and $I - mP_{\mathbf{T}}$ are connected via the identity

$$(aP_{\mathbf{T}} + Q_{\mathbf{T}})(h_0P_{\mathbf{T}} + Q_{\mathbf{T}}) = I - mP_{\mathbf{T}}$$

and are invertible. Thus the operator $h_0P_{\mathbf{T}} + Q_{\mathbf{T}}$ is also invertible, and this implies that $h_0^{-1} \in L_\infty^+(\mathbf{T})$. Set $h = h_0^{-1}$ and $g = 1 - m$. These functions satisfy the conditions of the theorem, and $a = gh$. ■

Theorem 2.1. *Let* Γ *be a simple closed Lyapunov curve,* $0 \in F_\Gamma^+$, *and* $a \in L_\infty(\Gamma)$. *If the operator* $A = aP_\Gamma + Q_\Gamma$ *is a* Φ*-operator in* $L_2(\Gamma)$ *then the function* a *is factorizable into* $a(t) = h(t)t^\kappa g(t)$ *with*

$$h^{\pm 1} \in L_\infty^+(\Gamma), \quad \kappa \in \mathbf{Z}, \quad \textit{and} \quad \operatorname{Re} g \geq \delta > 0. \tag{2.2}$$

Proof. Since $A \in \Phi(L_2(\Gamma))$ we can find a number $\kappa \in \mathbf{Z}$ such that the operator $B = at^{-\kappa}P_\Gamma + Q_\Gamma$ is invertible in $L_2(\Gamma)$. Set $b(t) = t^{-\kappa}a(t)$, and let $t = \beta(z)$ be a conformal mapping from the unit disc onto the region F_Γ^+. Further define $(Vf)(z) = f(\beta(z))$. Since $VBV^{-1} = \widetilde{b}P_\mathbf{T} + Q_\mathbf{T} + T$ with $\widetilde{b}(z) = b(\beta(z))$ and $T \in \mathcal{T}(L_2(\mathbf{T}))$, the operator $\widetilde{b}P_\mathbf{T} + Q_\mathbf{T}$ is a Φ–operator with index zero. Theorem 5.1 of Chapter 7 implies that this operator is even invertible and, by Corollary 2.2 of Chapter 8, the operator $cP_\mathbf{T} + Q_\mathbf{T}$ with $c(z) = \widetilde{b}(z)/|\widetilde{b}(z)|$ is invertible, too. Now Lemma 2.2 implies that the function c can be written as $c(z) = \widetilde{h}(z)\widetilde{g}(z)$ with $\widetilde{h}^{\pm 1} \in L_\infty^+(\mathbf{T})$ and $\operatorname{Re} \widetilde{g}(z) \geq \delta > 0$. Introducing the functions $h(t) = \widetilde{h}(\alpha(t))$ and $g(t) = \widetilde{g}(\alpha(t))\, |\widetilde{b}(\alpha(t))|$ where $\beta(\alpha(t)) = t$, we arrive at the equality $a(t) = h(t)t^\kappa g(t)$, all factors of which are subject to the corresponding conditions in (2.2). ∎

Let us remark an important consequence.

Corollary 2.1. *Let Γ be a simple closed Lyapunov curve, $0 \in F_\Gamma^+$, and $a \in L_\infty(\Gamma)$. The following three conditions are equivalent:*

(1°) $aP_\Gamma + Q_\Gamma \in \Phi(L_2(\Gamma))$.

(2°) $a \in \mathcal{S}_2(\Gamma)$.

(3°) *The function a admits a representation in the form*

$$a = ht^\kappa g \tag{2.3}$$

with $h^{\pm 1} \in L_\infty^+(\Gamma)$, $\kappa \in \mathbf{Z}$, *and* $\operatorname{Re} g \geq \delta > 0$.

In other words: For each function $a \in L_\infty(\Gamma)$, the local condition $a \in \mathcal{S}_2(\Gamma)$ coincides with the global condition (2.3).

12.3 Lemmas

Let Γ be a closed not necessarily simple curve.

Theorem 3.1. *If the range of the function $a \in GL_\infty(\Gamma)$ is contained in a sector with vertex at the origin and with an opening angle less than $2\pi/\max(p,q)$ $(1 < p < \infty,\ p^{-1} + q^{-1} = 1)$ then the operator $aP_\Gamma + Q_\Gamma$ is invertible in the space $L_p(\Gamma)$.*

The proof is based on the following lemma.

Lemma 3.1. *Let φ be a real-valued measurable function on the unit circle $\mathbf{T}$ satisfying the condition*

$$\operatorname*{ess\,sup}_{|t|=1} |\varphi(t)| < \pi/p \quad (p \geq 2). \tag{3.1}$$

Then the function $\exp(i\varphi)$ *admits a factorization* $\exp(i\varphi) = a_- a_+$ *in the spaces* $L_p(\mathbf{T})$ *and* $L_q(\mathbf{T})$ $(p^{-1} + q^{-1} = 1)$ *with the factors*

$$a_+ = \exp(iP_{\mathbf{T}}\varphi) \quad \text{and} \quad a_- = \exp(iQ_{\mathbf{T}}\varphi)\,. \tag{3.2}$$

Proof. Denote by $w(z)$ the Schwarz integral

$$w(z) := \frac{1}{2\pi i}\int_{\mathbf{T}} \frac{\zeta + z}{\zeta - z}\,\frac{\varphi(\zeta)}{\zeta}\,d\zeta \quad (|z| < 1)\,.$$

We claim that the functions $g_\pm(z) := \exp(\pm\frac{i}{2}w(z))$ belong to the Hardy space H_p $(= L_p^+(\mathbf{T}))$. Let $u = \frac{1}{2}\mathrm{Re}\, w$ and $v = \frac{1}{2}\mathrm{Im}\, w$. Since $v(0) = 0$ we have

$$\frac{1}{2\pi i}\int_{|\zeta|=r} e^{\pm ip(u+iv)(\zeta)}\,\frac{d\zeta}{\zeta} = e^{\pm ipu(0)} \tag{3.3}$$

for each r between 0 and 1.

On the other hand,

$$\frac{1}{2\pi i}\int_{|\zeta|=r} e^{\pm ip(u+iv)(\zeta)}\,\frac{d\zeta}{\zeta} = \frac{1}{2\pi}\int_0^{2\pi} |e^{\pm ip(u+iv)}|(\cos pu \pm i\sin pu)\,d\theta\,. \tag{3.4}$$

The identities (3.3) and (3.4) imply

$$\int_0^{2\pi} |g_\pm(re^{i\theta})|^p \cos(pu(re^{i\theta}))\,d\theta = \cos(pu(0))\,.$$

The function $u(\zeta)$ is harmonic in the disc $|\zeta| < 1$, and its boundary values on $\mathbf{T}$ coincide with $\frac{1}{2}\varphi(\zeta)$ almost everywhere (see HOFFMAN[1]). Thus,

$$|u(z)| \le \frac{1}{2}\sup_{|\zeta|=1}|\varphi(\zeta)| = M < \frac{\pi}{2p}\,,$$

and it follows that

$$\int_0^{2\pi} |g_\pm(re^{i\theta})|^p\,d\theta \;\le\; \frac{1}{\cos(Mp)}\int_0^{2\pi} |g_\pm(re^{i\theta})|^p \cos(pu(re^{i\theta}))\,d\theta = \frac{\cos(pu(0))}{\cos(Mp)}\,.$$

This relation implies that $g_\pm \in L_p^+(\mathbf{T})$ (cf. HOFFMAN [1, p. 39]). Now we let $f(z)$ refer to the Cauchy type integral

$$f(z) := \frac{1}{2\pi i}\int_{\mathbf{T}} \frac{\varphi(\zeta)}{\zeta - z}\,d\zeta \quad (|z| < 1)\,,$$

and define $\psi_\pm := \exp(\pm if)$. Since

$$f(z) - \frac{1}{2}w(z) = \frac{1}{4\pi}\int_0^{2\pi} \varphi(e^{i\theta})\, d\theta\ ,$$

we have $|\psi_\pm(z)g_\pm^{-1}(z)| = 1$ and, consequently, $|\psi_\pm(z)| = |g_\pm(z)|$.

This shows that the functions $\psi_\pm$ also belong to the space $L_p^+(\mathbf{T})$. Thus, the functions $\exp(\pm iP_\mathbf{T}\varphi)$ being the boundary values almost everywhere of the functions $\psi_\pm(z)$ from the interior, belong to the space $L_p^+(\mathbf{T})$. In an analogous manner one verifies that $\exp(\pm iQ_\mathbf{T}\varphi) \in L_p^-(\mathbf{T})$. Since $p \geq 2$ we have $L_p^\pm(\mathbf{T}) \subset L_q^\pm(\mathbf{T})$, and this yields that $a_+, a_+^{-1} \in L_q^+(\mathbf{T})$ and $a_-, a_-^{-1} \in L_q^-(\mathbf{T})$.

We next prove the boundedness of the operator $a_+^{-1}P_\mathbf{T}a_-^{-1}I$. Let $c := \exp(i\alpha\varphi)$ where the number $\alpha > p/2$ is chosen such that

$$\operatorname*{ess\ sup}_{|t|=1} |\alpha\varphi(t)| < \frac{\pi}{2}\ .$$

By Theorem 1.1, the operator $cP_\mathbf{T} + Q_\mathbf{T}$ is invertible in the space $L_2(\mathbf{T})$, and Theorem 3.1 in Chapter 8 tells us that the function c admits a factorization $c = \widetilde{c}_-\widetilde{c}_+$ in the space $L_2(\mathbf{T})$. Furthermore, by what we have already seen, the function c is equal to $c = c_-c_+$ with $c_+ := \exp(i\alpha P_\mathbf{T}\,\varphi), c_- := \exp(i\alpha Q_\mathbf{T}\varphi)$, and $c_+^{\pm 1} \in L_2^+(\mathbf{T})$, $c_-^{\pm 1} \in L_2^-(\mathbf{T})$. The identity $c_-c_+ = \widetilde{c}_-\widetilde{c}_+$ immediately gives that $\widetilde{c}_- = \mu c_-$ and $\widetilde{c}_+ = \mu^{-1}c_+$ for a certain $\mu \in \mathbf{C}$. The operator $\widetilde{c}_+^{-1}P_\mathbf{T}\widetilde{c}_-^{-1}I$ is bounded on $L_2(\mathbf{T})$. Thus, the operator $c_+^{-1}P_\mathbf{T}c_-^{-1}I$ is also bounded in the same space.

Introducing $h := |\exp(\frac{1}{2}S_\mathbf{T}\varphi)|$ we clearly have $|c_+| = h^\alpha,\ |c_-| = h^{-\alpha}$, and, hence, the operator $h^{-\alpha}P_\mathbf{T}h^\alpha$ is bounded in $L_2(\mathbf{T})$. This leads to the inequality

$$\|h^{-\alpha}P_\mathbf{T}f\|_{L_2(\mathbf{T})} \leq k_1\|h^{-\alpha}f\|_{L_2(\mathbf{T})} \tag{3.5}$$

which holds for each function f which is continuous on $\mathbf{T}$.

Now let $p_1 := 2p(\alpha-1)/(2\alpha-p)$ and $t = 1/\alpha$. Then

$$\frac{1}{p} = \frac{t}{2} + \frac{1-t}{p_1}\ .$$

The operator $P_\mathbf{T}$ is bounded in $L_{p_1}(\mathbf{T})$, that is

$$\|P_\mathbf{T}f\|_{L_{p_1}(\mathbf{T})} \leq k_2\|f\|_{L_{p_1}(\mathbf{T})} \tag{3.6}$$

for a certain constant k_2 .

The relations (3.5), (3.6) and (3.7) in combination with the interpolation theorem [GK 1, 1.1] yield

$$\|h^{-1}P_\mathbf{T}f\|_{L_p(\mathbf{T})} \leq k\|h^{-1}f\|_{L_p(\mathbf{T})}\ .$$

Hence, the operator $h^{-1}P_{\mathbf{T}}hI$ is bounded in $L_p(\mathbf{T})$, and since $|a_+| = h$ and $|a_-| = h^{-1}$ with $a_\pm$ denoting the functions defined in (3.2), we conclude that the operator $a_+^{-1}P_{\mathbf{T}}a_-^{-1}I$ is bounded in $L_p(\mathbf{T})$. The boundedness of this operator on $L_q(\mathbf{T})$ follows by taking adjoints. ■

Proof of Theorem 3.1. To start with suppose that $\Gamma = \mathbf{T}$. The assumptions of the theorem imply that the function a is representable in the form $a = \beta|a|\exp(i\varphi)$, with the function φ fulfilling the conditions of Lemma 3.1 and β a constant. This fact, in combination with Theorem 2.2 of Chapter 8, yields that the function a admits the factorization $a = a_-a_+$ in the space $L_p(\mathbf{T})$, and by Theorem 4.1 in Chapter 8, the operator $aP_{\mathbf{T}} + Q_{\mathbf{T}}$ is invertible on $L_p(\mathbf{T})$.

Now let Γ be a simple closed curve, and let $\nu : \mathbf{T} \to \Gamma$ be a mapping for which the derivative $\nu'(z)$ does not vanish and satisfies a Hölder condition on $\mathbf{T}$. Then (confer [GK 1, 1.4),

$$B(aP_\Gamma + Q_\Gamma)B^{-1} = a_0P_{\mathbf{T}} + Q_{\mathbf{T}} + T\ ,$$

T a compact operator, $(B\varphi)(\zeta) := \varphi(\nu(\zeta))$, and $a_0(\zeta) := a(\nu(\zeta))$. We conclude by the above that $aP_\Gamma + Q_\Gamma$ is a Fredholm operator with index zero. Invoking Theorem 1.2 of Chapter 7 one can generalize this assertion to the case of an arbitrary closed composed curve Γ . Finally, Theorem 5.4 of Chapter 7 states the invertibility of the operator $aP_\Gamma + Q_\Gamma$.■

12.4 Singular operators with coefficients in $\mathcal{S}_p(\Gamma)$. Sufficient conditions

Let Γ be a closed curve. We denote by $\mathcal{S}_p(\Gamma)$ $(1 < p < \infty)$ the class of all functions $a \in L_\infty(\Gamma)$ subject to the following conditions:

(1) $\operatorname{ess\,inf}_{t\in\Gamma} |a(t)| > 0$;

(2) for each point $\tau \in \Gamma$ there exist an open arc $l(\tau)$ $(\subset \Gamma)$ containing the point τ and functions $g_\infty^\pm$ which, together with their inverses $1/g_\tau^\pm$, belong to $L_\infty^\pm(\Gamma)$, such that the range of the restriction of the function $g_\tau^+ a g_\tau^-$ onto $l(\tau)$ is contained in a sector $\Lambda_\tau(p)$ with vertex coinciding to origin and opening angle less than $2\pi/\max(p,q)$ $(p^{-1}+q^{-1} = 1)$.

Theorem 4.1. *Let $a \in \mathcal{S}_p(\Gamma)$. Then the operator $aP_\Gamma + Q_\Gamma$ is Fredholm in $L_p(\Gamma)$.* [1]

Proof. We prove this theorem by means of the local principle. Let τ be an arbitrary point of the curve Γ, $g_\tau^\pm \in L_\infty^\pm(\Gamma)$, and let $l(\tau)$ be a neighbourhood of the point τ such

[1] In Section 12.7 we shall discuss a wider class of functions a for which $aP_\Gamma + Q_\Gamma$ is a Fredholm operator in $L_p(\Gamma)$.

that the range of the restriction of the function $ag_\tau^+ g_\tau^-$ onto $l(\tau)$ is contained in the sector $\Lambda_\tau(p)$. Define

$$a_\tau(t) := \begin{cases} a(t) & \text{if } t \in l(\tau) \\ \alpha(g_\tau^+(t)g_\tau^-(t))^{-1} & \text{if } t \in \Gamma \backslash l(\tau), \end{cases}$$

with $\alpha \in \Lambda_\tau(p)$ a suitably chosen complex number. By Theorem 1.1, the operator $bP_\Gamma + Q_\Gamma$ with $b := a_\tau g_\tau^+ g_\tau^-$ is invertible in $L_p(\Gamma)$, and the identity

$$bP_\Gamma + Q_\Gamma = g_\tau^-(a_\tau P_\Gamma + Q_\Gamma)(g_\tau^+ P_\Gamma + (g_\tau^-)^{-1} Q_\Gamma)$$

yields the invertibility of the operator $a_\tau P_\Gamma + Q_\Gamma$ in $L_p(\Gamma)$. Since the function a is equivalent to the function a_τ for $\tau \in \Gamma$ and since the operator $a_\tau P_\Gamma + Q_\Gamma$ is invertible in $L_p(\Gamma)$, the local principle gives that the operator A is Fredholm. ■

As a consequence of Theorem 4.1 and Theorem 3.1 of Chapter 8 we obtain

Corollary 4.1. *Each function $a \in \mathcal{S}_p(\Gamma)$ admits a factorization $a = a_- t^\kappa a_+$ in the space $L_p(\Gamma)$ with $\kappa = \operatorname{ind} a|L_p(\Gamma) = -\operatorname{Ind}(aP_\Gamma + Q_\Gamma)$.*

We let $\mathcal{S}_p^\circ(\Gamma)$ refer to the set of all functions $a \in \mathcal{S}_p(\Gamma)$ satisfying the following condition: To each point $\tau \in \Gamma$ there exists an open arc $l(\tau)$ containing the point τ such that the range of the restriction of the function a onto $l(\tau)$ is contained in the sector $\Lambda_\tau(p)$. Recall that the class $\mathcal{S}_2^\circ(\Gamma)$ was introduced in Section 12.1.

If a is a function in $\mathcal{S}_p^\circ(\Gamma)$ then its index, $\operatorname{ind} a|L_p(\Gamma)$, admits a geometrical interpretation. For let Γ be a simple closed curve and $a \in \mathcal{S}_p^\circ(\Gamma)$. Since $\mathcal{S}_p^\circ(\Gamma) \subset \mathcal{S}_2^\circ(\Gamma)$ we can associate to the function a the number $(\varphi_a(t^+) - \varphi_a(t^-))/2\pi$ defined in Section 12.1. We claim that

$$\operatorname{ind} a|L_p(\Gamma) = \frac{1}{2\pi}(\varphi_a(t^+) - \varphi_a(t^-)) . \tag{4.1}$$

To prove our claim we need the following lemma.

Lemma 4.1. *Each function $a \in \mathcal{S}_p^\circ(\Gamma)$ can be written as $a = bg$ where $b \in GL_\infty(\Gamma)$ is a function whose range lies in a sector with vertex coinciding with the origin and an opening angle less than $2\pi/\max(p,q)$, and where $g \in GC(\Gamma)$ is a function with index* $\operatorname{ind} g = (\varphi_a(t^+) - \varphi_a(t^-))/2\pi$.

Proof. For simplicity, let $p \geq q$. The proof of the lemma will follow once we have constructed a real-valued function ψ which is continuous at each point $t \neq t_0$ of the curve Γ and fulfills the conditions

$$\sup |\varphi_a(t) - \psi(t)| < \pi/p, \quad \psi(t^+) - \psi(t^-) = \varphi_a(t^+) - \varphi_a(t^-) . \tag{4.2}$$

Then the functions g and b will we defined by

$$g := \exp(i\psi), \quad b := |a| \exp i(\varphi_a - \psi) .$$

In order to determine such a function ψ we divide the curve Γ into finitely many arcs γ_k with end points t_{k-1} and t_k $(k = 1, \ldots, n;\ t_0 = t^-, t_n = t^+)$ in such a manner that

$$\sup_{t\in\gamma_k} \varphi_a(t) - \inf_{t\in\gamma_k} \varphi_a(t) < \frac{2\pi}{p},$$

which is obviously possible. Now abbreviate

$$\lambda_k := \inf_{t\in\gamma_k} \varphi_a(t), \quad \mu_k := \sup_{t\in\gamma_k} \varphi_a(t),$$

and define

$$\begin{aligned} \psi(t_k) &:= \frac{1}{2}(\max(\mu_k, \mu_{k+1}) + \min(\lambda_k, \lambda_{k+1})) \quad (k = 1, \ldots, n-1) \\ \psi(t^-) &:= \frac{1}{2}(\max(\mu_1, \mu_n - m) + \min(\lambda_1, \lambda_n - m)), \end{aligned}$$

where $m := \varphi(t^+) - \varphi(t^-)$. Set $\psi(t^+) := \psi(t^-) + m$. At the other points $t \in \Gamma$ we define the function ψ by linear interpolation along the arcs γ_k . Then the inequalities

$$\sup_{t\in\gamma_k} |\psi(t) - \lambda_k| < \frac{\pi}{p} \quad \text{and} \quad \sup_{t\in\gamma_k} |\psi(t) - \mu_k| < \frac{\pi}{p}$$

follow easily. But these inequalities also hold for the function φ in place of ψ , and since $0 \le \mu_k - \lambda_k < 2\pi/p$, we arrive at the first of the conditions (4.2). The second condition in (4.2) is immediate from the definition of the function ψ . ■

The preceding lemma implies that the operator $aP_\Gamma + Q_\Gamma$ can be rewritten as $(bP_\Gamma + Q_\Gamma)(gP_\Gamma + Q_\Gamma) + T$ for some operator $T \in \mathcal{T}(L_p(\Gamma))$. The operator $gP_\Gamma + Q_\Gamma$ is Fredholm and, by what we have just seen, its index is given by

$$\mathrm{Ind}\,(gP_\Gamma + Q_\Gamma) = -\mathrm{ind}\, g = -(\varphi_a(t^+) - \varphi_a(t^-))/2\pi .$$

Furthermore, by Theorem 3.1, the operator $bP_\Gamma + Q_\Gamma$ is invertible which entails that $\mathrm{Ind}\,(aP_\Gamma + Q_\Gamma) = -(\varphi(t^+) - \varphi(t^-))/2\pi$, and hence

$$\mathrm{ind}\, a|L_p(\Gamma) = \frac{1}{2\pi}(\varphi(t^+) - \varphi(t^-)) .$$

Till now we have only considered a simple closed curve Γ . If Γ is an arbitrary closed curve consisting of the simple closed curves $\Gamma_1, \ldots \Gamma_n$ then we have $\mathrm{ind}\, a|L_p(\Gamma) = \sum_{k=1}^n \mathrm{ind}\, a|L_p(\Gamma_k)$ (see Section 8.1), and this equality moreover allows to give a geometric interpretation of the index $\mathrm{ind}\, a|L_p(\Gamma)$.

Finally, it is possible to define the class $\mathcal{S}_p(\Gamma)$ for an arbitrary composed curve Γ . This can be done in the same way as in Section 12.1 for the case $p = 2$. Notice that, in case Γ is a closed composed curve consisting of the simple closed curves $\Gamma_1 \ldots, \Gamma_n$, the function a belongs to $\mathcal{S}_p(\Gamma)$ if and only if the restrictions $a|\Gamma_k$ belong to $\mathcal{S}_p(\Gamma_k)$ for each $k = 1, \ldots, n$. Obviously, Theorem 4.1 remains valid if Γ is replaced by an arbitrary composed curve. The Theorems 4.1 and 4.1 of Chapter 9 combine to give

Corollary 4.2. *Let $a \in \mathcal{S}_p(\Gamma)$ and $\kappa = \operatorname{ind} a|L_p(\Gamma)$. Then the operator $aP_\Gamma + Q_\Gamma$ is invertible, left–invertible, or right–invertible on the space $L_p(\Gamma)$ depending on whether the number κ is equal to zero, positive, or negative.*

12.5 The Helson–Szegö theorem and its generalization

It will be convenient in this section to consider methods of defining a norm in the space L_p with weight. Let us agree upon the following notations:

$\mathcal{L}_p(\Gamma, \rho)$ – the weighted space L_p provided with the norm

$$\|\varphi\|_{\mathcal{L}_p(\Gamma,\rho)} = \|\varphi\rho\|_{L_p(\Gamma)}, \tag{5.1}$$

$\mathcal{W}_p(\Gamma)$ – the collection of all weight functions ρ such that the operator S_Γ becomes bounded on the space $\mathcal{L}_p(\Gamma, \rho)$.

Notice that

$$\rho \in \mathcal{W}_p(\Gamma) \iff \rho^p \in W_p(\Gamma). \tag{5.2}$$

Using this notation, the Hunt–Muckenhoupt–Wheeden criterion concerning the boundedness of the operator S_Γ on the space $\mathcal{L}_p(\Gamma, \rho)$ (see [GK 1, Ch. 1, Comments and references]) can be stated as follows:

$$\rho \in \mathcal{W}_p(\Gamma) \iff \left(\int_l \rho^p(t)|dt|\right)^{1/p} \left(\int_l \rho^{-q}(t)|dt|\right)^{1/q} \leq A(p)\,|l| \tag{5.3}$$

where $l(\subset \Gamma)$ is an arbitrary arc of length $|l|$, and the constant $A(p)$ is independent of l.

Theorem 5.1. (Helson–Szegö) *The operator $S_{\mathbf{T}}$ is bounded on $L_2(\mathbf{T}, \rho)$ if and only if the weight ρ is of the form $\rho(t) = \exp(u(t) + \widetilde{v}(t))$, where u and v are real-valued functions in $L_\infty(\mathbf{T})$ with $\|v\|_\infty < \pi/2$ and*

$$\widetilde{v}(e^{i\theta}) := \frac{1}{2\pi} \int_{-\pi}^{\pi} \cot \frac{\sigma - \theta}{2} v(e^{i\sigma})\, d\sigma. \tag{5.4}$$

Proof. Let $\rho(t) = \exp(u(t) + \widetilde{v}(t))$ and define $a(t) = \exp(-iv(t))$. Because $\|v\|_\infty < \pi/2$, the function a admits a factorization defined by (3.2) with

$$a_+ = \exp(-iP_{\mathbf{T}}v) = \exp\left(-i\frac{v}{2} - i\frac{S_{\mathbf{T}}v}{2}\right) .$$

Since $\widetilde{v} = iS_{\mathbf{T}}v + c$ for some constant c , we have $|a_+^{-2}| = \gamma \exp \widetilde{v}$. The general definition of the factorizability in the space $L_2(\mathbf{T})$ implies that $|a_+^{-2}| \in W_2(\mathbf{T})$ and, hence, $\exp \widetilde{v} \in W_2(\mathbf{T})$. Finally, since $\exp(\pm u) \in L_\infty(\mathbf{T})$ we arrive at $\rho \in W_2(\mathbf{T})$.

Conversely, let $\rho \in W_2(\mathbf{T})$. Then, by Theorem 4.5 in [GK 1, Chapter 1], both ρ and ρ^{-1} are in $L_1(\mathbf{T})$, and since $|\ln \rho| \le \rho + \rho^{-1}$ we conclude that $\ln \rho \in L_1(\mathbf{T})$. Define $b = b_+ b_-$ with $b_+ = \exp(-P_{\mathbf{T}} \ln \rho)$ and $b_- = \overline{b}_+^{-1}$. It is well-known (see HOFFMAN [1]) that for each non-negative function ρ on $\mathbf{T}$ with $\ln \rho \in L_1(\mathbf{T})$, the (outer) function

$$F(z) = \exp\left[\frac{1}{2\pi} \int\limits_{\mathbf{T}} \frac{t+z}{t-z} \ln \rho(t)\, |dt|\right]$$

belongs to the space $L_1^+(\mathbf{T})$ and $|F(t)| = \rho(t)$ almost everywhere on $\mathbf{T}$. This shows that $b_+^{\pm 1} \in L_1^+(\mathbf{T})$ and $b_-^{\pm 1} \in L_1^-(\mathbf{T})$. Since $F(z) = c\ \exp(2P_{\mathbf{T}} \ln \rho)$, we have $|b_+|^{-2} = c_1 \rho \in W_2(\mathbf{T})$, and consequently, the equality $b = b_+ b_-$ is actually a factorization of the function b in the space $L_2(\mathbf{T})$. Now, by Theorem 1.1, the operator $B = bP_{\mathbf{T}} + Q_{\mathbf{T}}$ is invertible in $L_2(\mathbf{T})$, and from Lemma 2.2 we conclude that the function b can be factored into $b(t) = h(t)g(t)$ with $h^{\pm 1} \in L_\infty(\mathbf{T})$ and $g(t) = |g(t)| \exp(iv(t))$, where $\|v\|_\infty < \pi/2$. The function $|g(t)|$ admits a factorization $g_+(t)g_-(t)$ with $g_+^{\pm 1} \in L_\infty^+(\mathbf{T})$ and $g_-^{\pm 1} \in L_\infty^-(\mathbf{T})$ (cf. Theorem 2.1 in Chapter 8), and the function $\exp(iv(t))$ is factorizable into $\exp(iv(t)) = a_+(t)a_-(t)$ with $|a_+|^{-2} = \gamma \exp \widetilde{v}$ (see the first part of this proof). The identity $b_+ b_- = hg_+ g_- a_+ a_-$ implies that $b_+ = \lambda h g_+ a_+$, λ a constant. This means that $|b_+|^{-2} = |\lambda h g_+| \gamma e^v = e^{u+\widetilde{v}}$ with $\|v\|_\infty < \pi/2$. ■

Remark 5.1. *Since $\rho^2 \in W_2(\mathbf{T})$ if and only if $\rho \in \mathcal{W}_2(\mathbf{T})$, we have*

$$\rho \in \mathcal{W}_2(\mathbf{T}) \iff \rho = \exp(u_1 + \widetilde{v}_1) \tag{5.5}$$

where u_1 and v_1 are real-valued functions in $L_\infty(\mathbf{T})$ with

$$\|v_1\|_\infty < \pi/4 . \tag{5.6}$$

As a generalization of this result we formulate

Theorem 5.2. *Let $2 \le p < \infty$ and $p^{-1} + q^{-1} = 1$. Then $\rho \in \mathcal{W}_p(\mathbf{T}) \cap \mathcal{W}_q(\mathbf{T})$ if and only if*

$$\rho(t) = \exp(u(t) + \widetilde{v}(t)) \tag{5.7}$$

with real-valued functions u *and* v *in* $L_\infty(\mathbf{T})$ *satisfying* $\|v\|_\infty < \pi/2p$.

For the proof of this theorem we need

Lemma 5.1. *Let* $2 \le p < \infty$ *and* $p^{-1} + q^{-1} = 1$. *Then the implication*

$$\rho \in \mathcal{W}_p(\mathbf{T}) \cap \mathcal{W}_q(\mathbf{T}) \implies \rho^{p/2} \in \mathcal{W}_2(\mathbf{T}) \tag{5.8}$$

holds.

Proof. Let $\rho \in \mathcal{W}_p(\mathbf{T}) \cap \mathcal{W}_q(\mathbf{T})$ and $l(\subset \mathbf{T})$ be an arbitrary arc. Then, by (5.3),

$$\left(\int_l \rho^p(t)\,|dt|\right)^{1/p} \left(\int_l \rho^{-q}(t)|dt|\right)^{1/q} \le A(p)\,|l| \tag{5.9}$$

and

$$\left(\int_l \rho^q(t)\,|dt|\right)^{1/q} \left(\int_l \rho^{-p}(t)|dt|\right)^{1/p} \le A(q)\,|l|\,. \tag{5.10}$$

The estimates (5.9) and (5.10) in combination with the Cauchy–Schwarz inequality

$$|l|^2 \le \int_l \rho^q(t)|dt| \int_l \rho^{-q}(t)|dt|$$

give

$$\left(\int_l \rho^p(t)|dt|\right)^{1/2} \left(\int_l \rho^{-p}(t)|dt|\right)^{1/2} \le |l|\,(A(p)A(q))^{p/2}\,.$$

Hence, by criterion (5.3), it follows that $\rho^{p/2} \in \mathcal{W}_2(\mathbf{T})$. ■

Proof of the theorem. Let $\rho \in \mathcal{W}_p(\mathbf{T}) \cap \mathcal{W}_q(\mathbf{T})$. Since $\rho^{p/2} \in \mathcal{W}_2(\mathbf{T})$ we know from (5.5) that $\rho = \exp(u_1 + \widetilde{v}_1)$ with $\|v_1\|_\infty < \pi/4$. Upon defining $u = \frac{2}{p}u_1$ and $v = \frac{2}{p}v_1$ we obtain (5.7) with $\|v\|_\infty < \pi/2p$.

Conversely, let $\rho = \exp(u + \widetilde{v})$ with $\|v\|_\infty < \pi/2p$. Then the function $a(t) = \exp(-2iv)$ admits a factorization $a = a_+a_-$ both in $L_p(\mathbf{T})$ and in $L_q(\mathbf{T})$ with factors being given by (3.2). In the same way as in the proof of Theorem 5.1, one can show that $|a_+^{-1}| = \gamma \exp \widetilde{v}$ for some constant γ . Further, the definition of factorizability in the spaces $L_p(\mathbf{T})$ and $L_q(\mathbf{T})$ implies that $|a_+^{-1}|^p \in W_p(\mathbf{T})$ and $|a_+^{-1}|^q \in W_q(\mathbf{T})$. Consequently, $|a_+^{-1}| \in \mathcal{W}_p(\mathbf{T}) \cap \mathcal{W}_q(\mathbf{T})$,and it follows that $\exp(u + \widetilde{v}) \in \mathcal{W}_p(\mathbf{T}) \cap \mathcal{W}_q(\mathbf{T})$. ■

Corollary 5.1. *Let* $p \ge 2$ *and* $p^{-1} + q^{-1} = 1$. *Then the inclusions* $\rho^p \in W_p(\mathbf{T})$ *and* $\rho^q \in W_q(\mathbf{T})$ *hold simultaneously if and only if the weight* ρ *is of the form (5.7).*

12.6 On the necessity of the condition $a \in \mathcal{S}_p$

The condition $a \in \mathcal{S}_p(\Gamma)$ which, as we have seen in Section 12.4, is sufficient for the Fredholmness of the operator $A = aP_\Gamma + Q_\Gamma$ in $L_p(\Gamma)$, is not necessary. To understand this, notice that the definition of the class $\mathcal{S}_p(\Gamma)$ implies that if $a \in \mathcal{S}_p(\Gamma)$ for any number $p > 2$, then $a \in \mathcal{S}_r(\Gamma)$ for all $r \in [q,p]$ with $q^{-1}+p^{-1} = 1$. Thus, if $a \in \mathcal{S}_p(\Gamma)$ then $A \in \Phi(L_r(\Gamma))$ for all $r \in [q,p]$, and in particular, $A \in \Phi(L_2(\Gamma))$. On the other hand, the operator $A = t^{1/2}P_{\mathbf{T}} + Q_{\mathbf{T}}$ is Fredholm in all spaces $L_p(\mathbf{T})$ with $p \neq 2$ but it fails to be a Fredholm operator in $L_2(\mathbf{T})$. Consequently, the function $a(t) = t^{1/2}$ is not contained in $\mathcal{S}_p(\mathbf{T})$ for any $p > 1$.

In the present section we shall show that the condition $a \in \mathcal{S}_p(\Gamma)$ is necessary and sufficient for the operator A to belong to $\Phi(L_r(\Gamma))$ for all $r \in [q,p]$.

First we establish the following lemma.

Lemma 6.1. *Let Γ be a simple closed Lyapunov curve. Then the operator $A = aP_\Gamma + Q_\Gamma$ is invertible both in $L_p(\Gamma)$ and in $L_q(\Gamma)$ $(p^{-1}+q^{-1} = 1,\ p \geq 2)$ if and only if the function a can be represented in the form*

$$a(t) = h(t)g(t) \tag{6.1}$$

with $h^{\pm 1} \in L^+_\infty(\Gamma),\ g \in GL_\infty(\Gamma)$, and

$$\operatorname*{ess\,sup}_{t\in\Gamma} |\arg g(t)| < \pi/p\ . \tag{6.2}$$

Proof. The sufficiency of the conditions (6.1) and (6.2) is a consequence of Theorem 4.1. We are going to prove their necessity. As in Theorem 2.1 we can confine ourselves to the situation when $\Gamma = \mathbf{T}$ and $|a(t)| = 1$. Then the invertibility of the operator $aP_{\mathbf{T}} + Q_{\mathbf{T}}$ both in $L_p(\mathbf{T})$ and in $L_q(\mathbf{T})$ implies that the function a admits a factorization $a = a_- a_+$ with $a_+^{\pm} \in L_p^+(\mathbf{T}),\ a_-^{\pm 1} \in L_p^-(\mathbf{T})$, and $|a_+|^{-1} \in \mathcal{W}_p(\mathbf{T}) \cap \mathcal{W}_q(\mathbf{T})$. As in the proof of Theorem 5.1 (and noting the inclusion $\ln|a_+| \in L_1(\mathbf{T})$) we can write the outer function a_+ in the form

$$a_+ = c\exp(2P_{\mathbf{T}} \ln|a_+|) \quad (c = \text{const.})\ ,$$

whereas Theorem 5.2 gives that $|a_+|^{-1} = \exp(u+\widetilde{v})$ with $\|v\|_\infty < \pi/2p$. Consequently [1] ,

$$a_+ = c_1 \exp(2P_{\mathbf{T}}(u+\widetilde{v}))\ ,$$

c_1 a constant.

---

[1]The c_m's stand for certain constants appearing in the process of the transformation.

Since $|a(t)| = 1$ we have $a_+ a_- \overline{a}_+ \overline{a}_- = 1$, from which it follows $a_- = c_2 \overline{a}_+^{-1}$, that is,

$$\begin{aligned} a &= a_+ a_- = c_3 \exp(2P_{\mathbf{T}}(u + \widetilde{v}) - \overline{2P_{\mathbf{T}}(u + \widetilde{v})}) \\ &= c_4 \exp(2S_{\mathbf{T}}(u + \widetilde{v})) \\ &= c_5 \exp(2iv) \exp(4P_{\mathbf{T}}u - 2u) \\ &= h(t)g(t) , \end{aligned}$$

with $h(t) = c_5 \exp(4P_{\mathbf{T}}u)$ and $g(t) = \exp(-2u + 2iv)$.

Theorem 2.1 of Chapter 8 implies that $h^{\pm 1} \in L^+_\infty(\mathbf{T})$, and hence, the functions h and g are subject to the desired conditions. ■

Theorem 6.1. *Let Γ be a simple closed Lyapunov curve and $a \in L_\infty(\Gamma)$. Then the operator $A = aP_\Gamma + Q_\Gamma$ is a Φ-operator both in $L_p(\Gamma)$ and $L_q(\Gamma)$ $(p \geq 2,\ p^{-1} + q^{-1} = 1)$ with* Ind $A|L_p(\Gamma) =$ Ind $A|L_q(\Gamma)$ $(= \kappa)$ *if and only if the function a admits a factorization $a(t) = g(t)h(t)t^\kappa$ with $h^{\pm 1} \in L^+_\infty(\Gamma)$, $g \in GL_\infty(\Gamma)$ and* ess sup $|\arg g(t)| < \pi/p$.

Proof. Apply Lemma 6.1 to the operator $B = t^{-\kappa} a P_\Gamma + Q_\Gamma$. ■

Theorem 6.2 *Let Γ be a simple closed Lyapunov curve, $0 \in F^+_\Gamma$, $a \in L_\infty(\Gamma)$, $A = aP_\Gamma + Q_\Gamma$, $p \geq 2$, and $q^{-1} + p^{-1} = 1$. Then the following assertions are equivalent:*

(1°) $a \in \mathcal{S}_p(\Gamma)$

(2°) $A \in \Phi(L_r(\Gamma))$ *for all* $r \in [q, p]$

(3°) $A \in \Phi(L_p(\Gamma)) \cap \Phi(L_q(\Gamma))$, *and* Ind $A|L_p(\Gamma) =$ Ind $A|L_q(\Gamma)$

(4°) *the function a may be factored as*

$$a(t) = g(t)h(t)t^\kappa$$

with $h^{\pm 1} \in L^+_\infty(\Gamma)$, $g \in GL_\infty(\Gamma)$, ess sup $|\arg g(t)| < \pi/p$, *and* $\kappa = -$Ind A .

Proof. The implication $(1°) \Longrightarrow (2°)$ follows from Theorem 4.1, $(3°) \Longrightarrow (4°)$ from Theorem 6.1, and $(4°) \Longrightarrow (1°)$ is evident. It remains to verify the implication $(3°) \Longrightarrow (4°)$. Its proof results from the following lemma:

Lemma 6.2. *Let $a \in L_\infty(\Gamma)$ and $r \in (1, \infty)$. If the operator $A = aP_\Gamma + Q_\Gamma$ is a Φ-operator on the space $L_r(\Gamma)$, then there is a number $\delta > 0$ such that $A \in \Phi(L_p(\Gamma))$ for all $p \in (r - \delta, r + \delta)$ and*

$$\text{Ind } A|L_p(\Gamma) = \text{Ind } A|L_r(\Gamma) .$$

Proof of the lemma. Let $\operatorname{Ind} A|L_r(\Gamma) =: -\kappa$. Then the function a is factorizable into $a(t) = a_-(t)t^\kappa a_+(t)$ with $a_+^{\pm 1} \in L_1^+(\Gamma)$, $a_-^{\pm 1} \in L_1^-(\Gamma)$, and $|a_+|^{-1} \in \mathcal{W}_r(\Gamma)$. In HUNT-MUCKENHOUPT-WHEEDEN [1] it is shown that if $\rho \in \mathcal{W}_r(\Gamma)$, then there exists a neighbourhood $(r-\delta,\, r+\delta)$ of r such that $\rho \in \mathcal{W}_p(\Gamma)$ whenever $p \in (r-\delta,\, r+\delta)$. For these values of p , the equality $a(t) = a_-(t)t^\kappa a_+(t)$ gives the factorization of the function a in $L_p(\Gamma)$. Consequently, $A \in \Phi(L_p(\Gamma))$, and $\operatorname{Ind} A|L_p(\Gamma) = -\kappa$. ■

12.7 Extension of the class of coefficients

Denote by $\mathcal{N}_p(\Gamma)$ the set of all p–non–singular functions in $PC(\Gamma)$ (see Sections 9.1 and 9.3). None of the function classes $\mathcal{N}_p(\Gamma)$ and $\mathcal{S}_p(\Gamma)$ is contained in the other, and each of the conditions $a \in \mathcal{N}_p(\Gamma)$ and $a \in \mathcal{S}_p(\Gamma)$ is sufficient for the Fredholmness of the operator $A = aP_\Gamma + Q_\Gamma$ in the space $L_p(\Gamma)$. In the present section we examine a larger class of coefficients, $\mathcal{M}_p(\Gamma)$, which is distinguished by the inclusion $\mathcal{N}_p(\Gamma) \cup \mathcal{S}_p(\Gamma) \subset \mathcal{M}_p(\Gamma)$ and by the implication

$$a \in \mathcal{M}_p(\Gamma) \Longrightarrow aP_\Gamma + Q_\Gamma \in \Phi(L_p(\Gamma)) .$$

We let $\mathcal{M}_p(\Gamma)$ refer to the set of all functions $f \in GL_\infty(\Gamma)$ which are representable in the form

$$f = ag , \tag{7.1}$$

where $a \in \mathcal{S}_p(\Gamma)$ and for which the function g possesses one–sided limits $g(t \pm 0)$ at each point $t \in \Gamma$ satisfying the condition

$$\frac{g(t-0)}{g(t+0)} = \exp(2\pi i \gamma(t))$$

with

$$\min\left(0, \frac{2}{p} - 1\right) \leq \operatorname{Re} \gamma(t) \leq \max\left(0, \frac{2}{p} - 1\right) . \tag{7.2}$$

It is immediate from the definition of the class $\mathcal{M}_p(\Gamma)$ that $\mathcal{S}_p(\Gamma) \subset \mathcal{M}_p(\Gamma)$. Let us show that $\mathcal{N}_p(\Gamma) \subset \mathcal{M}_p(\Gamma)$, as well.

For let f be a p–non–singular function, $\tau_1, \ldots, \tau_m$ its points of discontinuity, and

$$\frac{f(\tau_k - 0)}{f(\tau_k + 0)} = \exp(2\pi i \lambda_k) \quad \left(\frac{1}{p} - 1 < \operatorname{Re} \lambda_k < \frac{1}{p}\right) .$$

Assume, for definiteness, $p \geq 2$, and let δ_k be an arbitrary number satisfying

$$\max\left(\alpha_k - \frac{1}{p}, \frac{2}{p} - 1\right) < \delta_k < \min\left(0, \alpha_k + \frac{1}{p}\right) , \tag{7.3}$$

where $\alpha_k = \operatorname{Re} \gamma_k$.

Define $\Psi = \Psi_{\tau_1,\delta_1}\Psi_{\tau_2,\delta_2}\cdot\ldots\cdot\Psi_{\tau_m,\delta_m}$ (see Section 9.2), and set $a = f/\Psi$. Then it is easy to see that

$$\frac{\Psi(\tau_k - 0)}{\Psi(\tau_k + 0)} = \exp(2\pi i\delta_k)$$

and

$$\frac{a(\tau_k - 0)}{a(\tau_k + 0)} = \exp(2\pi i(\lambda_k - \delta_k)) .$$

The relation (7.3) yields that $-1/p < \operatorname{Re}\lambda_k - \delta_k < 1/p$, and it follows that $a \in \mathcal{S}_p(\Gamma)$. Furthermore, the relation (7.3) shows that the numbers δ_k satisfy condition (7.2). Thus, the function $f = a\Psi$ is in $\mathcal{M}_p(\Gamma)$.

Theorem 7.1. *If $f \in \mathcal{M}_p(\Gamma)$ then $fP_\Gamma + Q_\Gamma \in \Phi(L_p(\Gamma))$.*

The proof of this theorem will be given in Section 13.4 after establishing criteria for the Fredholmness of singular operators in the spaces $L_p(\Gamma, \rho)$.

12.8 Exercises

12.1. Represent the function

$$r(t) = \frac{(t-2)(3t-1)}{(2t-1)(t-3)}$$

in the form $h(t)g(t)$, with $h^{\pm 1} \in C^+(\mathbf{T})$ and $|\arg g(t)| < \pi/2$ for all $t \in \mathbf{T}$.

12.2. Let $\varepsilon > 0$ be an arbitrary number and

$$r(t) = \frac{(t-\alpha_1)\ldots(t-\alpha_p)(t-\beta_1)\ldots(t-\beta_q)}{(t-\gamma_1)\ldots(t-\gamma_m)(t-\delta_1)\ldots(t-\delta_n)},$$

with $|\alpha_j| > 1$, $|\beta_j| < 1$, $|\gamma_j| > 1$, $|\delta_j| < 1$. Describe the factorization of the function r in the form $r = hgt^\kappa$ where $h^{\pm 1} \in L_\infty^+(\mathbf{T})$, $\kappa \in \mathbf{Z}$, and $|\arg g(t)| < \varepsilon$ for all $t \in \mathbf{T}$.

12.3. Let $a \in C(\mathbf{T})$ and $\operatorname{ind} a = 0$. Show that the function a can be factored as $a = hg$ where $h \in GC^+(\mathbf{T})$ and $|\arg g(t)| < \varepsilon$.

12.4. Let a be a 2–non–singular function with index $\operatorname{ind} a|L_2(\Gamma) = 0$. Prove that the function a can be written in the form $a = gh$ with $h \in GC^+(\Gamma)$ and $|\arg g| < \pi/2$.

12.5. Is the operator $\sqrt[4]{t^4}P_\mathbf{T} + Q_\mathbf{T}$ a Φ–operator in $L_2(\mathbf{T})$? Verify the existence of three functions $a_1, a_2, a_3 \in L_\infty(\mathbf{T})$ such that $a_k^4(t) = t^4$ $(k = 1, 2, 3)$ the operator $a_1P_\mathbf{T} + Q_\mathbf{T}$ is invertible on $L_2(\mathbf{T})$, the operator $a_2P_\mathbf{T} + Q_\mathbf{T}$ is a Φ–operator but not invertible, and the operator $a_3P_\mathbf{T} + Q_\mathbf{T}$ is not a Φ–operator.

12.6. Let $a\ (\in L_\infty(\mathbf{T}))$ be a function taking only two values. Prove that the operator $aP_{\mathbf{T}} + Q_{\mathbf{T}}$ is invertible on $L_2(\mathbf{T})$ whenever it is Fredholm.

12.7. Does the preceding assertion hold for spaces $L_p(\mathbf{T})$ with $p \neq 2$?

12.8. Formulate an analogue to Theorem 2.1 for the space $L_2[0,1]$.

12.9. Let $g \in PC(\Gamma)$ and $g(t \pm 0) \neq 0$ for all $t \in \Gamma$. Prove that the operator $agP_\Gamma + Q_\Gamma$ is Fredholm on $L_p(\Gamma)$ for each function $a \in \mathcal{S}_p(\Gamma)$ if and only if at each point $t \in \Gamma$, the quotient $g(t-0)/g(t+0)$ can be written as

$$\frac{g(t-0)}{g(t+0)} = \exp(2\pi i \gamma(t)) ,$$

with

$$\min\left(0, \frac{2}{p} - 1\right) \leq \gamma(t) \leq \max(0, \frac{2}{p} - 1) .$$

12.10. Denote by $\overline{\mathcal{M}_p}(\Gamma)$ the "local closure" of the class $\mathcal{M}_p(\Gamma)$: We say that $a \in \overline{\mathcal{M}_p}(\Gamma)$ if for each point $\tau \in \Gamma$ and each $\varepsilon > 0$ there are a neighbourhood $U(\tau)$ of τ and a function $a_\tau \in \mathcal{M}_p(\Gamma)$ such that

$$\operatorname{ess} \sup_{t \in U(\tau)} |a(t) - a_\tau(t)| < \varepsilon .$$

Prove that if $a \in \overline{\mathcal{M}_p}(\Gamma)$ then $aP_\Gamma + Q_\Gamma \in \Phi(L_p(\Gamma))$.

Comments and references

12.1, 12.3, 12.4. The results of these sections go back to SIMONENKO [2-6]. He introduced the classes $\mathcal{S}_p(\Gamma)$ and investigated them by means of his local principle (see SIMONENKO [2-6]).

12.2. Necessary conditions for the invertibility of singular operators on the space $L_2(\Gamma)$ were obtained in the papers HELSON/SZEGÖ[1], DEVINATZ [1], SIMONENKO [3], and criteria for the one-sided invertibility of singular operators on $L_2(\mathbf{T})$ were achieved by LEE/SARASON [1], DOUGLAS/SARASON [1] and DOUGLAS [1].

12.5, 12.6. Theorem 5.1 was established by HELSON/SZEGÖ [1] whereas Theorems 5.2, 6.1 and 6.2 are due to KRUPNIK [2].

12.7. These results were obtained by FROLOV [1, 2].

Chapter 13

Exact constants in theorems on the boundedness of singular operators

In the ninth chapter we derived upper estimates for the norm and the quotient norm of singular operators on spaces $L_p(\Gamma, \rho)$. Now we are going to show that these estimates are exact for the operator S_Γ in the spaces $L_p(\Gamma, \rho)$ as well as for the operators $aI + bS_\Gamma$ in $L_2(\Gamma, \rho)$. These results will be used to obtain sufficient conditions for the Fredholmness of singular integral operators in $L_p(\Gamma, \rho)$.

Throughout this chapter we denote the quotient norm of an operator A acting in $\mathcal{B}$ by $|A|$,

$$|A| = \inf_{T \in \mathcal{T}(\mathcal{B})} \|A + T\| .$$

13.1 Norm and quotient norm of the operator of singular integration

This section is devoted to the computation of the norm and the quotient norm of the operator of singular integration. We start with the following

Theorem 1.1. *Let*

$$f(t) = \sum_{k=-n}^{n} f_k t^k \qquad (|t| = 1)$$

be a trigonometric polynomial with coefficients $f_k \in \mathbf{C}$ *satisfying* $f_{-k} = \overline{f}_k$. *Then the function*

$$g(t) := i \left(\sum_{k=-n}^{-1} - \sum_{k=1}^{n} \right) f_k t^k \tag{1.1}$$

is subject to the estimate

$$\|g\|_{L_p(\mathbf{T})} \leq \tan \frac{\pi}{2p} \|f\|_{L_p(\mathbf{T})} \tag{1.2}$$

for $1 < p \leq 2$.

The proof of this theorem is based on the following two lemmas.

Lemma 1.1. *If* $|x| < \pi/2$ *and* $1 < p \leq 2$, *then*

$$|\sin x|^p \leq \tan^p \frac{\pi}{2p} \cos^p x - \beta(p) \cos px \ , \tag{1.3}$$

with

$$\beta(p) := \frac{\sin^{p-1}\left(\frac{\pi}{2p}\right)}{\cos\left(\frac{\pi}{2p}\right)} \ .$$

Proof. Consider the function

$$F(x) := \frac{\sin^p x + \beta(p) \cos px}{\cos^p x}$$

on the interval $0 < x < \pi/2$. Obviously,

$$F'(x) = p \, \frac{\sin^{p-1} x}{\cos^{p-1} x} \, g(x) \ ,$$

with $g(x) := 1 - \beta(p)(\sin(p-1)x)/\sin^{p-1} x$. Since $g'(x) = -\beta(p)(p-1)\sin(2-p)x \sin^{-p} x < 0$, the function g is decreasing. Moreover, $g(\pi/2p) = 0$, and thus $F'(x) > 0$ if $0 < x < \pi/2p$ and $F'(x) < 0$ if $\pi/2p < x < \pi/2$. This shows that $F(x) \leq F(\pi/2p) = \tan^p(\pi/2p)$; $0 < x < \pi/2$, whence the inequality (1.3) follows. ∎

Let D be a region in the complex plane and Γ its boundary. Remember that a continuous real-valued function u defined on D is called subharmonic if the inequality

$$u(z_0) \leq \frac{1}{2\pi} \int_0^{2\pi} u(z_0 + re^{i\theta}) \, d\theta \ . \tag{1.4}$$

holds for each closed disc $Q \subset D$ with centre z_0 and radius r . It is well-known (see TIMAN/TROFIMOV[1]) that a continuous real-valued function u is subharmonic in the region D if and only if the inequality (1.4) holds for each point $z_0 \in D$ and for each sufficiently small radius r .

Lemma 1.2. *The function* $g_0 : \mathbf{C} \to \mathbf{R}$ *defined by*

$$g_0(z) := |z|^p \cos(p\alpha(z))$$

with $\alpha(x+iy) := \arctan(y/|x|)$ *and* $1 < p \leq 2$ *is subharmonic in* $\mathbf{C}$.

Proof. Since g_0 is a continuous function in $\mathbf{C}$ it suffices to verify the inequality (1.4) for each point $z_0 \in \mathbf{C}$ and for each sufficiently small r .

The function g_0 coincides with the harmonic function $\text{Re } z^p$ $(|\arg z| < \pi/2)$ in the right half plane $\text{Re } z > 0$ and with the harmonic function $\text{Re } (-z)^p$ $(|\arg(-z)| < \pi/2)$ in the left half plane. Thus the estimate (1.4) holds for all z_0 with $\text{Re } z_0 \neq 0$ whenever r is sufficiently small. Now let $z_0 = 0$. Then, for each positive r ,

$$\frac{1}{2\pi}\int_{-\pi}^{\pi} g_0(re^{i\theta})\, d\theta = \frac{1}{\pi}\int_{-\pi/2}^{\pi/2} r^p \cos(p\theta)\, d\theta = \frac{2r^p}{\pi p}\sin\frac{p\pi}{2} \geq 0 = g_0(0)$$

and, hence, the inequality (1.4) also holds for $z_0 = 0$.

It remains to consider the points $z_0 = iy$ with y being an arbitrary real number different from zero. Define $h_0(z) := \text{Re } z^p, z \neq 0, |\arg z| < \pi$. The functions g_0 and h_0 coincide in the right half plane $\text{Re } z \geq 0$. We claim that the difference $g_0 - h_0$ is non-negative at those points z in the left half plane which satisfy $\text{Im } z \neq 0$. Let $z = ae^{ix}$ $(a > 0,\ \pi/2 \leq x < \pi)$. Then

$$g_0(z) - h_0(z) = a^p(\cos p(x-\pi) - \cos px) = 2a^p \sin p(x - \frac{\pi}{2}) \sin\frac{\pi p}{2} \geq 0 .$$

On the other hand, if $z = ae^{ix}$ $(a > 0, -\pi < x \leq -\pi/2)$, then

$$g_0(z) - h_0(z) = a^p(\cos p(x+\pi) - \cos px) = -2a^p \sin((x + \frac{\pi}{2})p) \sin\frac{\pi p}{2} \geq 0 .$$

Thus, $g_0 - h_0 \geq 0$. Since, moreover, the function h_0 is harmonic in the complex plane with a cut along the negative real half-axis, we obtain for each number $r,\ \ 0 < r < |y|$,

$$g_0(iy) = h_0(iy) = \frac{1}{2\pi}\int_{-\pi}^{\pi} h_0(iy + re^{i\theta})\, d\theta \leq \frac{1}{2\pi}\int_{-\pi}^{\pi} g_0(iy + re^{i\theta})\, d\theta .$$

Hence, (1.4) is valid. ■

Proof of Theorem 1.1. Define

$$h(z) := \sum_{k=0}^{n} f_k z^k .$$

It is easy to see that $f = \text{Re } h$, and $g = \text{Im } h$ on $\mathbf{T}$. Denote by F the function $F(z) = g_0(h(z))$, where g_0 is the subharmonic function appearing in Lemma 1.2. Since the function h is holomorphic in $\mathbf{C}$ and the function g_0 is subharmonic, the function F is subharmonic as well.

Let $\psi(z) := \alpha(h(z))$ with $\alpha(x+iy) = \text{arc}\ \tan(y/|x|)$; the function ψ being defined at all points z with $h(z) \neq 0$. At those points z which are zeros of the polynomial h we define ψ in an arbitrary manner. The function F can be written in the form $F(z) = |h| \cos(p\psi)$, and a little thought shows that, for $|z| = 1$, the equalities

$$|f(z)| = |h(z)| \cos \psi(z), \quad |g(z)| = |h(z) \sin \psi(z)| \tag{1.5}$$

hold. Lemma 1.1 implies that

$$\begin{aligned} \int_{\mathbf{T}} |h(z)|^p \, |\sin \psi(z)|^p \, |dz| \ &\leq \ \tan^p \frac{\pi}{2p} \int_{\mathbf{T}} |h(z)|^p \cos^p \psi(z) |dz| \\ &- \ \beta(p) \int_{\mathbf{T}} |h(z)|^p \cos(p\psi(z)) \, |dz| \, . \end{aligned} \tag{1.6}$$

We are going to show that the second integral on the right hand side of inequality (1.6) is non-negative. Since f is a real-valued, we have $\text{Im}\, h(0) = 0$. Thus, if $h(0) \neq 0$, then $\psi(0) = 0$, and hence $F(0) = |h(0)|^p$. Furthermore, since the function F is subharmonic,

$$0 \leq F(0) \leq \frac{1}{2\pi} \int_0^{2\pi} F(e^{i\theta}) \, d\theta = \frac{1}{2\pi} \int_{\mathbf{T}} |h(z)|^p \cos(p\psi(z)) |dz| \tag{1.7}$$

From (1.5), (1.6), and (1.7) we conclude that

$$\int_{\mathbf{T}} |g(t)|^p |dt| \leq \tan^p \frac{\pi}{2p} \int_{\mathbf{T}} |f(t)|^p |dt| \, .$$

■

The basic result of this section is the following theorem.

Theorem 1.2. *Let Γ be an arbitrary composed curve. Then*

$$\inf_{T \in \mathcal{T}} \|S_\Gamma + T\|_p = \begin{cases} \tan \frac{\pi}{2p} & \text{if} \quad 1 < p \leq 2 \\ \cot \frac{\pi}{2p} & \text{if} \quad 2 \leq p < \infty \, . \end{cases} \tag{1.8}$$

In order to prove Theorem 1.2 we need the following lemma.

Lemma 1.3. *Let $L_p^{\mathbf{R}}(\Gamma)$ denote the real L_p–space on the curve Γ , and let $A \in L(L_p^{\mathbf{R}}(\Gamma))$. If $\widetilde{A} \in L(L_p(\Gamma))$ is an operator satisfying $\widetilde{A} f = Af$ for all $f \in L_p^{\mathbf{R}}(\Gamma)$, then*

$$\|\widetilde{A}\|_{L(L_p(\Gamma))} = \|A\|_{L(L_p^{\mathbf{R}}(\Gamma))} \, .$$

Proof. Obviously, $\|A\|_{L(L_p^{\mathbf{R}}(\Gamma))} \leq \|\widetilde{A}\|_{L(L_p(\Gamma))}$. Let us show the reverse inequality. For let f_1, f_2 be functions in $L_p^{\mathbf{R}}(\Gamma)$ and α a number such that $0 \leq \alpha \leq 2\pi$. Then

$$\int_\Gamma |(Af_1)(t)\cos\alpha + (Af_2)(t)\sin\alpha|^p |dt| \leq \|A\|^p_{L(L_p^{\mathbf{R}}(\Gamma))} \int_\Gamma |f_1(t)\cos\alpha + f_2(t)\sin\alpha|^p |dt|,$$

and consequently,

$$\int_\Gamma \int_0^{2\pi} |(Af_1)(t)\cos\alpha + (Af_2)(t)\sin\alpha|^p d\alpha |dt|$$

$$\leq \|A\|^p_{L(L_p^{\mathbf{R}})} \int_\Gamma \int_0^{2\pi} |f_1(t)\cos\alpha + f_2(t)\sin\alpha|^p \, d\alpha |dt| .$$

Taking into account that

$$\int_0^{2\pi} |a\cos\alpha + b\sin\alpha|^p \, d\alpha = C_p(a^2+b^2)^{p/2}$$

for a certain constant C_p which is independent of a and b, one immediately obtains

$$\int_\Gamma |(Af_1)^2 + (Af_2)^2|^{p/2} \, |dt| \leq \|A\|^p_{L(L_p^{\mathbf{R}}(\Gamma))} \int_\Gamma |f_1^2 + f_2^2|^{p/2} \, |dt| .$$

Defining the vector f by $f = f_1 + if_2$, we can rewrite this inequality as

$$\|\widetilde{A}f\| \leq \|A\|_{L(L_p^{\mathbf{R}}(\Gamma))} \|f\| ,$$

and the lemma is proved. ■

Proof of Theorem 1.2. Let H stand for the operator which is defined for real trigonometric polynomials by

$$H\left(\sum_{k=-n}^{n} f_k t^k\right) = i\left(\sum_{k=-n}^{-1} - \sum_{k=1}^{n}\right) f_k t^k \quad (|t|=1) .$$

This operator H differs from the operator $-iS_\Gamma$ by only a rank one operator, K_0, and so Theorem 1.1 implies that

$$\|iS_{\mathbf{T}} - K_0\|_{L(L_p^{\mathbf{R}}(\mathbf{T}))} = \|H\|_{L(L_p^{\mathbf{R}}(\mathbf{T}))} \leq \tan\frac{\pi}{2p} \quad (1 < p \leq 2) .$$

By Lemma 1.3,

$$\|iS_{\mathbf{T}} - K_0\|_{L(L_p(\mathbf{T}))} \leq \tan\frac{\pi}{2p}, \quad (1 < p \leq 2)$$

and so

$$\inf_{T\in\mathcal{T}} \|S_{\mathbf{T}} + T\| \le \tan\frac{\pi}{2p} \quad (1 < p \le 2).$$

Passing to the adjoint operator, we obtain

$$\inf_{T\in\mathcal{T}} \|S_{\mathbf{T}} + T\| \le \cot\frac{\pi}{2p} \quad (2 \le p < \infty)\,.$$

These estimates, in combination with (9.5) and (9.6) of Chapter 9, give the equality (1.8) in case Γ is a circle. By means of Theorem 2.1 of Chapter 7 one can generalize this equality to the case of an arbitrary composed curve. ■

For the sake of completeness we mention two other theorems concerning the norm of the operator S in L_p.

Theorem 1.3. *For the operator*

$$(S\varphi)(t) = \frac{1}{\pi i}\int_{-\infty}^{\infty} \frac{\varphi(\tau)}{\tau - t}\,d\tau \quad (-\infty < t < \infty)$$

the equality

$$\|S\|_{L(L_p(\mathbf{R}))} = \begin{cases} \cot\frac{\pi}{2p} & \text{if} \quad 2 \le p < \infty \\ \tan\frac{\pi}{2p} & \text{if} \quad 1 < p \le 2 \end{cases}$$

holds.

Proof. The upper estimate of the norm $\|S\|_{L(L_p(\mathbf{R}))}$ proceeds by means of Theorem 1.1, as in the proof of Theorem 3.1 in Chapter XVI of ZYGMUND [1]. The lower estimate was obtained in Section 9.9 of the present monograph. ■

Another consequence of Theorem 1.2 is the following result:

Theorem 1.4. *Let $S_{\mathbf{T}}$ be the singular integral operator on the unit circle $\mathbf{T}$ and $1 < p < \infty$. Then*

$$\inf_{T\in\mathcal{T}} \|S_{\mathbf{T}} + T\|_{L(L_p(\mathbf{T}))} = \|S_{\mathbf{T}}\|_{L(L_p)}\,.$$

Proof. Denote by P_N the projection operator acting via

$$P_N\left(\sum_j \varphi_j t^j\right) := \sum_{|k|<N} \varphi_k t^k \quad (|t| = 1)\,.$$

It is well–known that for each $1 < p < \infty$ and $\varphi \in L_p(\mathbf{T})$,

$$\lim_{N\to\infty} \|P_N\varphi - \varphi\| = 0\,.$$[1]

[1] All norms are taken in $L_p(\mathbf{T})$ or $L(L_p(\mathbf{T}))$, respectively.

Consider the operator B_N defined by

$$(B_N\varphi)(t) := \varphi(t^{2N})t^N \quad (|t| = 1) .$$

Obviously, B_N is an isometric operator, which satisfies the identities

$$S_{\mathbf{T}}B_N = B_N S_{\mathbf{T}} \quad \text{and} \quad P_N B_N = 0 .$$

Now let T refer to an arbitrary compact linear operator. Then

$$\|S_{\mathbf{T}} - TP_N\| \geq \|S_{\mathbf{T}}\| .$$

Indeed, if φ_0 is an vector such that $\|S_{\mathbf{T}}\varphi_0\| \geq (\|S_{\mathbf{T}}\| - \varepsilon)\|\varphi_0\|$, then

$$\|(S_{\mathbf{T}} - TP_N)B_N\varphi_0\| = \|S_{\mathbf{T}}B_N\varphi_0\| = \|B_N S_{\mathbf{T}}\varphi_0\| = \|S_{\mathbf{T}}\varphi_0\| ,$$

from which it follows that

$$\|(S_{\mathbf{T}} - TP_N)B_N\varphi_0\| \geq (\|S_{\mathbf{T}}\| - \varepsilon)\|B_N\varphi_0\| .$$

This inequality, in combination with the fact that

$$\lim_{N\to\infty} \|T - TP_N\| = \lim_{N\to\infty} \|T^* - P_N T^*\| = 0 ,$$

gives

$$\|S_{\mathbf{T}} - T\| \geq \|S_{\mathbf{T}}\| .$$

■

Finally we remark that Theorem 1.4 fails to be true if the circle $\mathbf{T}$ is replaced by another closed simple Lyapunov curve Γ , even in the Hilbert space case $L_2(\Gamma)$.

Indeed, if Γ is not a circle then, by Theorem 7.2 in [GK 1], Chapter 1, $S_\Gamma^* \neq S_\Gamma$. Since, on the other hand, $S_\Gamma^2 = I$, we have $S_\Gamma^* S_\Gamma \neq I$, and thus, the spectrum of $S_\Gamma^* S_\Gamma$ necessarily contains a point which is different from one. The identity $S_\Gamma^* S_\Gamma - \lambda I = S_\Gamma^*(I - \lambda S_\Gamma^* S_\Gamma)S_\Gamma$ and the invertibility of the operators S_Γ^* and S_Γ then imply that the spectrum of $S_\Gamma^* S_\Gamma$ contains a real number greater than one. Hence, $\|S_\Gamma\|_{L(L_2(\Gamma))} > 1$, but $\inf_{T\in\mathcal{T}} \|S_\Gamma + T\|_{L(L_2(\Gamma))} = 1$ by Theorem 2.1 of Chapter 7.

Furthermore, it is interesting to notice that there exist curves with corners (different from the circle) such that

$$\|S_\Gamma\| = \inf_{T\in\mathcal{T}} \|S_\Gamma + T\| .$$

For example one can take a curve consisting of two circular arcs or of one circular arc and one straight line (see Exercise 13.6).

13.2 A second proof of Theorem 4.1 of Chapter 12

Here we present a simple proof for the implication

$$a \in \mathcal{S}_p(\Gamma) \Longrightarrow aP_\Gamma + Q_\Gamma \in \Phi(L_p(\Gamma)) , \tag{2.1}$$

which is true for closed Lyapunov curves. Invoking the local principle established in [GK 1, 5.1] (see also Proposition 6.8.1 of Chapter 6), we can limit ourselves to the case when $a = h_- c h_+$ with $h_-^{\pm 1} \in L_\infty^-(\Gamma)$, $h_+^{\pm 1} \in L_\infty^+(\Gamma)$, $c(t) = |c(t)|^{i\varphi(t)}$ and $\|\varphi\|_\infty < \pi / \max(p, \rho)$. Moreover, by Corollary 2.2 of Chapter 8, it suffices to check the implication (2.1) for the operator $A_0 = \exp(i\varphi)P_\Gamma + Q_\Gamma$. Finally, we can assume without loss that $p \geq 2$. In that case, $\|\varphi\|_\infty < \pi/p$. The operator A_0 can be represented as

$$A_0 = \frac{a_0+1}{2}\left(I + \frac{a_0-1}{a_0+1}S_\Gamma\right) = \frac{a_0+1}{2}\left(I + \left(i\tan\frac{\varphi}{2}\right)S_\Gamma\right)$$

with $a_0(t) = \exp i\varphi(t)$.

Since ess sup $|\tan\frac{\varphi}{2}| < \tan\frac{\pi}{2p}$ and $\inf\|S_\Gamma + T\| = \cot\frac{\pi}{2p}$ we have

$$\inf\|i\tan\frac{\varphi}{2}S_\Gamma + T\| < 1 ,$$

and consequently, $A_0 \in \Phi(L_p(\Gamma))$. ■

13.3 Norm and quotient norm of the operator S_Γ on weighted spaces

In this section we compute the norm of the operator $S_\mathbf{T}$ in the spaces $L_p(\mathbf{T}, |t-t_0|^\alpha)$ $(t_0 \in \mathbf{T})$ for all pairs of numbers p, α which guarantee the boundedness of the operator $S_\mathbf{T}$. After this we apply these results to calculate the quotient norm of the operator S_Γ along an arbitrary simple closed Lyapunov curve Γ on the space $L_p(\Gamma, \rho)$, where

$$\rho(t) = \prod_{k=1}^{n} |t - t_k|^{\beta_k} \tag{3.1}$$

$(1 < p < \infty, -1 < \beta_k < p-1)$. Remember that the norm in $L_p(\Gamma, \rho)$ is given by

$$\|\varphi\|^p = \int_\Gamma |\varphi(t)|^p \rho(t) |dt| ,$$

and so it follows that

$$L_p(\Gamma, \rho)^* = L_q(\Gamma, \rho^{1-q}) .$$

In particular,

$$\|S_{\mathbf{T}}\|_{L_p(\mathbf{T},|t-t_0|^\alpha)} = \|S_{\mathbf{T}}\|_{L_q(\mathbf{T},|t-t_0|^{\alpha(1-q)})} \tag{3.2}$$

$(1 < p < \infty,\ -1 < \alpha < p-1,\ q^{-1} + p^{-1} = 1)$.

The following lemmas are useful for the problem of computing the norm of singular integral operators in the spaces $L_p(\Gamma, \rho)$.

Lemma 3.1. *Let $x_1, \ldots, x_n$ be pairwise distinct real numbers, $t_k = (x_k + i)\cdot$ $\cdot(x_k - i)^{-1}$ $(k = 1, \ldots, n)$, $t_0 = 1$, and $p, \beta_0, \beta_1, \ldots, \beta_n$ be certain real numbers satisfying the conditions*

$$1 < p < \infty,\ -1 < \beta_k < p-1 \quad (k = 0, 1, \ldots, n).$$

Define

$$\rho(t) = \prod_{k=0}^{n} |t - t_k|^{\beta_k}, \quad \rho_1(x) = |x - i|^{\beta} \prod_{k=1}^{n} |x - x_k|^{\beta_k},$$

where $\beta = p - 2 - \beta_0 - \sum_{k=1}^{n} \beta_k$. Then

$$\|S_{\mathbf{R}}\|_{L_p(\mathbf{R},\rho_1)} = \|S_{\mathbf{T}}\|_{L_p(\mathbf{T},\rho)} \tag{3.3}$$

and

$$|S_{\mathbf{R}}|_{L_p(\mathbf{R},\rho_1)} = |S_{\mathbf{T}}|_{L_p(\mathbf{T},\rho)}. \tag{3.4}$$

Proof. The operator B defined by

$$(B\varphi)(t) = \frac{1}{t-1}\varphi\left(i\frac{t+1}{t-1}\right)$$

is a bounded linear operator acting from $L_p(\mathbf{R}, \rho_1)$ into $L_p(\mathbf{T}, \rho)$. Moreover, $S_{\mathbf{R}} = -B^{-1} S_{\mathbf{T}} B$, and

$$\|B_\varphi\|_{L_p(\mathbf{T},\rho)} = \delta \|\varphi\|_{L_p(\mathbf{R},\rho_1)}$$

with a constant δ being independent of φ (see [GK 1, Theorem 5.1 of Chapter 1]). This easily implies (3.3) and (3.4). ∎

In particular, choosing $\beta = \beta_1 = \ldots = \beta_n = 0$, we obtain

$$\|S_{\mathbf{T}}\|_{L_p(\mathbf{T},|t-1|^{p-2})} = \|S_{\mathbf{R}}\|_{L_p(\mathbf{R})} =: \nu_p. \tag{3.5}$$

By invoking the interpolation theorem 1.4 of Chapter 1 of [GK 1] (see also 6.9), we find that

$$\|S_{\mathbf{T}}\|_{L_p(\mathbf{T},|t-1|^\alpha)} = \nu_p \quad (\min(0, p-2) \le \alpha \le \max(0, p-2)).$$

Let $t_0 \in \mathbf{T}$ be an arbitrary point and define $(V\varphi)(t) = \varphi(t_0 t)$. Because $V^{-1} S_{\mathbf{T}} V = S_{\mathbf{T}}$ and $\|V\varphi\|_{L_p(\mathbf{T},|t-1|^\alpha)} = \|\varphi\|_{L_p(\mathbf{T},|t-t_0|^\alpha)}$, the following lemma holds:

Lemma 3.2. *Let* $t_0 \in \mathbf{T}$, $1 < p < \infty$, *and* $\min(0, p-2) \leq \alpha \leq \max(0, p-2)$. *Then*

$$\|S_{\mathbf{T}}\|_{L_p(\mathbf{T},|t-t_0|^\alpha)} = \nu_p . \tag{3.6}$$

It is interesting to remark that in the case $\min(0,p-2) \leq \max(0,p-2)$ the norm of the operator $S_{\mathbf{T}}$ actually depends on α , as the next theorem shows.

Theorem 3.1. *Let* $t_0 \in \mathbf{T}$ *and* $-1 < \alpha < p-1$. *Then*

$$\|S_{\mathbf{T}}\|_{L_p(\mathbf{T},|t-t_0|^\alpha)} = \nu(p,\alpha) , \tag{3.7}$$

with $\nu(p,\alpha)$ *defined by*

$$\nu(p,\alpha) = \begin{cases} \cot \frac{\pi(p-1-\alpha)}{2p} & \text{if } \; p-2 < \alpha < p-1 \\ \cot \frac{\pi}{2p} & \text{if } \; 0 \leq \alpha \leq p-2 \\ \cot \frac{\pi(1+\alpha)}{2p} & \text{if } \; -1 < \alpha < 0 \end{cases} \tag{3.8}$$

in the case $2 \leq p < \infty$, *and by*

$$\nu(p,\alpha) = \nu(q, \alpha(1-q)) \quad (q^{-1} + p^{-1} = 1)$$

in the case $1 < p \leq 2$.

Proof. To start with, let $2 \leq p < \infty$ and $p-2 < \alpha < p-1$. Put $r = p(p-1-\alpha)^{-1}$ and $s = r(r-1)^{-1}$. Then $s < p < r$, that is the number $1/p$ can be written as $p^{-1} = r^{-1}(1-\theta) + s^{-1}\theta$ for some $0 < \theta < 1$. A straightforward computation shows that

$$(1-\theta)(r-2) = \frac{pr - p - r}{p} = \frac{\alpha r}{p} ,$$

and it follows that

$$|t - t_0|^{\alpha/p} = |t - t_0|^{(1-\theta)\frac{r-2}{r}} .$$

Let us abbreviate the norm $\|S_{\mathbf{T}}\|_{L_p(\mathbf{T},|t-t_0|^\alpha)}$ by $\|S_{\mathbf{T}}\|_{p,\alpha}$. The latter identity, in combination with the interpolation theorem 9.1 of Chapter 6, yields that

$$\|S_{\mathbf{T}}\|_{p,\alpha} \leq \|S_{\mathbf{T}}\|_{r,r-2}^{1-\theta} \, \|S_{\mathbf{T}}\|_s^\theta . \tag{3.9}$$

From Lemma 3.2 we know that $\|S_{\mathbf{T}}\|_{r,r-2} = \|S_{\mathbf{T}}\|_s = \nu_p$. Thus, since $r > 2$,

$$\|S_{\mathbf{T}}\|_{p,\alpha} \leq \cot \frac{\pi}{2r} = \nu(p,\alpha) . \tag{3.10}$$

This estimate implies in particular that $\|S_{\mathbf{T}}\|_{2,\alpha} \leq \nu(2,\alpha)$ $(0 < \alpha < 1)$, and by means of equality (3.2), we get that $\|S_{\mathbf{T}}\|_{2,\alpha} \leq \nu(2,\alpha)$ for $\alpha \in (-1,0)$.

Next let us consider the case $2 < p < \infty$ and $-1 < \alpha < 0$. Define $r = p(1+\alpha)^{-1}$ and $\beta = (2+2\alpha-p)p^{-1}$. It is immediate that $2 < p < r$ and $-1 < \beta < 0$, and so we can write $1/p$ in the form $p^{-1} = (1-\theta)/2 + \theta/r$ with $\theta = (p-2)(p-2-2\alpha)^{-1} \in (0,1)$. Since

$$\beta = \frac{2-r}{r},\ 1-\theta = \frac{2(r-p)}{p(r-2)} \quad \text{and} \quad \alpha = \frac{p-r}{r}$$

we have $\frac{(1-\theta)\beta}{2} = \frac{\alpha}{p}$ and, again by the interpolation theorem 9.1 of Chapter 6,

$$\|S_{\mathbf{T}}\|_{p,\alpha} \le \|S_{\mathbf{T}}\|_{2,\beta}^{1-\theta} \|S_{\mathbf{T}}\|_r^{\theta} . \tag{3.11}$$

As we have already seen,

$$\|S_{\mathbf{T}}\|_{2,p} \le \nu(2,\beta) = \cot \frac{\pi(1+\alpha)}{2p}$$

and

$$\|S_{\mathbf{T}}\|_r = \cot \frac{\pi}{2r} = \cot \frac{\pi(1+\alpha)}{2p} .$$

Substituting these estimates into (3.11), we obtain

$$\|S_{\mathbf{T}}\|_{p,\alpha} \le \nu(p,\alpha) \qquad (p > 2 \quad \text{and} \quad -1 < \alpha < 0) .$$

Since the case $p \ge 2,\ 0 \le \alpha \le p-2$ has been already considered in Lemma 3.2, we conclude that the estimate (3.10) holds for all values $p \ge 2$ and $-1 < \alpha < p-1$. The reverse inequality has been verified in Section 9.9. Consequently,

$$\|S_{\mathbf{T}}\|_{p,\alpha} = \nu(p,\alpha) \quad (p \ge 2, -1 < \alpha < p-1) .$$

In case that $p < 2$ the assertion follows by passing to the adjoints (cf. equality (3.2)).■

Analogously one proves

Theorem 3.2. *Let $1 < p < \infty$, $-1 < \alpha < p-1$ and $t_0 \in \mathbf{R}$. Then*

$$\|S_{\mathbf{R}}\|_{p,\alpha} = \nu(p,\alpha) , \tag{3.12}$$

where $\|\cdot\|_{p,\alpha}$ here denotes the norm in $L_p(\mathbf{R}, |t-t_0|^\alpha)$.

Theorem 3.1 implies that the norm of the operator $S_{\mathbf{T}}$ in the space $L_p(\mathbf{T}, |t-t_0|^\alpha)$ does not depend on the position of the point t_0 on the curve $\mathbf{T}$. On the other hand, if $\rho(t) = \prod_{k=1}^n |t-t_k|^{\beta_k}$ and $n \ge 2$ then it turns out that the norm of the operator $S_{\mathbf{T}}$ depends essentially on the location of the points $t_1, \dots, t_k$ on $\mathbf{T}$ (see Exercise 13.9). In Section 7.2. we saw that $\|S_\Gamma\|$ depends on the form of the curve Γ. However the quotient norm $|S_\Gamma|$ of the operator S_Γ depends neither on the form of the Lyapunov curve Γ nor on the location of the points $t_1, \dots, t_n$ on Γ if these points are only pairwise distinct. Indeed, Theorem 2.1 of Chapter 7, Example 4.2 of [GK 1, Chapter 5], and Theorem 3.1 proved above, combine to give the following assertion.

Theorem 3.3. *Let the curve Γ consist of a finite number of pairwise disjoint simple Lyapunov curves, and let $t_1, \dots, t_n$ be different points on Γ and $\rho(t)$ the weight function defined by (3.1). Then, in the space $L_p(\Gamma, \rho)$,*

$$|S_\Gamma| = \max_{k=1,\dots,n} \nu(p, \beta_k) \tag{3.13}$$

with $\nu(p, \alpha)$ defined in (3.8).

Remark 3.1. Theorem 3.3 fails if the curve Γ has corner points (see, e.g. AVENDANIO/KRUPNIK [1]).

13.4 Conditions for Fredholmness of singular operators in spaces $L_p(\Gamma, \rho)$

Let Γ be a closed curve, $t_1, \dots, t_n$ be different points on Γ, and

$$\rho(t) = \prod_{k=1}^{n} |t - t_k|^{\beta_k} \quad (1 < p < \infty, \ -1 < \beta_k < p - 1) .$$

We denote by $\mathcal{S}_p(\Gamma, \rho)$ the class of all functions a in $\mathcal{S}_p(\Gamma)$ (see Section 12.4) which are subject to the following additional condition: For each point t_k $(k = 1, \dots, n)$, there exist an open arc $l(t_k)$ $(t_k \in l(t_k) \subset \Gamma)$ and functions $g_k^{\pm} \in L_\infty^{\pm}(\Gamma)$ such that $(g_k^{\pm})^{-1} \in L_\infty^{\pm}(\Gamma)$ and that the range of the restriction of the function $g_k^+ g_k^- a$ onto $l(t_k)$ is located in the interior of a sector with vertex $z = 0$ and with an opening angle less than $\theta_k := 2\pi / \max(p_k, q_k)$, $p_k = p(1 + \beta_k)^{-1}$ and $q_k^{-1} + p_k^{-1} = 1$. In other words,

$$\theta_k = 2\pi (\max(p, \frac{p}{p-1}, \frac{p}{1+\beta_k}, \frac{p}{p-1-\beta_k}))^{-1} .$$

Theorem 4.1 *Let $a \in \mathcal{S}_p(\Gamma, \rho)$. Then the operator $aP_\Gamma + Q_\Gamma$ belongs to $\Phi(L_p(\Gamma, \rho))$.*

Proof. Invoking the local principle in the same way as in the proofs to Theorems 1.1 and 4.1 of Chapter 12, we can restrict ourselves to the situation where $\Gamma = \mathbf{T}, \rho(t) = |t-1|^\beta$ $(-1 < \beta < p-1)$, $a(t) = \exp(i\varphi(t))$, and the sector which contains the range of the function a lies symmetrically with respect to the axis $[0, +\infty)$. Furthermore, we assume without loss of generality,that $p \geq 2$ (otherwise we take adjoints and apply Theorem 6.2 of Chapter 7). Under these restrictions, the conditions of the theorem imply that

$$\operatorname*{ess\,sup}_{t\in\mathbf{T}} |\varphi(t)| < \pi \left(\max \left(p, \frac{p}{1+\beta}, \frac{p}{p-1-\beta} \right) \right)^{-1} . \tag{4.1}$$

As in the proof of Theorem 2.1, the operator $A = aP_\Gamma + Q_\Gamma$ can be written in the form

$$A = \frac{a+1}{2}(I + i\tan\frac{\varphi}{2}S_{\mathbf{T}}) . \tag{4.2}$$

If $p \geq 2$ then

$$\max \left(p, \frac{p}{1+\beta}, \frac{p}{p-1-\beta} \right) = \begin{cases} \frac{p}{p-1-\beta} & \text{if } p-2 < \beta < p-1 \\ p & \text{if } 0 \leq \beta \leq p-2 \\ \frac{p}{1+\beta} & \text{if } -1 < \beta < 0 . \end{cases}$$

Consequently,

$$\text{ess sup}\, |\tan\frac{\varphi}{2}| < \begin{cases} \tan\frac{\pi(p-1-\beta)}{2p} & \text{if } p-2 < \beta < p-1 \\ \tan\frac{\pi}{2p} & \text{if } 0 \leq \beta \leq p-2 \\ \tan\frac{\pi(\beta+1)}{2p)} & \text{if } -1 < \beta < 0 , \end{cases}$$

and since

$$\|S_{\mathbf{T}}\|_{L_p(\mathbf{T},|t-1|^\beta)} = \begin{cases} \cot\frac{\pi(p-1-\beta)}{2p} & \text{if } p-2 < \beta < p-1 \\ \cot\frac{\pi}{2p} & \text{if } 0 \leq \beta \leq p-2 \\ \cot\frac{\pi(\beta+1)}{2p)} & \text{if } -1 < \beta < 0 , \end{cases}$$

we conclude that the operator $I + i\tan\frac{\varphi}{2}S_{\mathbf{T}}$ is invertible. ■

The subsequent theorem will prove useful for the derivation of Fredholm criteria for singular operators in several $L_p(\Gamma,\rho)$–spaces.

Theorem 4.2. *Let* $d \in W_p(\Gamma)$, $b \in L_\infty(\Gamma)$ *, and suppose the function* $c(\in L_\infty(\Gamma))$ *admits a factorization* $c = c_+c_-$ *in the space* $L_p(\Gamma,d)$ *. For* $\mathcal{B}_1 = L_p(\Gamma,d)$ *and* $\mathcal{B}_2 = L_p(\Gamma,|c_+|^{-p}d)$ *the following is true:*

$$bP_\Gamma + Q_\Gamma \in \Phi(\mathcal{B}_2) \Longleftrightarrow bcP_\Gamma + Q_\Gamma \in \Phi(\mathcal{B}_1) . \tag{4.3}$$

Proof. The operator R defined by $(R\varphi)(t) = c_+^{-1}(t)\varphi(t)$, and thought of as acting from $\mathcal{B}_2$ into $\mathcal{B}_1$, is invertible. The conditions of the theorem imply that the operator $cP_\Gamma + Q_\Gamma$ is invertible in $L_p(\Gamma,d)$. Set $B = cR$ and $D = B^{-1}(cP_\Gamma + Q_\Gamma)$. By what we have just remarked, the operators $B \in L(\mathcal{B}_2,\mathcal{B}_1)$ and $D \in L(\mathcal{B}_1,\mathcal{B}_2)$ are invertible. The identity

$$B(bP_\Gamma + Q_\Gamma)D = bcP_\Gamma + Q_\Gamma ,$$

which can be readily checked, immediately yields (4.3). ∎

As a consequence of this theorem we obtain the following

Proof of Theorem 7.1 of Chapter 12. Following our standard scheme, we confine ourselves to the case when $\Gamma = \mathbf{T}$ and $g(t) = t^\gamma$, with

$$\min(0, 2-p) \le p \operatorname{Re}\gamma \le \max(0, 2-\mathrm{p}) . \tag{4.4}$$

Choose $d(t) \equiv 1$, $b(t) = a(t)$ and $c(t) = g(t)$ in Theorem 4.2. Then $\mathcal{B}_2 = L_p(\mathbf{T}, |t-1|^{-\beta \operatorname{Re}\gamma})$ with $\beta = -p\operatorname{Re}\gamma$, and $\mathcal{B}_1 = L_p(\mathbf{T})$. We shall show by means of Theorem 4.1 that $aP_\mathbf{T} + Q_\mathbf{T} \in \Phi(\mathcal{B}_2)$. For, since $a \in \mathcal{S}_p(\mathbf{T})$, it suffices to check that $\mathcal{S}(\mathbf{T}, |t-1|^\beta) = \mathcal{S}_p(\mathbf{T})$ whenever $\beta = -p\operatorname{Re}\gamma$ and the condition (4.4) is satisfied. For definiteness, assume $p \ge 2$. Then (4.4) entails that $2 - p \le p\operatorname{Re}\gamma \le 0$; that is $0 \le \beta \le p-2$. For these values of β, we have

$$\max\left(p, \frac{p}{1+\beta}, \frac{p}{p-1-\beta}\right) = p$$

(see the proof of Theorem 4.1), and this leads to the desired equality $\mathcal{S}_p(\mathbf{T}, |t-1|^\beta) = \mathcal{S}_p(\mathbf{T})$. Now it remains to apply Theorem 4.2 which gives

$$aP_\mathbf{T} + Q_\mathbf{T} \in \Phi(L_p(\mathbf{T}, |t-1|^\beta)) \Longrightarrow agP_\mathbf{T} + Q_\mathbf{T} \in \Phi(L_p(\mathbf{T})) .$$

∎

As a further corollary we shall derive a Fredholm criterion for the operator $A = aP_\mathbf{T} + Q_\mathbf{T}$ in the space $L_2(\mathbf{T}, \rho)$, where a is an arbitrary function in $L_\infty(\mathbf{T})$ and ρ is an arbitrary weight in $W_2(\mathbf{T})$.

In the course of the proof of Theorem 5.1 of Chapter 12 we saw that if $\rho \in W_2(\mathbf{T})$, then the equality $b = b_- b_+$ with $b_+ = \exp(-P_\mathbf{T} \ln \rho)$ and $b_- = \overline{b}_+^{-1}$ describes a factorization of a given function b in the space $L_2(\mathbf{T})$, and that $|b_+|^{-2} = c_1 \rho$. Furthermore, Theorem 4.2 gives

$$aP_\mathbf{T} + Q_\mathbf{T} \in \Phi(L_2(\mathbf{T}, \rho)) \Longleftrightarrow acP_\mathbf{T} + Q_\mathbf{T} \in \Phi(L_2(\mathbf{T})) ,$$

and one can immediately check that $c(t) = \gamma \exp(i(\widetilde{\ln \rho}))$. [1] In accordance with the Fredholm criterion for the operator $acP_\mathbf{T} + Q_\mathbf{T}$ in $L_2(\mathbf{T})$ (cf. Theorem 2.1) we obtain the following result:

Theorem 4.3. *Let $a \in L_\infty(\mathbf{T})$ and $\rho \in W_2(\mathbf{T})$. Then the operator $A = aP_\mathbf{T} + Q_\mathbf{T}$ is a Φ-operator in $L_2(\mathbf{T}, \rho)$ if and only if the function a admits a representation in the form*

$$a(t) = \exp(-i\, \widetilde{\ln \rho}(t))\, h(t) g(t) t^\kappa , \tag{4.5}$$

[1] Here and in what follows, below the tilde denotes the conjugate harmonic function.

with $\kappa \in \mathbf{Z}$, $h^{\pm 1} \in L^+_\infty(\mathbf{T})$, *and*

$$\operatorname*{ess\,sup}_{t\in\mathbf{T}} |\arg g(t)| < \pi/2 \,.$$

In particular, for the Fredholmness of the operator $aP_{\mathbf{T}}+Q_{\mathbf{T}}$ in the space $L_2(\mathbf{T}, |t-1|^\beta)$, it is necessary and sufficient that the function $a(t)t^{-\beta/2}$ is representable in the form

$$a(t)t^{-\beta/2} = h(t)g(t)t^\kappa \,,$$

with the functions g and h being subject to the conditions (4.5). In other words,

$$aP_{\mathbf{T}} + Q_{\mathbf{T}} \in \Phi(L_2(\mathbf{T}, |t-1|^\beta) \Longleftrightarrow a = h(t)g_1(t)t^\kappa \,,$$

with

$$h^{\pm 1} \in L^+_\infty(\mathbf{T}), \kappa \in \mathbf{Z} \quad \text{and} \quad \operatorname{ess\,sup} |\arg(g(t)t^{\beta/2})| < \frac{\pi}{2} \,. \tag{4.6}$$

Here we did not directly apply Theorem 4.3 but instead used the factorization $t^\alpha = (t-1)^\alpha\,(1-t^{-1})^{-\alpha}$ $(\alpha = -\beta/2)$ in combination with Theorem 4.2. It is also possible to invoke Theorem 4.3 directly. For this one considers the identity $(\ln 2|\sin\frac{\theta}{2}|)^\sim = \frac{\pi-\theta}{2}$ (see, e.g., GAKHOV [1, p. 89] Problem 14.c), which implies that

$$\exp(-i(\ln|t-1|^\beta)^\sim) = \gamma t^{\beta/2} \quad (\gamma = \text{const.}\,)\,.$$

13.5 Norm and quotient norm of the operator $aI + bS_\Gamma$

In this section we shall derive some formulas for the computation of the norms of the operators $aI + bS_\Gamma$. The needed restrictions for the curve Γ , the weight ρ , and the coefficients a, b will be formulated in context.

Theorem 5.1. *Let* $a, b \in \mathbf{C}$. *Then*

$$\|aI + bS_\Gamma\|_{L_2(\Gamma,\rho)} = \mu(a, b, \|S_\Gamma\|_{L_2(\Gamma,\rho)}) \tag{5.1}$$

and

$$|aI + bS_\Gamma|_{L_2(\Gamma,\rho)} = \mu(a, b, |S_\Gamma|_{L_2(\Gamma,\rho)})\,, \tag{5.2}$$

where Γ *is an arbitrary curve and* ρ *an arbitrary weight such that*

$$S_\Gamma \in L(L_2(\Gamma,\rho)) \quad \text{and} \quad S_\Gamma^2 = I\,,$$

and

$$\mu(a,b,c) = \left[|b|^2\left(\frac{c^2-1}{2c}\right)^2 + \left(\frac{|a+b|+|a-b|}{2}\right)^2\right]^{1/2} + \left[|b|^2\left(\frac{c^2-1}{2c}\right)^2 + \left(\frac{|a+b|-|a-b|}{2}\right)^2\right]^{1/2}. \qquad (5.3)$$

In particular,

$$\mu(a,b,\cot\theta) = \left[|b|^2\cot^2 2\theta + \left(\frac{|a+b|+|a-b|}{2}\right)^2\right]^{1/2} + \left[|b|^2\cot^2 2\theta + \left(\frac{|a+b|-|a-b|}{2}\right)^2\right]^{1/2}. \qquad (5.4)$$

In order to prove Theorem 5.1 we establish some lemmas which work in the case of an arbitrary Hilbert space H.

If the space H is decomposed in the orthogonal sum of its subspaces H_1 and H_2, then it is convenient to write the operators in $L(H)$ in block form

$$A = \begin{pmatrix} A_{11} & A_{12} \\ A_{21} & A_{22} \end{pmatrix},$$

where $A_{jk} \in L(H_k, H_j)$.

The following identity is well-known and can be checked immediately:

$$\begin{pmatrix} I & B \\ C & D \end{pmatrix} = \begin{pmatrix} I & 0 \\ C & I \end{pmatrix}\begin{pmatrix} I & 0 \\ 0 & D-CB \end{pmatrix}\begin{pmatrix} I & B \\ 0 & I \end{pmatrix}. \qquad (5.5)$$

Lemma 5.1. *Let a be a non-zero complex number. Then the operator*

$$\begin{pmatrix} aI & B \\ C & D \end{pmatrix}$$

is invertible in $L(H)$ if and only if the operator $aD - CB$ is invertible in $L(H_2)$.

This lemma follows immediately from (5.5).

Lemma 5.2. *If $X \in L(H_2, H_1), \alpha, \beta \in \mathbf{C}$, and*

$$A = \begin{pmatrix} \alpha I & X \\ 0 & \beta I \end{pmatrix} \quad (\in L(H)) \qquad (5.6)$$

then

$$\|A\| = \frac{1}{2}((\|X\|^2 + (|\alpha| + |\beta|)^2)^{1/2} + (\|X\|^2 + (|\alpha| - |\beta|)^2)^{1/2}) \,. \tag{5.7}$$

Proof. It is easy to verify that

$$A^*A - \lambda I = \begin{pmatrix} (|\alpha|^2 - \lambda)I & \overline{\alpha}X \\ \alpha X^* & X^*X + (|\beta|^2 - \lambda)I \end{pmatrix} .$$

By Lemma 5.1, the operator $A^*A - \lambda I$ ($\lambda \neq |\alpha|^2$) is invertible if and only if the operator $(|\alpha|^2 - \lambda)(X^*X + (|\beta|^2 - \lambda)I) - |\alpha|^2 X^*X = (|\alpha|^2 - \lambda)(|\beta|^2 - \lambda)I - \lambda X^*X$ is invertible. Consequently, λ ($\neq 0, |\alpha|^2$) belongs to $\sigma(A^*A)$ if and only if $(|\alpha|^2 - \lambda)(|\beta|^2 - \lambda)\lambda^{-1}$ belongs to $\sigma(X^*X)$. It follows from (5.7) that $\|A\| \geq \max(|\alpha|, |\beta|)$. Since the function

$$f(\lambda) = \frac{(|\alpha|^2 - \lambda)\,(|\beta|^2 - \lambda)}{\lambda}$$

increases for $\lambda \geq \max(|\alpha|^2, |\beta|^2)$, the maximal point μ_0 of the spectrum of the operator X^*X and the maximal point of the spectrum of the operator A^*A are connected by the relation $\mu_0 = f(\lambda_0)$. In view of the equality

$$\|A\|^2 = \max\{\lambda : \lambda \in \sigma(A^*A)\}$$

this means that $\|X\|^2 = f(\|A\|^2)$, that is,

$$\|X\|^2 = \frac{(|\alpha|^2 - \|A\|^2)\,(|\beta|^2 - \|A\|^2)}{\|A\|^2} .$$

The number $\|A\|^2$ coincides with the maximal root of the equation

$$\|X\|^2 = \frac{(|\alpha|^2 - t)\,(|\beta|^2 - t)}{t} ,$$

or in other words,

$$\|A\|^2 = \frac{1}{2}(|\alpha|^2 + |\beta|^2 + \|X\|^2 + ((|\alpha|^2 + |\beta|^2 + \|X\|^2)^2 - 4|\alpha\beta|^2)^{1/2}) \,. \tag{5.8}$$

In view of the elementary equality

$$(\frac{1}{2}(a + b^{\frac{1}{2}}))^{\frac{1}{2}} = \frac{1}{2}((a + (a^2 - b)^{\frac{1}{2}})^{\frac{1}{2}} + (a - (a^2 - b)^{\frac{1}{2}})^{\frac{1}{2}}) \quad (a^2 \geq b > 0) ,$$

we can rewrite equality (5.8) in the form (5.7). ■

Lemma 5.3. *If P ($\neq 0, I$) is a projection in H, $Q = I - P$ is the complementary projection, and $\alpha, \beta \in \mathbf{C}$ then*

$$\begin{aligned} \|\alpha P + \beta Q\| &= \frac{1}{2}((|\alpha - \beta|^2(\|P\|^2 - 1) + (|\alpha| + |\beta|)^2)^{1/2} \\ &+ (|\alpha - \beta|^2(\|P\|^2 - 1) + (|\alpha| - |\beta|)^2)^{1/2}) \,. \end{aligned} \tag{5.9}$$

Proof. Let us denote by P_0 the orthogonal projection onto the subspace $\operatorname{im} P$. With respect to the decomposition $H = \operatorname{im} P_0 \oplus \ker P_0$, the operator P has the form

$$P = \begin{pmatrix} I & T \\ 0 & 0 \end{pmatrix} \tag{5.10}$$

where $T \in L(\ker P_0, \operatorname{im} P_0)$. Therefore, the operator $A = \alpha P + \beta Q$ has the form

$$A = \begin{pmatrix} \alpha I & (\alpha - \beta)T \\ 0 & \beta I \end{pmatrix} . \tag{5.11}$$

By virtue of Lemma 5.2,

$$\begin{aligned} \|A\| &= \frac{1}{2}((|\alpha - \beta|^2 \|T\|^2 + (|\alpha| + |\beta|)^2)^{1/2} \\ &\quad + (|\alpha - \beta|^2 \|T\|^2 + (|\alpha| - |\beta|)^2)^{1/2}) . \end{aligned} \tag{5.12}$$

For $\alpha = 1,\ \beta = 0$ this equality yields

$$\|P\|^2 = \|T\|^2 + 1 \tag{5.13}$$

Now equality (5.9) follows from (5.12) and (5.13). ∎

The equality (5.9) involves the following relation between the norms of the operators $S_\Gamma, P_\Gamma, Q_\Gamma$ in the space $L_2(\Gamma, \rho)$, under the assumption that the space $L_2(\Gamma, \rho)$ satisfies the conditions of Theorem 5.1:

$$\|P_\Gamma\| = \|Q_\Gamma\| = \frac{\|S_\Gamma\| + \|S_\Gamma\|^{-1}}{2} . \tag{5.14}$$

Proof of Theorem 5.1. The identity (5.1) follows from (5.9) by setting $\alpha = a + b, \beta = a - b$ and observing that

$$\|P_\Gamma\|^2 - 1 = \frac{1}{4}(\|S_\Gamma\| - \|S_\Gamma\|^{-1})^2 .$$

For a proof of (5.2) we need that the quotient algebra $\widehat{L} = L/T$ is a Banach C^*-algebra, which is, by the Gelfand–Naimark theorem (see GELFAND, RAIKOV, SHILOV [1], Applications, §3), isometrically isomorphic to a subalgebra of the algebra $L(H)$ of all linear bounded operators acting on a certain Hilbert space H.

Since both projections P_Γ and Q_Γ have infinite-dimensional kernel and range, their cosets $\widehat{P}_\Gamma$ and $\widehat{Q}_\Gamma$ in the algebra $\widehat{L}$ are neither the zero nor the identity element. Thus, the projection P and Q which are associated to $\widehat{P}_\Gamma$ and $\widehat{Q}_\Gamma$ by the Gelfand–Naimark theorem are subject to the conditions in Lemma 5.3, and the equality (5.9) for P and Q implies the identity (5.2). ∎

Theorem 5.2. *Let $a, b \in \mathbf{C}$, $t_0 \in \mathbf{T}$, and $-1 < \alpha < 1$. Then*

$$\begin{aligned}\|aI + bS_{\mathbf{T}}\|_{L_2(\mathbf{T},|t-t_0|^\alpha)} &= \left[|b|^2 \tan^2 \frac{\pi\alpha}{2} + \left(\frac{|a+b| - |a-b|}{2}\right)^2\right]^{1/2} \\ &+ \left[|b|^2 \tan^2 \frac{\pi\alpha}{2} + \left(\frac{|a+b| + |a-b|}{2}\right)^2\right]^{1/2}. \qquad (5.15)\end{aligned}$$

Proof. Theorem 3.1 tells us that $\|S_{\mathbf{T}}\|_{2,\alpha} = \nu(2,\alpha) = \cot \frac{\pi(1-|\alpha|)}{4}$, and so it remains to apply the equalities (5.1) and (5.4) to finish the proof. ∎

Up to now we have considered the space $L_2(\mathbf{T}, |t-t_0|^\alpha)$. For the spaces $L_p(\mathbf{T}, |t-t_0|^\alpha)$ with $p \neq 2$, the computation of the norm $\|S_{\mathbf{T}}\|$ was pursued in Theorem 3.1. The general case remains an open problem. There is only a lower estimate which we present now.

Theorem 5.3. *Let $a, b \in \mathbf{C}, t_0 \in \mathbf{T}$, and $-1 < \alpha < p-1$. Then*

$$\begin{aligned}\|aI + bS_{\mathbf{T}}\|_{L_p(\mathbf{T},|t-t_0|^\alpha)} &\geq \left[|b|^2 \cot^2 \frac{\pi(1+\beta)}{p} + \left(\frac{|a+b| - |a-b|}{2}\right)^2\right]^{1/2} \\ &+ \left[|b|^2 \cot \frac{\pi(1+\beta)}{p} + \left(\frac{|a+b| + |a-b|}{2}\right)^2\right]^{1/2} \qquad (5.16)\end{aligned}$$

where

$$\beta = \begin{cases} \alpha & \text{if } \ |\alpha| > |p-2| \\ 0 & \text{if } \ |\alpha| \leq |p-2| \,. \end{cases} \qquad (5.17)$$

Notice that the right hand sides of (5.16) and (5.15) coincide in case that $p = 2$.

Proof. The idea of the proof is the same as the one for Theorem 9.1 in Section 9. So we shall only sketch the proof here.

For simplicity, let $\alpha = 0$ and $p > 2$ (in the general case the proof proceeds analogously; one only has to take into account the interval containing the number α).

Define

$$c = \frac{a+b}{a-b}, \quad A_0 = cP_{\mathbf{T}} + Q_{\mathbf{T}}, \quad A_1 = gP_{\mathbf{T}} + Q_{\mathbf{T}}, \qquad (5.18)$$

where $g \in PC(\mathbf{T})$ is a function which takes only two values, z_1 and z_2. These values will be specified later on, here we only suppose the angle $(z_1, 0, z_2)$ to be equal to $2\pi/p$.

In this situation, A_1 is not a Φ-operator (see Section 9.3) in $L_p(\mathbf{T})$. We represent this operator in the form

$$A_1 = \frac{c-g}{c-1}\left(I + \frac{g-1}{c-g}A_0\right) . \tag{5.19}$$

Since $A_1 \notin \Phi(L_p(\mathbf{T}))$, we have

$$\left|\frac{g-1}{c-g}A_0\right| \geq 1 . \tag{5.20}$$

Now choose z_1 and z_2 in such a way that the quotient $|g-1|/|c-g|$ is constant, that is, such that these points lie on a circle

$$\left|\frac{z-c}{z-1}\right| = \delta \tag{5.21}$$

(the circle of Apollonius), with a certain constant δ. Then

$$|A_0|_{L_p(\mathbf{T})} \geq \delta , \tag{5.22}$$

and the estimate

$$|aI + bS_{\mathbf{T}}|_{L_p(\mathbf{T})} \geq |a-b|\delta \tag{5.23}$$

follows.

It remains to show that this construction will give the estimate (5.16). For this we set

$$\begin{aligned} 2\delta &= \left[|c-1|^2 \cot\frac{\pi}{p} + (|c|-1)^2\right]^{1/2} \\ &\quad + \left[|c-1|^2 \cot\frac{\pi}{p} + (|c|+1)^2\right]^{1/2} , \end{aligned} \tag{5.24}$$

and consider the circle defined by (5.21). Its center is the point $z_0 = (\delta^2 - c)(\delta^2-1)^{-1}$, and its radius R equals $\delta|c-1|(\delta^2-1)^{-1}$. Further we consider the tangents drawn from the coordinate origin to this circle. The choice of the number δ in (5.24) ensures that the angle between these tangents is equal to $2\pi/p$. Now specify z_1 and z_2 to be the points where the tangents meet the circle. Then we obtain from (5.23) and (5.24) the estimate

$$|aI + bS_{\mathbf{T}}|_{L_p(\mathbf{T})} \geq \mu\left(a, b, \cot\frac{\pi}{2p}\right) , \tag{5.25}$$

and consequently,

$$\|aI + bS_{\mathbf{T}}\|_{L_p(\mathbf{T})} \geq \mu\left(a, b, \cot\frac{\pi}{2p}\right) . \tag{5.26}$$

■

By means of the local principle for the quotient norm of singular integral operators (see 8.1 of Chapter 6) one can generalize the estimate (5.25) to the case of singular integral operators with arbitrary continuous coefficients:

Let Γ be a closed Lyapunov curve and $a, b \in C(\Gamma)$. Then

$$\begin{aligned} |aI + bS_\Gamma|_{L_p(\Gamma)} &= \max_{\tau\in\Gamma} |a(\tau)I + b(\tau)S_\Gamma|_{L_p(\Gamma)} \\ &= \max_{\tau\in\Gamma} |a(\tau)I + b(\tau)S_{\mathbf{T}}|_{L_p(\mathbf{T})} \\ &\geq \max_{\tau\in\Gamma} \mu\left(a(\tau), b(\tau), \cot\frac{\pi}{2p}\right) . \end{aligned} \tag{5.27}$$

Here we used the observation that the quotient norm of a singular operator with constant coefficients does not depend on the form of the Lyapunov curve Γ (cf. Lemma 2.2 of Chapter 7).

In case $p = 2$ in (5.27) equality holds, as the following result indicates.

Theorem 5.4. *Let Γ be a closed Lyapunov curve, $a, b \in C(\Gamma)$, and $\rho(t) = \prod_{k=1}^{n} |t - t_k|^{\beta_k}$. Then*

$$|aI + bS_\Gamma|_{L_2(\Gamma,\rho)} = \max_{\tau\in\Gamma} \mu(a(\tau), b(\tau), \tan\theta(\tau)) ,$$

where $\theta(t_k) = \frac{\pi\beta_k}{4}$ and $\theta(\tau) = 0$ at all other points.

Proof. Set $A = aI + bS_\Gamma$ and $A_\tau = a(\tau)I + b(\tau)S_\Gamma$. From identity (8.3) of Chapter 6 we conclude that $|A| = \max_{\tau\in\Gamma} |A_\tau|$. Theorem 5.1 entails that

$$|A_\tau| = \mu(a(\tau), b(\tau), |S_\Gamma|_{L_2(\Gamma,\rho_\tau)})$$

(remember the definition of ρ_τ given in equality 8.3 of Chapter 6), and the quotient norm $|S_\Gamma|_{L_2(\Gamma,\rho_\tau)}$ has been computed in Theorem 3.3. ■

13.6 Exercises

13.1. Let A be an operator in $L(\mathcal{B})$ and $\{B_n\}$ be a sequence of operators in $L(\mathcal{B})$ such that

1. $\|B_n\varphi\| = \|\varphi\|$ $(\varphi \in \mathcal{B})$,

2. $AB_n = B_nA$,

3. the sequence $\{B_n\}$ converges weakly to zero.

Prove that then

$$\inf_{T\in\mathcal{T}(\mathcal{B})} \|A+T\| = \|A\| \ . \tag{6.1}$$

Hint. Consult the proof of Theorem 4.1.

13.2. Let $\mathcal{B} = L_p(\mathbf{R}, |t|^\beta)$ $(-1<\beta<p-1)$, $a,b\in\mathbf{C}$, $A = aI + bS_{\mathbf{R}}$, and

$$(B_n\varphi)(t) = n^{\frac{1+\beta}{p}}\varphi(nt) \ .$$

Show that the operators A and B_n satisfy the conditions of the preceding exercise.

13.3. Let Γ be a curve composed by two circular arcs (or by one circular arc and one straight line) with the common points ± 1 . Define

$$(B_n\varphi)(z) = \frac{2\sqrt{n}}{n+1+z(n-1)}\,\varphi\left(\frac{(n+1)z+n-1}{n+1+z(n-1)}\right) .$$

Show that these operators and the operator S_Γ in place of A satisfy the conditions in Exercise 13.1 in case the underlying space is $L_2(\Gamma)$.

13.4. Is it possible to define a circle as a simple closed Lyapunov curve Γ for which

$$\|S_\Gamma\|_{L_2(\Gamma)} = |S_\Gamma|_{L_2(\Gamma)} \ ? \tag{6.2}$$

13.5. Is it possible to define a circle as a simple piecewise Lyapunov curve for which (6.2) holds?

13.6. Let Γ be a curve composed by two circular arcs or by one circular arc and one straight line. Verify that then

$$\|S_\Gamma\|_{L_2(\Gamma)} = |S_\Gamma|_{L_2(\Gamma)} \ . \tag{6.3}$$

13.7. Does equality (6.3) remain valid for arbitrary simple piecewise Lyapunov curves consisting of circular arcs and straight lines?

Hint. Consider a Lyapunov curve (different from a circle) which consists of two semi–circles and two straight lines.

13.8.* Does equality (6.3) hold if Γ is the boundary of a square?[1]

13.9. Demonstrate that the norm

$$\|S_{\mathbf{T}}\|_{L_2(\mathbf{T},|t-t_1|^{1/2}|t-t_2|^{1/2})}$$

depends on the position of the points t_1 and t_2 on $\mathbf{T}$.

[1]The answer to this question is unknown to the authors.

Hint. Let t_1 tend to t_2 .

13.10. Let $P \in L(H)$, $P^2 = P$, $P \neq 0$, $P \neq I$, and $Q = I - P$. Prove that the equality

$$\|\alpha P + \beta Q\| = \|\alpha Q + \beta P\| \tag{6.4}$$

holds for each pair α, β of complex numbers.

13.11. Verify that the assertion in Exercise 13.10 generally fails to be true if H is a Banach space.

13.12. Let $1 < p < \infty$. Show that, for each choice of complex numbers α, β , the equality

$$\|\alpha P_{\mathbf{T}} + \beta Q_{\mathbf{T}}\|_{L_p(\mathbf{T})} = \|\beta P_{\mathbf{T}} + \alpha Q_{\mathbf{T}}\|_{L_p(\mathbf{T})}$$

holds.

13.13. Prove that, under the assumptions of Exercise 13.10,

a) $\|\alpha P + \beta Q\| \geq \max(|\alpha|, |\beta|)$.

b) If the equality

$$\|\alpha P + \beta Q\| = \max(|\alpha|, |\beta|) \tag{6.5}$$

holds for at least one pair α, β (with $\alpha \neq \beta$) of complex numbers then $P^* = P$, and equality (6.5) is valid for each pair of complex numbers.

Comments and references

13.1. The equalities

$$\begin{aligned} \|S_{\mathbf{T}}\|_{L_p(\mathbf{T})} &= \|S_{\mathbf{R}}\|_{L_p(\mathbf{R})} \\ &= \inf_{T \in \mathcal{T}} \|S_\Gamma + T\|_{L_p(\Gamma)} \\ &= \begin{cases} \cot \frac{\pi}{2p} & (p \geq 2) \\ \tan \frac{\pi}{2p} & (p \leq 2) \end{cases} \end{aligned}$$

were obtained by GOHBERG/KRUPNIK [2,3] for $p = 2^n$ and $p = 2^n(2^n - 1)^{-1}$. In these papers were also derived lower estimates for the norm and quotient norm of singular operators for all p in $(1, \infty)$.

An important role for the derivation of upper estimates for all $p \in (1, \infty)$ is played by Theorem 1.1, obtained by PICHORIDES [1]. Theorem 1.4 is due to KRUPNIK and POLONSKII [1].

13.2. The second proof of Theorem 4.1 of Chapter 12 which is presented in this section is GOHBERG and KRUPNIK's (see [6], Ch. XIII, §4).

13.3, 13.4 The Theorems 3.1–3.3 and 4.1 were obtained by KRUPNIK and VERBITSKII [1] whereas Lemma 3.2 and Theorem 4.2 were found by KRUPNIK and NYAGA [1]. Another proof of Theorem 4.1 goes back to FROLOV [1,2] (see also GOHBERG and KRUPNIK [6]).

13.5 Theorem 5.1 is contained in FELDMAN, KRUPNIK and MARKUS' paper [1]. The connection between the norms of the operators S_Γ and P_Γ cited in (5.14) was remarked earlier by SPITKOVSKII [1]. Theorem 5.2 was obtained by KRUPNIK ([1, p. 57]) in another way, and the Theorems 5.3 and 5.4 go back to AVENDANIO and KRUPNIK [1].

References

AVENDANIO, R.E., KRUPNIK, N. YA.

[1] Local principle of computation of the factor-norms of singular integral operators. Funct. Anal. Appl. **22** (1988), 130-131.

BABENKO, K. I.

[1] On adjoint functions (Russian). Doklady AN SSSR **62** (1948), 2, 157-160.

BÖTTCHER, A., ROCH, S., SILBERMANN, B. SPITKOVSKIĬ, I. M.

[1] A Gohberg-Krupnik-Sarason symbol calculus for algebras of Toeplitz, Hankel, Cauchy, and Carleman operators. In: Operator Theory: Advances and Appl., vol. 48, Birkhäuser Verlag Basel 1990, 189-234.

BÖTTCHER, A., SILBERMANN, B., SPITKOVSKIĬ, I. M.

[1] Toeplitz operators with piecewise quasisectorial symbols. Bull. London Math. Soc. **22** (1990), 281-286.

BOURBAKI, N.

[1] Eléments de Mathématique, Livre IX, Fonctions d'une variable réelle (Théorie élémentaire), Hermann & Cie, Paris 1949.

[2] Eléments de mathématique, Livre XXXII, Théories spectrales, Hermann & Cie, Paris 1967.

BOYD, D.

[1] The Hilbert transform of rearrangement spaces. Canadian J. Math. **19** (1967) ,3, 599-616.

CHEBOTAREV, G. N.

[1] On a first kind equation of convolution type (Russian). Izv. vusov (matem.) **2** (1967), 80-92.

COBURN, L. A.

[1] Weyl's theorem for non-normal operators. Michigan Math. J. **13** (1966), 285-286.

COBURN, L. A., DOUGLAS, R. G.

[1] Translation operators on the half-line. Proc. Nat. Acad. Sci U.S.A. **62** (1969), 1010-1013.

COSTABEL, M.

[1] An inverse for the Gohberg-Krupnik symbol map. Proc. Royal Soc. of Edinburgh **87A** (1980), 153-165.

[2] Singular integral operators on curves with corners. Integral Equ. and Operator Theory **3** (1980), 323-349

DEVINATZ, A.

[1] Toeplitz operators on H_2–space. Trans. Amer. Math. Soc. **112** (1964), 2, 304-317.

DOUGLAS, R. G.

[1] Banach algebra techniques in operator theory. Academic Press, New York 1972.

DOUGLAS, R.G., SARASON, D. E.

[1] Fredholm Toeplitz operators. Proc. Amer. Math. Soc. **26** (1970), 117-120.

DUDUCHAVA, R. V.

[1] On singular integral operators on weighted Hölder spaces. (Russian). Doklady AN SSSR **191** (1970)1, 16-19.

[2] On the boundedness of the operator of singular integration in weighted Hölder spaces (Russian). Matem. issled. Kishinev **5** (1970), 1, 56-76.

[3] Singular integral equations in weighted Hölder spaces, 1. Hölder continuous coefficients (Russian). Matem issled. Kishinev **5** (1970), 2, 104-124.

[4] Singular integral equations in weighted Hölder spaces, 2. Piecewise Hölder continuous coefficients (Russian). Matem. issled. Kishinev **5** (1970), 3, 58-82.

[5] Discrete Wiener–Hopf equations in weighted l_p–spaces (Russian). Soobshzh. AN Gruz. SSR **67** (1972), 1, 17-20.

[6] Discrete Wiener–Hopf equations generated by the Fourier coefficients of piecewise Wiener functions (Russian). Doklady AN SSSR **208** (1972), 1, 19-22.

[7] Wiener–Hopf integral operators with discontinuous symbols. Soviet Math. Dokl. **14** (1973), 4, 1001-1005.

[8] On convolution integral operators with discontinuous coefficients. Soviet Math. Dokl. **15** (1974), 5, 1302-1306.

DYBIN, V. B.

[1] Normalization of Wiener–Hopf operators (Russian). Doklady AN SSSR **191** (1970), 4, 759-762.

DYBIN, V. B., KARAPETYANC, N. K.

[1] The application of the method of normalization to a class of systems of linear algebraic equations (Russian). Izv. vusov (matem.) **10**(1967), 39-49.

ESKIN, G. I.

[1] Systems of elliptic pseudodifferential equations (Russian). Institute for Problems in Mechanics of the AN SSSR, Moscow 1972 (Preprint Nr. 11).

FREIDKIN, S. A.

[1] On unbounded singular operators (Russian). Materialy nauchn. konferencii Kishinevskogo universiteta (sektsia estestv. i eksperim. nauk), Kishinev 1970, 84-86.

FROLOV, V. D.

[1] On the theory of integral equations with measurable coefficients (Russian). Doklady AN SSSR **189** (1969), 6, 1185-1188.

[2] On singular integral equations with measurable coefficients on weighted spaces (Russian). Matem. Issled. Kishinev **5** (1970), 1, 141-151.

GAKHOV, F. D.

[1] Boundary Value Problems. Pergamon Press, London 1966.

GELFAND, I. M., RAIKOV, D. A., SHILOV, G. E.

[1] Kommutative normierte Algebren. Deutscher Verlag der Wissenschaften, Berlin 1966.

GOHBERG, I. C.

[1] On the number of the solutions of homogenous singular equations with continuous coefficients (Russian). Doklady AN SSSR **122** (1958), 3, 327-330.

GOHBERG, I. C. FELDMAN, I. A.

[1] Faltungsgleichungen und Projektionsmethoden zu ihrer Lösung. Birkhäuser–Verlag, Basel–Stuttgart 1974. (English transl.: Amer. Math. Soc. Transl. of Math. Monographs 41, Providence, R. I., 1974.)

GOHBERG, I. C., KRUPNIK, N. YA.

[1] The spectrum of one–dimensional singular integral operators with piecewise continuous coefficients (Russian). Matem. Issled. Kishinev **3** (1968), 1, 16-30. (English transl.: Amer. Math. Soc. Trans. Ser. 2, **103** (1974), 181-193.)

[2] On the spectrum of singular integral operators on the space L_p (Russian). Studia Mathematica **31** (1968), 347–362.

[3] On the norm of the Hilbert transform on the space L_p (Russian). Funkc. analiz i ego prilozh. **2** (1968), 2, 91–92.

[4] On the spectrum of singular integral operators on weighted L_p–spaces (Russian). Doklady AN SSSR **185** (1969), 4, 745–749.

[5] On the quotient norm of singular integral operators (Russian). Matem. Issled. Kishinev **4** (1969), 3, 136–139.

[6] Introduction to the theory of one–dimensional singular integral operators (Russian). Ştiintsa, Kishinev 1973.

[7] On singular integral operators on a non–simple curve (Russian). Soobshzh. AN Gruz. SSR **64** (1971) 1, 21–24.

[8] On a local principle and on algebras generated by Toeplitz matrices (Russian). Annalele ştiintifice ale univ. "Al. I. Cuza Iaşi, Sect. I. a, Matematica, 19, 1, 1973.

[9] Singular integral equations with continuous coefficients on a composed curve (Russian). Matem. issled. Kishinev **5** (1970), 2, 89-103.

[10] Einführung in die Theorie der eindimensionalen singulären Integraloperatoren. Birkhäuser Verlag, Basel–Boston–Stuttgart 1979.

[11] One–dimensional linear singular integral equations I. Introduction. Birkhäuser Verlag, Basel–Boston–Stuttgart 1991.

GOHBERG, I. C. SEMENCUL, A. A.

[1] Toeplitz matrices generated by the Fourier coefficients of functions having discontinuities of almost periodic type (Russian). Matem. issled. Kishinev **5** (1970) 4, 63-83.

GORDADSE, E.

[1] On a Riemann–Privalov boundary value problem in case of a non–smooth boundary (Russian). Trudy Tbilisk. matem. instituta im. A. M. Rasmadse **33** (1967), 25–31.

HARDY, G. H., LITTLEWOOD, J. E., POLYA, G.

[1] Inequalities. University Press. London–Cambridge 1951.

HOFFMAN, K.

[1] Banach spaces of analytic functions. Prentice Hall, Englewood Cliffs. New Jersey 1962.

HUNT, R., MUCKENHOUPT, B., WHEEDEN, R.

[1] Weighted norm inequalities for the conjugate function and Hilbert transform. Trans. Amer. Math. Soc. **176** (1973), 227–251.

KHAIKIN, M. I.

[1] On first kind integral equations of convolution type (Russian). Izv. vusov (matem.) **3**(1967), 105–116.

[2] On a special singular integral operator with piecewise continuous coefficients (Russian). Matem. issled. Kishinev **7** (1972), 3, 194–209.

[3] On the kernel of a singular integral operator with piecewise continuous coefficients in a special case (Russian). Mathem. issled. Kishinev **8** (1973), 1, 171–179.

[4] On the image of a singular integral operator in the degenerated case (Russian). Izv. vusov (matem.) **165** (1976), 2, 119–122.

KHVEDELIDSE, B. V.

[1] Linear discontinuous boundary value problems of function theory, singular integral equations, and some of their applications (Russian). Trudy Tbilisk. matem. instituta AN Gruz. SSR **23** (1956), 3–158.

[2] Remarks to my paper: "Linear discontinuous boundary value problems ..." (Russian). Soobshzh. AN Gruz. SSR **21**(1958), 2, 129–130.

KRASNOSELSKIĬ, M. A.

[1] On a theorem of M. Riesz (Russian). Doklady AN SSR **131** (1960), 2, 246–248.

KRUPNIK, N. YA.

[1] Banach Algebras with Symbol and Singular Integral Operators. Operator Theory: Advances and Applications Vol. 26, Birkhäuser Verlag, Basel–Boston–Berlin 1987.

[2] Some consequences of a theorem of Hunt, Muckenhoupt, Wheeden (Russian). Matem. Issled. Kishinev **47** (1978), 64–70.

KRUPNIK, N. YA., MARKUS, A. S., FELDMAN, I. A.

[1] On the norm of a linear combination of projections in Hilbert space. Funct. Anal. Appl. **23** (1989), 85–86.

KRUPNIK, N: YA., NYAGA, V. I.

[1] On singular integral operators on weighted L_p–spaces (Russian). Matem. issled. Kishinev **9** (1974), 3, 206–210.

KRUPNIK, N. YA., POLONSKIĬ, E. P.

[1] On the norm of the operator of singular integration (Russian). Funkc. analiz i ego prilozh. **9** (1975), 4, 73–75.

KRUPNIK, N. YA, VERBITSKIĬ, I. E.

[1] Exact constants in the theorems of K. I. Babenko and B. V. Khvedelidze on the boundedness of singular operators (Russian). Soobshzh. AN Gruz. SSR **85** (1977), 1, 21–24.

[2] Exact constants in theorems on the boundedness of singular operators on weighted spaces, and their applications (Russian). Matem. issled. Kishinev **54** (1980), 21–35.

LEE, M, SARASON, D.

[1] The spectra of some Toeplitz operators. J. of Math. Analysis and Applic. **33** (1971), 529–543.

LEVITAN, B. M.

[1] Almost periodic functions (Russian). Gostekhizdat, Moscow 1953.

LITVINCHUK, G. S., SPITKOVSKIĬ, I. M.

[1] Factorization of Measurable Matrix Functions. Operator Theory: Advances and Applications Vol. 25, Birkhäuser Verlag, Basel–Boston–Stuttgart 1987.

MEISTER, E.

[1] Das Riemannsche Randwertproblem, Ergebnisse und Anwendungen. Überblicke Mathematik **6** (1973), 113–178.

[2] Ein Überblick über analytische Methoden zur Lösung singulärer Integralgleichungen. ZAMM **57** (1977) 5, T 23 – T 35.

MUSKHELISHWILI, N. I.

[1] Singuläre Integralgleichungen. Akademie–Verlag, Berlin 1965.

PICHORIDES, S. K.

[1] On the best values of the constants in the theorems of M. Riesz, Zygmund and Kolmogorov. Studia Mathematica **19** (1972), 2, 165–179.

PRIVALOV, I. I.

[1] Randeigenschaften analytischer Funktionen. Deutscher Verlag der Wissenschaften, Berlin 1956.

PRÖSSDORF, S.

[1] Über eine Klasse singulärer Integralgleichungen nichtnormalen Typs. Math. Ann. **183** (1969), 130–150.

[2] Einige Klassen singulärer Gleichungen. Birkhäuser Verlag, Basel–Stuttgart 1974. (English transl.: North Holland Publishing 1978.)

ROCH, S., SILBERMANN, B.

[1] Algebras generated by idempotents and the symbol calculus for singular integral operators. Integral Equ. and Operator Theory **11** (1988), 385–419.

[2] The Calkin image of algebras of singular integral operators. Integral Equ. and Operator Theory **12** (1989), 855–897.

RYSHIK, I. M., GRADSTEIN, I. S.

[1] Summen-, Produkt- und Integraltafeln. Deutscher Verlag der Wissenschaften, Berlin 1963.

SARASON, D.

[1] Toeplitz operators with semi-almost periodic symbols. Duke Math. J. **44** (1977), 2, 357–364.

SCHLEIFF, M.

[1] Über eine lineare Integrodifferentialgleichung mit Zusatzkern. Wiss. Beiträge Univ. Halle, Beiträge zur Analysis und Angewandten Mathematik, 1968/9 (M 1), 145–154.

SCHÜPPEL, B.

[1] Regularisierung singulärer nicht elliptischer Integralgleichungen mit unstetigen Koeffizienten. Inaug.-Diss. Univ. München 1973.

[2] Regularisierung singulärer Integralgleichungen vom nicht normalen Typ mit stückweise stetigen Koeffizienten. In: "Function–theoretic methods for partial differential equations", Springer Lecture Notes in Math., vol. 561, Berlin 1976, 430-442.

SCHWARTZ, J.

[1] Some results on the spectra and spectral resolutions of a class of singular integral operators. Communications on pure and applied mathematics **15** (1962), 1, 75-90.

SEMENOV, E. M.

[1] A new interpolation theorem (Russian). Funkc. analiz. i ego prilozh. **2** (1968)2, 68–80.

[2] Interpolation of linear operators on symmetric spaces (Habilitation Thesis) (Russian). Woronezh 1968.

SEMENCUL, A. A.

[1] On singular integral equations with coefficients having discontinuities of almost periodic type (Russian). Matem. issled. Kishinev **6** (1971), 3, 92–114.

SHAMIR, E.

[1] The solution of a Riemann–Hilbert system with piecewise continuous coefficients (Russian). Doklady AN SSSR **167** (1966), 5, 1000-1003.

SHILOV, G. E.

[1] On locally analytic functions (Russian). Uspekhi matem. nauk **21** (1966), 6.

SIMONENKO, I. B.

[1] The Riemann boundary value problem with continuous coefficients (Russian). Doklady AN SSSR **124** (1959), 2, 278–281.

[2] The Riemann boundary value problem with measurable coefficients (Transl. from Russian). Soviet Math. Doklady **1** (1960), 1295–1298 (1961).

[3] The Riemann boundary value problem for an n–tuple of function pairs with measurable coefficients and its application to the investigation of singular integrals on weighted L_p-spaces (Russian). Izv. AN SSSR (seria matem) **28** (1964), 2, 277–306.

[4] A new general method to study linear operator equations of the singular integral equation type, part 1 (Russian). Izv. AN SSSR (seria matem.) **29** (1965), 3, 567-586.

[5] A new general method to study linear operator equations of the singular integral equation type, part 2 (Russian). Izv. AN SSSR (seria matem.) **29** (1965), 4, 757–782.

[6] Some general problems in the theory of the Riemann boundary value problem (Russian). Izv. AN SSSR (seria matem.) **32** (1968), 5, 1138–1146.

SPITKOVSKIĬ, I. M.

[1] On partial indices of continuous matrix-valued functions. Soviet Math. Doklady **17** (1976), 1155–1159.

STEIN, E. M.

[1] Interpolation of linear operators. Trans. Amer. Math. Soc. **83** (1956), 222–234.

TIMAN, A. F., TROFIMOV, V. N.

[1] Introduction to the theory of harmonic functions (Russian). Nauka, Moscow 1968.

TRIEBEL, H.

[1] Interpolation theory, function spaces, differential operators. North-Holland Publishing, Amsterdam 1978.

WIDOM, H.

[1] Singular integral equations in L_p. Trans. Amer. Math. Soc. **97** (1960), 1, 131–160.

[2] On the spectra of a Toeplitz operator. Pacific J. Math. **14** (1964), 365–375.

ZYGMUND, A.

[1] Trigonometric Series I, II. Cambridge Univ. Press, 1959.

Subject Index